T0183989

Graduate Texts in Mathematics

Series Editors:

Sheldon Axler
San Francisco State University, San Francisco, CA, USA

Kenneth Ribet
University of California, Berkeley, CA, USA

Advisory Board:

Alejandro Adem, *University of British Columbia*
David Eisenbud, *MSRI & University of California, Berkeley*
Brian C. Hall, *University of Notre Dame*
Patricia Hersh, *North Carolina State University*
J.F. Jardine, *University of Western Ontario*
Jeffrey C. Lagarias, *University of Michigan*
Ken Ono, *Emory University*
Jeremy Quastel, *University of Toronto*
Fadil Santosa, *University of Minnesota*
Barry Simon, *California Institute of Technology*
Ravi Vakil, *Stanford University*
Steven H. Weintraub, *Lehigh University*
Melanie Matchett Wood, *University of California, Berkeley*

Graduate Texts in Mathematics bridge the gap between passive study and creative understanding, offering graduate-level introductions to advanced topics in mathematics. The volumes are carefully written as teaching aids and highlight characteristic features of the theory. Although these books are frequently used as textbooks in graduate courses, they are also suitable for individual study.

More information about this series at http://www.springer.com/series/136

Laurenţiu G. Maxim

Intersection Homology & Perverse Sheaves

with Applications to Singularities

 Springer

Laurenţiu G. Maxim
Department of Mathematics
University of Wisconsin–Madison
Madison, WI, USA

ISSN 0072-5285 ISSN 2197-5612 (electronic)
Graduate Texts in Mathematics
ISBN 978-3-030-27646-1 ISBN 978-3-030-27644-7 (eBook)
https://doi.org/10.1007/978-3-030-27644-7

Mathematics Subject Classification: 55N33, 32S60, 32S35, 32S05, 32S20

© Springer Nature Switzerland AG 2019
This work is subject to copyright. All rights are reserved by the Publisher, whether the whole or part of the material is concerned, specifically the rights of translation, reprinting, reuse of illustrations, recitation, broadcasting, reproduction on microfilms or in any other physical way, and transmission or information storage and retrieval, electronic adaptation, computer software, or by similar or dissimilar methodology now known or hereafter developed.
The use of general descriptive names, registered names, trademarks, service marks, etc. in this publication does not imply, even in the absence of a specific statement, that such names are exempt from the relevant protective laws and regulations and therefore free for general use.
The publisher, the authors, and the editors are safe to assume that the advice and information in this book are believed to be true and accurate at the date of publication. Neither the publisher nor the authors or the editors give a warranty, express or implied, with respect to the material contained herein or for any errors or omissions that may have been made. The publisher remains neutral with regard to jurisdictional claims in published maps and institutional affiliations.

This Springer imprint is published by the registered company Springer Nature Switzerland AG.
The registered company address is: Gewerbestrasse 11, 6330 Cham, Switzerland

Dedicated to my family:
Bridget, Juliana, and Alex.

Preface

In recent years, intersection homology and perverse sheaves have become indispensable tools for studying the topology of singular spaces. This book provides a gentle introduction to these concepts, with an emphasis on geometric examples and applications.

Part of the motivation for the development of intersection homology is that the main results and properties of manifolds (such as Poincaré duality, existence of multiplicative characteristic class theories, Lefschetz-type theorems and Hodge theory for complex algebraic varieties, Morse theory, etc.) fail to be true for singular spaces when considering usual homology. Intersection homology was introduced by M. Goresky and R. MacPherson in 1974 for the purpose of recovering such properties and results when dealing with singular spaces.[1]

The first part of these notes provides an elementary introduction of intersection homology and some of its applications. We first recall the results and main properties in the manifold case and then show to what extent the intersection homology allows to recover these properties in the singular case. The guiding principle of these notes is to provide an explicit and geometric introduction of the mathematical objects that are defined in this context, as well as to present some of the most significant examples.

The basic idea of intersection homology is that if one wants to recover classical properties of homology (e.g., Poincaré duality) in the case of singular spaces, one has to consider only cycles that meet the singularities with a controlled defect of transversality (encoded by a perversity function). This approach is explained in Chapter 2. As an application of Poincaré duality, in Chapter 3, we explain how the duality pairing on the middle-perversity intersection homology groups and its associated signature invariant can be used to construct characteristic L-classes in the singular setting.

---

[1]In fact, according to [129], Goresky and MacPherson were initially seeking a theory of characteristic numbers for complex analytic varieties and other singular spaces.

One of the most important (and intriguing) properties of intersection homology is the local calculus. In fact, this is what distinguishes intersection homology theory from classical homology theory, in the sense that intersection homology is not a homotopy invariant. The local calculus also provides the transition to sheaf theory and motivates the second part of these notes, which presents a sheaf-theoretic description of intersection homology.

A second definition of intersection homology makes use of sheaf theory and homological algebra, and it was introduced by Goresky and MacPherson in [83], following a suggestion of Deligne. In Chapters 4 and 5, we develop the necessary background on sheaves needed to define Deligne's intersection cohomology complex, whose (hyper)cohomology computes the intersection homology groups. This complex of sheaves, introduced in Chapter 6, can be described axiomatically in a way that is independent of the stratification or any additional geometric structure (such as a piecewise linear structure), leading to a proof of the topological invariance of intersection homology groups.

In complex algebraic geometry, the middle-perversity Deligne intersection cohomology complex is a basic example of a perverse sheaf. Perverse sheaves are fundamental objects at the crossroads of topology, algebraic geometry, analysis, and differential equations, with notable applications in number theory, algebra, and representation theory. For instance, perverse sheaves have seen striking applications in representation theory (proof of the Kazhdan–Lusztig conjecture, proof of the fundamental lemma in the Langlands program, etc.), and in geometry and topology (the BBDG decomposition theorem). They also form the backbone of Saito's mixed Hodge module theory. However, despite their fundamental importance, perverse sheaves remain rather mysterious objects. After a quick introduction of the theory of constructible sheaves in complex algebraic geometry (Chapter 7), we present a down-to-earth treatment of the deep theory of perverse sheaves (Chapter 8), with an emphasis on basic geometric examples. Of particular importance here is Artin's vanishing theorem for perverse sheaves on complex affine varieties, which plays an essential role in proving the Lefschetz hyperplane section theorem for intersection homology in the subsequent chapter.

Chapter 9 is devoted to what is usually referred to as the "decomposition package," consisting of Lefschetz-type results for perverse sheaves and intersection homology (Sections 9.1 and 9.2), as well as the BBDG decomposition theorem (Section 9.3). The Beilinson–Bernstein–Deligne–Gabber decomposition theorem is one of the most important results of the theory of perverse sheaves, and it contains as special cases some of the deepest homological and topological properties of algebraic maps. Since its proof in 1981, the decomposition theorem has found spectacular applications in algebraic topology and geometry, number theory, representation theory, and combinatorics. In Section 9.3, we give a brief overview of the motivation and the main ideas of its proof and discuss some of its immediate consequences. Furthermore, in Section 9.4, we sample several of the numerous applications of the decomposition package. We begin with a computation of topological invariants of Hilbert schemes of points on a surface and then move to combinatorial applications and overview Stanley's proof of McMullen's conjecture (about a complete characterization of face vectors of simplicial polytopes) as well as Huh–Wang's recent resolution of the Dowling–Wilson top-heavy conjecture (on the enumeration of subspaces of a projective space generated by a finite set of points).

In Chapter 10, we indicate several applications of perverse sheaves to the study of local and global topology of complex hypersurface singularities. In Section 10.1, we give a brief overview of the local topological structure of hypersurface singularities. Global topological aspects of complex hypersurfaces and of their complements are discussed in Section 10.2 by means of Alexander-type invariants inspired by knot theory. The nearby and vanishing cycle functors, introduced in Section 10.3, are used to glue the local topological data around singularities into constructible complexes of sheaves. We also discuss here the interplay between nearby/vanishing cycles and perverse sheaves. Very concrete applications of the nearby and vanishing cycles are presented in Section 10.4 (to the computation of Euler characteristics of complex projective hypersurfaces), in Section 10.5 (for obtaining generalized Riemann–Hurwitz-type formulae), and in Section 10.6 (for deriving homological connectivity statements for the local topology of complex singularities).

Chapter 11 gives a quick introduction of Saito's theory of mixed Hodge modules, with an emphasis on concrete applications to Hodge-theoretic aspects of intersection homology. Mixed Hodge modules are extensions in the singular context of variations of mixed Hodge structures and can be regarded, informally, as sheaves of mixed Hodge structures. Section 11.1 reviews some of the main concepts and results from the classical mixed Hodge theory, due to Deligne. In Section 11.2, we discuss the basic calculus of mixed Hodge modules and discuss some basic examples. In Section 11.3, we explain how to use Saito's mixed Hodge module theory to construct mixed Hodge structures on the intersection cohomology groups of complex algebraic varieties and, respectively, of links of closed subvarieties. We also show that the generalized Poincaré duality isomorphism in intersection homology is compatible with these mixed Hodge structures.

Each of the main actors of these notes, namely, intersection homology, perverse sheaves, and mixed Hodge modules, is at the center of a large and growing subject, touching on many aspects of modern mathematics. As a consequence, there is a vast research literature. In the *Epilogue*, we provide a succinct summary of (and references for) some of the recent applications of these theories (other than those already discussed in earlier chapters) in various research fields such as topology, algebraic geometry, representation theory, and number theory. This list of applications is by no means exhaustive but rather reflects the author's own research interests and mathematical taste. While the discussion will be limited to a small fraction of the possible routes the interested reader might explore, it should nevertheless serve as a starting point for those interested in aspects of intersection homology and perverse sheaves in other areas than those already considered in the text.

This book is intended as a broadly accessible first introduction to intersection homology and perverse sheaves, and it is far from comprehensive. In order to keep the size of the material within a reasonable level, many important results are stated without proof (whenever this is the case, a reference is provided), while some of their applications are emphasized. The goal of these notes is not necessarily to introduce the readers to the general abstract theory but to provide them with a taste of the subject by presenting concrete examples and applications that motivate the theory. At the end of this journey, readers should feel comfortable enough to delve further into more specialized topics and to explore problems of current research. For more complete details and further reading, the interested reader is advised to consult standard references such as [6, 15, 61, 75, 122, 214]. For a nice account on the history of intersection homology and its connections with various problems in mathematics, see [129]. For excellent overviews of perverse sheaves and their many applications, the two ICM addresses [146] and [149], as well as the more recent [51], are highly recommended. While the text presented here has a sizable (and unavoidable) overlap with some of the above-mentioned references (especially on background material and classical aspects of the theory), it also complements them in terms of the range of applications and/or the level of detail.

Throughout these notes, we assume the reader has a certain familiarity with the basic concepts of algebraic topology and algebraic geometry. While many of the relevant notions are still defined in the text (often in the form of footnotes), the novice reader is expected to consult standard textbooks on these subjects, such as [91, 97, 98].

Acknowledgments These notes grew out of lectures given by the author at the University of Illinois at Chicago; the University of Wisconsin-Madison; the University of Science and Technology of China (USTC) in Hefei, China; and the Chinese University of Hong Kong (CUHK). In particular, I would like to thank my hosts, Xiuxiong Chen at USTC and Conan Leung at CUHK, for their excellent working conditions. I also thank my students, colleagues, and collaborators for their valuable feedback, and I thank Mr. Okan Akalin for helping me with drawing the pictures. Special thanks go to Jörg Schürmann, Julius Shaneson, Sylvain Cappell, Alex Dimca, Markus Banagl, Botong Wang, Yongqiang Liu, Greg Friedmann, Shoji Yokura, Mark Andrea de Cataldo, and David Massey for many enlightening discussions and for reading parts of earlier versions of the manuscript.

During the writing of this book I was partially supported by the Simons Foundation Collaboration Grant #567077 and by the Romanian Ministry of National Education, CNCS-UEFISCDI, grant PN-III-P4-ID-PCE-2016-0030.

Madison, WI, USA   Laurenţiu G. Maxim
November 2018

Contents

List of Figures

Chapter 1
Topology of Singular Spaces: Motivation, Overview

In this chapter, we overview the main results and properties of the (co)homology of manifolds, and show in examples that these results fail to be true for singular spaces. This motivates the use of intersection homology, which recovers the corresponding results in the singular context.

Complex algebraic (or analytic) varieties are major examples of singular spaces, and provide a convenient testing ground for topological theories. Most examples considered here are complex algebraic/analytic varieties, regarded as topological spaces with respect to their complex analytic topology.

1.1 Poincaré Duality

Manifolds have an amazing hidden symmetry, called *Poincaré Duality*, which ultimately is reflected in the equality of ranks of (co)homology groups in complementary degrees. As we shall see in the examples below, singular spaces do not possess such symmetry in general.

Before recalling the statement of Poincaré duality, we make a few definitions.

Definition 1.1.1 A *topological n-manifold* (without boundary) is a Hausdorff space X such that for every $x \in X$, there is a neighborhood U of x homeomorphic to an open ball in $\mathbb{R}^n$. A compact topological manifold with no boundary is said to be *closed*.

For an n-manifold, the *local homology groups* $H_i(X, X - x; \mathbb{Z})$ at $x \in X$ are computed by excising the complement of a small neighborhood of x as follows:

$$H_i(X, X - x; \mathbb{Z}) \cong H_i(\mathbb{R}^n, \mathbb{R}^n - 0; \mathbb{Z})$$

$$\cong \tilde{H}_{i-1}(\mathbb{R}^n - 0; \mathbb{Z})$$

© Springer Nature Switzerland AG 2019
L. G. Maxim, *Intersection Homology & Perverse Sheaves*, Graduate Texts in Mathematics 281, https://doi.org/10.1007/978-3-030-27644-7_1

$$\cong \tilde{H}_{i-1}(S^{n-1}; \mathbb{Z})$$

$$\cong \begin{cases} \mathbb{Z}, & i = n, \\ 0, & i \neq n. \end{cases}$$

Definition 1.1.2 An *orientation* of a topological n-manifold X is a continuous choice of a generator of $H_n(X, X - x; \mathbb{Z})$, as x varies through X. If an orientation exists, X is said to be *oriented*.

It is known that if X is a closed, oriented, connected topological n-manifold, then $H_n(X; \mathbb{Z}) = \mathbb{Z}$. The generator of this group is called the *fundamental class* of X, and it is denoted by $[X]$.

For stating the Poincaré duality isomorphism, we also need to recall the notion of *cap product* on an n-manifold X, that is,

$$C^i(X) \times C_n(X) \overset{\frown}{\longrightarrow} C_{n-i}(X),$$

where C_i and C^i denote the (simplicial/singular) i-(co)chains on X with $\mathbb{Z}$-coefficients. The cap product is defined as follows: if $a \in C^{n-i}(X)$, $b \in C^i(X)$, and $\sigma \in C_n(X)$ then

$$a(b \frown \sigma) = (a \smile b)(\sigma).$$

The cap product is compatible with the boundary maps, thus it descends to a map

$$H^i(X; \mathbb{Z}) \times H_n(X; \mathbb{Z}) \overset{\frown}{\longrightarrow} H_{n-i}(X; \mathbb{Z}).$$

The following statement lies at the heart of algebraic and geometric topology. For a modern proof see, e.g., [98, Section 3.3]:

Theorem 1.1.3 (Poincaré Duality) *Let X be a closed, connected, oriented topological n-manifold with fundamental class $[X]$. Then capping with $[X]$ gives an isomorphism*

$$H^i(X; \mathbb{Z}) \overset{\cong}{\longrightarrow} H_{n-i}(X; \mathbb{Z}),$$

for all integers i.

As a consequence of Theorem 1.1.3 one gets a non-degenerate pairing

$$H_i(X, \mathbb{C}) \otimes H_{n-i}(X; \mathbb{C}) \longrightarrow \mathbb{C}.$$

In particular, the Betti numbers of X in complementary degrees coincide, i.e.,

$$\dim_{\mathbb{C}} H_i(X; \mathbb{C}) = \dim_{\mathbb{C}} H_{n-i}(X; \mathbb{C}). \tag{1.1}$$

Poincaré's original formulation of his duality statement was an equality of complementary Betti numbers. His proof, involving dual triangulation, was given in [196, 197]. Poincaré duality did not take on its modern form of Theorem 1.1.3 until the advent of cohomology in the 1930s, when Čech and Whitney invented the cup and cap products and formulated Poincaré duality in these new terms; see, for example, [58].

However, Poincaré duality fails in general for singular spaces, as the next example shows.

Example 1.1.4 Let

$$X = \{x_0 x_1 = 0\} \subset \mathbb{C}P^2 = \{[x_0 : x_1 : x_2]\}.$$

Topologically,

$$X = \{x_0 = 0\} \cup \{x_1 = 0\}$$

is the union of two copies of $\mathbb{C}P^1 \cong S^2$ intersecting at the point $p = [0 : 0 : 1]$. The point p is the unique singular point of X (this can be seen both algebraically by solving for the critical points of the defining equation, as well as topologically by noting that the local homology group $H_2(X, X - p; \mathbb{Z})$ at p is not $\mathbb{Z}$; in fact, $H_2(X, X - p; \mathbb{Z}) \cong \mathbb{Z} \oplus \mathbb{Z}$). Poincaré duality does not hold for X since, in the notations of Figure 1.1, we get by a simple Mayer–Vietoris argument that:

$$H_0(X; \mathbb{C}) = \mathbb{C} = \langle [a] \rangle = \langle [b] \rangle, \quad H_1(X; \mathbb{C}) = 0, \quad H_2(X; \mathbb{C}) = \mathbb{C} \oplus \mathbb{C}.$$

Note that $[a] = [b]$ since there is a 1-chain $\delta \in C_1(X)$ whose boundary is $\partial \delta = b - a$.

Remark 1.1.5 One of the goals of this book is to introduce a new homology theory, the (middle-perversity) *intersection homology* IH_*, which is a theory of "allowable chains," so that in the above example one has $[a] \neq [b]$ in $IH_0(X; \mathbb{C})$. It will then follow that

$$IH_0(X; \mathbb{C}) = \mathbb{C} \oplus \mathbb{C} = \langle [a], [b] \rangle$$

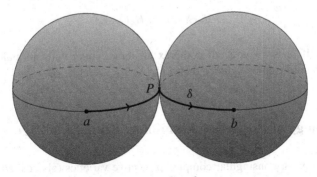

Fig. 1.1 $S^2 \vee S^2$

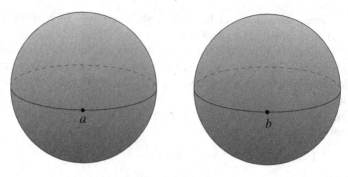

Fig. 1.2 $S^2 \sqcup S^2$

and

$$IH_2(X; \mathbb{C}) = \mathbb{C} \oplus \mathbb{C}$$

(so at least the above obstruction to Poincaré duality shall be removed). In this new theory, we do not "allow" 1-chains to pass through the singular point, so the 1-chain δ connecting the 0-cycles $[a]$ and $[b]$ will not be allowed in the new theory.

Alternatively, in the above example we can *normalize* X (see Section 2.4 for a definition), to get the normal space $\widetilde{X} = S^2 \sqcup S^2$, a disjoint union of two copies of S^2, with

$$H_0(\widetilde{X}; \mathbb{C}) = H_2(\widetilde{X}; \mathbb{C}) = \mathbb{C} \oplus \mathbb{C},$$

see Figure 1.2.

As it will be seen later on (in Chapter 2), in general the (middle-perversity) intersection homology theory satisfies the following properties:

(a) If $\widetilde{X}$ is a normalization of X, then

$$IH_*(X; \mathbb{Z}) \cong IH_*(\widetilde{X}; \mathbb{Z}).$$

(b) If X is a manifold (e.g., a nonsingular complex algebraic variety), then

$$IH_*(X; \mathbb{Z}) \cong H_*(X; \mathbb{Z}).$$

In particular, if X is a (reduced) complex algebraic curve with normalization $\widetilde{X}$ (which in this case is nonsingular), then $IH_*(X; \mathbb{Z}) \cong H_*(\widetilde{X}; \mathbb{Z})$.

1.2 Topology of Projective Manifolds: Kähler Package

The cohomology of nonsingular complex projective varieties (also called complex *projective manifolds*, in order to emphasize the use of complex topology) inherits additional properties from the complex algebraic structure. In addition to being a vector space, each complex cohomology group is endowed with a *Hodge structure*, which in turn imposes additional topological and analytical constraints on the topological space itself.

Hodge Decomposition

Let X be a nonsingular complex projective variety of (complex) dimension n. Then one has the *Hodge decomposition*, see [105] or [91]:

$$H^i(X;\mathbb{C}) \cong H^i_{DR}(X) \cong \bigoplus_{p+q=i} H^{p,q}(X),$$

where $H^i_{DR}(X)$ denotes the *de Rham cohomology* of X, $H^i(X;\mathbb{C}) = H^i(X;\mathbb{R}) \otimes \mathbb{C}$, and $H^{q,p}(X) = \overline{H^{p,q}(X)}$. More precisely, every element in the de Rham cohomology $H^i_{DR}(X)$ has a unique harmonic representative with respect to the induced Fubini–Study metric (coming from a projective embedding), which can be expressed as a sum of (p,q)-forms; in local coordinates $z_1,\ldots,z_n$ on X a (p,q)-form can be written as

$$\alpha \cdot dz_{i_1} \wedge dz_{i_2} \wedge \cdots \wedge dz_{i_p} \wedge d\bar{z}_{j_1} \wedge \cdots \wedge d\bar{z}_{j_q},$$

with α a smooth function and $i_1 < \cdots < i_p$, $j_1 < \cdots < j_q$. This Hodge decomposition of $H^i(X;\mathbb{C})$ is independent of the choice of the Fubini–Study metric, i.e., of the choice of a projective embedding. We say that $H^i(X;\mathbb{C})$ has a *pure Hodge structure of weight i*. (See Section 11.1 for more background on Hodge theory.)

An important consequence of the existence of Hodge structures on the cohomology of complex projective manifolds is the following:

Corollary 1.2.1 *The odd Betti numbers of a complex projective manifold are even.*

Example 1.2.2 Let X be the variety

$$X = \{x_0^3 + x_1^3 = x_0 x_1 x_2\} \subset \mathbb{C}P^2.$$

The singular locus of this complex projective variety is

$$\mathrm{Sing}(X) = \{P = [0:0:1]\}.$$

In the notations of Figure 1.3, we have

$$H_1(X;\mathbb{C}) = \mathbb{C} = \langle [\eta] \rangle,$$

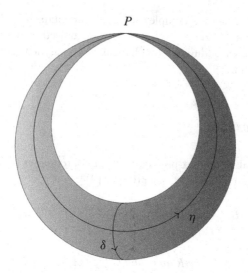

Fig. 1.3 X

where η is a longitude in X. Note that the meridian δ is a boundary in X, so its homology class vanishes. As the first Betti number of X is odd, there cannot exist a Hodge decomposition for $H^1(X; \mathbb{C})$.

Remark 1.2.3 In the new theory IH_* of allowable chains on the space X of Figure 1.3, 1-cycles are not allowed to go through the singular point P, but 2-cycles are allowed, so $[\delta] = 0$ in $IH_1(X; \mathbb{C})$. Therefore,

$$IH_1(X; \mathbb{C}) = 0.$$

In particular, the first intersection homology Betti number of X is even, so the above obstruction to the existence of a Hodge structure on $IH_1(X; \mathbb{C})$ is removed.

Remark 1.2.4 As it will be seen later on in Chapter 11, the (middle-perversity) intersection (co)homology groups of a (possibly singular) complex projective variety have pure Hodge structures. On the other hand, it is known that the usual cohomology of a singular complex algebraic variety carries Deligne's mixed Hodge structure.

Lefschetz Hyperplane Section Theorem

Let $X^n \subseteq \mathbb{C}P^N$ be a nonsingular complex projective variety of complex dimension n, and let H be a generic hyperplane in $\mathbb{C}P^N$. The *Lefschetz hyperplane section theorem* asserts that the homomorphism

$$H^i(X; \mathbb{C}) \longrightarrow H^i(X \cap H; \mathbb{C})$$

induced by restriction is an isomorphism for $i < n - 1$, and it is injective if $i = n - 1$. It was originally proved by Lefschetz [139], and recast by Andreotti–Frankel [2] using Morse theory (see also [178, Section 7]).

As the next example shows, the Lefschetz hyperplane section theorem fails to be true in general for singular complex projective varieties.

Example 1.2.5 Let $X = \mathbb{C}P^2 \cup \mathbb{C}P^2 \subset \mathbb{C}P^4 = \{[x_0 : x_1 : \cdots : x_4]\}$, where the two copies of $\mathbb{C}P^2$ in X meet at a point. Algebraically,

$$X = \{x_i x_j = 0 \mid i \in \{0, 1\}, j \in \{3, 4\}\},$$

with $\mathrm{Sing}(X) = \{[0 : 0 : 1 : 0 : 0]\}$. A simple computation based on the Mayer–Vietoris sequence yields that:

$$H^i(X; \mathbb{C}) = \begin{cases} \mathbb{C}, & i = 0, \\ 0, & i = 1, \\ \mathbb{C} \oplus \mathbb{C}, & i = 2, \\ 0, & i = 3, \\ \mathbb{C} \oplus \mathbb{C}, & i = 4. \end{cases}$$

If H is a generic hyperplane in $\mathbb{C}P^4$, then $X \cap H = \mathbb{C}P^1 \sqcup \mathbb{C}P^1$ consists of two disjoint copies of $\mathbb{C}P^1 \cong S^2$ (see Figure 1.4), hence

$$H^i(X \cap H; \mathbb{C}) = \begin{cases} \mathbb{C} \oplus \mathbb{C}, & i = 0, 2, \\ 0, & \text{otherwise.} \end{cases}$$

Notice that for $n = 2$ and $i = 0$, we have that:

$$H^0(X; \mathbb{C}) = \mathbb{C} \not\cong \mathbb{C} \oplus \mathbb{C} = H^0(X \cap H; \mathbb{C}),$$

so the Lefschetz hyperplane section theorem does not hold for the singular variety X of this example.

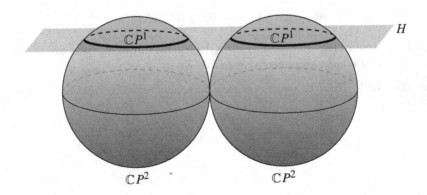

Fig. 1.4 X

Remark 1.2.6 The Lefschetz hyperplane section theorem implies the connectedness of a generic hyperplane section in dimension $n \geq 2$. The previous example shows that this property can also fail in the singular case.

Hard Lefschetz Theorem

Let X be a nonsingular complex projective variety of complex dimension n, and let H be a generic hyperplane. The intersection $X \cap H$ yields a homology class $[X \cap H] \in H_{2n-2}(X; \mathbb{Z})$, and its Poincaré dual is a degree-two cohomology class, denoted by $[H] \in H^2(X; \mathbb{Z})$. The *Lefschetz operator* is the map

$$L : H^i(X; \mathbb{C}) \xrightarrow{\smile [H]} H^{i+2}(X; \mathbb{C})$$

defined by taking the cup product with $[H]$. Then the following important result holds:

Theorem 1.2.7 (Hard Lefschetz Theorem) *The map*

$$L^i : H^{n-i}(X; \mathbb{C}) \xrightarrow{\smile [H]^i} H^{n+i}(X; \mathbb{C})$$

is an isomorphism, for all integers $i \geq 0$.

The Hard Lefschetz theorem was initially formulated in [139] (first published in 1924), but so far no one has succeeded in making Lefschetz's intuitive arguments precise. The first complete proof of the Hard Lefschetz theorem has been given by Hodge in his book [105] using harmonic differential forms (see also [91, Page 122]).

A simple consequence of Theorem 1.2.7 is the following symmetry of Betti numbers (which can also be deduced from Poincaré duality):

Corollary 1.2.8 *Let X be an n-dimensional nonsingular complex projective variety. Then, for all $i \geq 0$, we have*

$$\dim_{\mathbb{C}} H^{n-i}(X; \mathbb{C}) = \dim_{\mathbb{C}} H^{n+i}(X; \mathbb{C}).$$

Another application of Theorem 1.2.7 is the *unimodality* of the Betti numbers of an n-dimensional complex projective manifold X:

Corollary 1.2.9 *The Betti numbers of a n-dimensional complex projective manifold satisfy*

$$\dim_{\mathbb{C}} H^{i-2}(X; \mathbb{C}) \leq \dim_{\mathbb{C}} H^i(X; \mathbb{C})$$

for $i \leq n/2$.

Proof Indeed, Theorem 1.2.7 implies that the map $L : H^{i-2}(X; \mathbb{C}) \to H^i(X; \mathbb{C})$ is injective in the desired range. □

Example 1.2.10 In the previous Example 1.2.5, let $n = 2$ and $i = 2$. Then

$$H^0(X; \mathbb{C}) = \mathbb{C} \ncong \mathbb{C} \oplus \mathbb{C} = H^4(X; \mathbb{C}),$$

so the Hard Lefschetz theorem fails for the singular space X. Poincaré duality also fails for Example 1.2.5, as it can be easily seen from the above calculation of its cohomology.

As it will be seen later on in Chapters 9 and 11, all these results continue to hold true for singular complex projective varieties provided that cohomology is replaced by the (middle-perversity) intersection cohomology.

In order to emphasize the importance of the Kähler package, we conclude this introductory chapter with an interesting combinatorial application of the unimodality of Betti numbers of complex projective manifolds (Corollary 1.2.9); see [222] for the relevant references. Let $X = \mathbf{G}_d(\mathbb{C}^n)$ be the Grassmann variety of d-planes in $\mathbb{C}^n$; this is a complex projective manifold of complex dimension $d(n-d)$. It was shown by Ehresmann (see also [91, Chapter 1, Section 5]) that the odd Betti numbers of X are all zero (in fact, X has an algebraic cell decomposition by complex affine spaces, so all of its cells appear in even real dimensions), whereas the even Betti numbers are computed as

$$\dim_{\mathbb{C}} H^{2i}(X; \mathbb{C}) = p(i, d, n-d),$$

where $p(i, d, n-d)$ is the number of partitions of the integer i into $\leq d$ parts, with largest part $\leq n-d$ (i.e., partitions whose Young diagrams fit inside a $d \times (n-d)$ box). In particular, Corollaries 1.2.8 and 1.2.9 imply that the sequence

$$p(0, d, n-d), p(1, d, n-d), \cdots, p(d(n-d), d, n-d)$$

is symmetric and unimodal. For more combinatorial applications of the unimodality of (intersection homology) Betti numbers in the singular context, the reader is referred to Section 9.4.

Chapter 2
Intersection Homology: Definition, Properties

In this chapter, we introduce intersection homology from a chain-theoretic perspective (as originally developed by Goresky–MacPherson [81]). For a more comprehensive account of this approach, the reader is advised to consult Friedman's book [75], as well as [15] and [6].

2.1 Topological Pseudomanifolds

Intersection homology can be defined for a wide class of singular spaces called *topological pseudomanifolds*. Let us begin with a few definitions.

Definition 2.1.1 If L is a compact Hausdorff space, then the *open cone* on L is defined as (Fig. 2.1)

$$\mathring{c}L = L \times [0, 1) / L \times \{0\}.$$

Definition 2.1.2 A *topologically stratified space* is defined by induction on dimension as follows:

(i) A 0-dimensional topologically stratified space is a countable set of points with discrete topology.
(ii) For $n > 0$, an n-dimensional topologically stratified space is a Hausdorff topological space with a filtration

$$X = X_n \supseteq X_{n-1} \supseteq \ldots \supseteq X_1 \supseteq X_0 \supseteq X_{-1} = \emptyset$$

by closed subspaces X_j, so that the following *local normal triviality* condition is satisfied: if $x \in X_j - X_{j-1}$, there is a neighborhood U_x of x in X and a compact $n - j - 1$ dimensional topologically stratified space L with a filtration

© Springer Nature Switzerland AG 2019
L. G. Maxim, *Intersection Homology & Perverse Sheaves*, Graduate Texts in Mathematics 281, https://doi.org/10.1007/978-3-030-27644-7_2

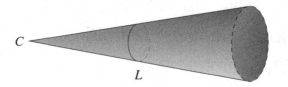

Fig. 2.1 $\mathring{c}L$

$$L = L_{n-j-1} \supseteq \ldots \supseteq L_1 \supseteq L_0 \supseteq \emptyset,$$

and a homeomorphism

$$\phi : U_x \to \mathbb{R}^j \times \mathring{c}L$$

such that ϕ takes $U_x \cap X_{j+i+1}$ homeomorphically onto $\mathbb{R}^j \times \mathring{c}L_i$ when $n - j - 1 \geq i \geq 0$, and $\phi : U_x \cap X_j \xrightarrow{\sim} \mathbb{R}^j \times \{x\}$ is a homeomorphism.

Definition 2.1.3 In the above definition:

(i) L is called the *link* of the connected component of $X_j - X_{j-1}$ containing x. (But see [75, Section 2.3] for a discussion around "the" link of a point.)
(ii) The connected components of $X_j - X_{j-1}$ are called j-dimensional *strata* of X.

As a consequence of Definition 2.1.2, one has the following:

Proposition 2.1.4 *Let X be a topologically stratified space. Then, in the notations of Definition 2.1.2, every nonempty $X_j - X_{j-1}$ is a j-dimensional topological manifold, and X is locally normal trivial along the strata.*

Remark 2.1.5 Up to homeomorphism, the link L of $x \in X$ depends only on the connected component of the stratum $X_j - X_{j-1}$ that contains x.

Remark 2.1.6 From a historical perspective, an old idea (already implicit in the notion of a simplicial complex) was to study a singular space by decomposing it into smooth pieces (the strata). Whitney [243, 242] was the first to point out that a good stratification should satisfy certain regularity conditions along strata (the famous "Whitney conditions (a) and (b)"). Topologically stratified spaces provide a purely topological setting for the study of singularities, analogous to the more differential-geometric theory of Whitney. They were introduced by Thom [232], who showed (by using his first isotopy lemma) that every Whitney stratified space was also a topologically stratified space, with the same strata. Another proof of this fact was given by Mather [156].

Definition 2.1.7 A *topological pseudomanifold* of dimension n is a topologically stratified space X with a filtration satisfying $X_{n-1} = X_{n-2}$ and so that $X - X_{n-2}$ is dense in X. The collection of strata of X will be denoted by $\mathfrak{X}$, and it will be referred to as a *pseudomanifold stratification* of X.

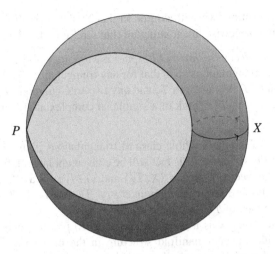

Fig. 2.2 Not a pseudomanifold

Example 2.1.8

1. Every manifold is a topological pseudomanifold.
2. The cone $\mathring{c}M$ of a manifold M is a topological pseudomanifold when the dimension of M is positive.
3. The wedge sum of two 2-spheres is a topological pseudomanifold.
4. Complex algebraic varieties admit *Whitney stratifications* (e.g., see [242, 233]) that, as already mentioned in Remark 2.1.6, make them into topologically stratified spaces. Moreover, complex algebraic varieties of pure dimension are topological pseudomanifolds (see [232, 156], and also [15, IV, Section 2]).
5. The open cone on three points is *not* a pseudomanifold.
6. A pinched torus with a central disc attached to it (see Figure 2.2) is *not* a pseudomanifold.

2.2 Borel–Moore Homology

Before introducing intersection homology, we recall the definition of the locally finite (Borel–Moore) homology, see [16]. This homology theory is relevant in the context of Poincaré duality for non-compact spaces. Indeed, in the non-compact case, Poincaré duality for an n-dimensional oriented manifold X yields isomorphisms

$$H_c^i(X; \mathbb{Z}) \cong H_{n-i}(X; \mathbb{Z})$$

and

$$H^i(X; \mathbb{Z}) \cong H_{n-i}^{BM}(X; \mathbb{Z}),$$

where $H_*^{BM}(X; \mathbb{Z})$ denotes the *Borel–Moore homology* of X, which will be defined below.

For ease of exposition, we will assume in this chapter that all spaces involved have an underlying piecewise linear structure (but see Remark 2.3.12 below).

Definition 2.2.1 A *piecewise linear (PL) space* is a topological space X with a class $\mathcal{T}$ of locally finite triangulations such that for any triangulation T of X in the class, every subdivision of T lies again in $\mathcal{T}$, and any two triangulations $T, T' \in \mathcal{T}$ have a common refinement in $\mathcal{T}$. (Think of a simplicial complex along with all possible refinements.)

The advantage of having a whole class of triangulations is that every open set $U \subset X$ inherits a PL-structure. This fact will be convenient later on for constructing sheaves. A map between PL-spaces $(X, \mathcal{T}_X)$ and $(Y, \mathcal{T}_Y)$, or a *PL-map*, is a map $X \to Y$ for which there are triangulations $T_X \in \mathcal{T}_X$ and $T_Y \in \mathcal{T}_Y$ such that the image of every simplex of T_X lies inside a simplex of T_Y.

Assume for now that X is a *piecewise linear (PL) pseudomanifold*, namely it is a PL-space with a pseudomanifold structure in the piecewise linear category (i.e., X has a triangulation such that each X_j is a union of simplices and all homeomorphisms in the definition of a pseudomanifold are piecewise linear homeomorphisms). An important class of examples of such PL pseudomanifolds is provided by complex quasi-projective varieties. Indeed, for any given Whitney stratification of a complex quasi-projective variety X, there is a triangulation of X compatible with the stratification (see, e.g., [80]).

If $(X, \mathcal{T})$ is a PL-space, the i-th homology group of X is defined by

$$H_i(X; \mathbb{Z}) := H_i(C_\bullet(X)),$$

where $C_\bullet(X)$ is the chain complex of piecewise linear (PL) chains on X, defined as

$$C_k(X) := \varinjlim_{T \in \mathcal{T}} C_k^T(X).$$

Here the limit is over triangulations T of X of the sets $C_k^T(X)$ of *finite* simplicial k-chains in such a triangulation T. Similarly, we define

$$C_k((X)) := \varinjlim_{T \in \mathcal{T}} C_k^T((X)),$$

where we now consider *locally finite* chains $\zeta = \sum_\sigma \zeta_\sigma \cdot \sigma \in C_k^T((X))$ with $\zeta_\sigma \in \mathbb{Z}$ and σ a k-simplex in T.[1] The *support* of a (locally finite) chain ζ is defined as

---

[1] Recall that a formal linear combination $\zeta = \sum_\sigma \zeta_\sigma \cdot \sigma$ of singular k-simplices in X is a *locally finite* k-chain if for each $x \in X$ there is an open neighborhood U_x of x in X such that the set

$$\{\zeta_\sigma \mid \zeta_\sigma \neq 0, \sigma^{-1}(U_x) \neq \emptyset\}$$

is finite.

$$|\xi| := \bigcup_{\xi_\sigma \neq 0} \sigma,$$

and it is a closed subset of X (by the local finiteness of T).

Definition 2.2.2 The *Borel–Moore homology* (or *homology with closed supports*) of X is defined as

$$H_i^{BM}(X; \mathbb{Z}) := H_i(C_\bullet((X))).$$

Remark 2.2.3 If X is compact, then $C_i^T(X) = C_i^T((X))$ for every triangulation T of X, so $H_i^{BM}(X; \mathbb{Z}) = H_i(X; \mathbb{Z})$ for all i.

Remark 2.2.4 The Universal Coefficient Theorem (with field coefficients) yields isomorphisms (e.g., see [98, Section 3.1])

$$H^i(X; \mathbb{C}) \cong H_i(X; \mathbb{C})^\vee$$

and

$$H_c^i(X; \mathbb{C}) \cong H_i^{BM}(X; \mathbb{C})^\vee,$$

where $-^\vee$ denotes the operation of taking the dual of a vector space.

Remark 2.2.5 If $\overline{X}$ is a compactification of X with $D = \overline{X} - X$, then it can be shown that

$$H_c^i(X; \mathbb{Z}) \cong H^i(\overline{X}, D; \mathbb{Z}),$$

see (5.24) below. In particular, if $\widehat{X}$ is a one-point compactification of X by $\{\infty\}$, then

$$H_c^i(X; \mathbb{Z}) \cong H^i(\widehat{X}, \{\infty\}; \mathbb{Z}).$$

Similarly, for such spaces X, the Borel–Moore homology can be computed by:

$$H_i^{BM}(X; \mathbb{Z}) \cong H_i(\widehat{X}, \{\infty\}; \mathbb{Z}).$$

Example 2.2.6

$$H_i^{BM}(\mathbb{R}^n; \mathbb{Z}) = H_i(S^n, \infty; \mathbb{Z}) = \widetilde{H}_i(S^n; \mathbb{Z}) = \begin{cases} \mathbb{Z}, & i = n, \\ 0, & i \neq n. \end{cases}$$

2.3   Intersection Homology via Chains

Let X be an n-dimensional PL pseudomanifold. In particular,

(i) X is filtered by closed PL subsets satisfying

$$X = X_n \supseteq X_{n-1} = X_{n-2} \supseteq \cdots \supseteq X_0 \supseteq X_{-1} = \emptyset.$$

(ii) $X - X_{n-2}$ is dense in X.
(iii) X has locally a conical structure (in the normal direction to strata).

Let $(C_\bullet((X)), \partial)$ be the chain complex of PL locally finite chains on X.

Definition 2.3.1 A PL i-chain ξ is *transverse* to the stratification of X if

$$\dim(|\xi| \cap X_{n-k}) = i + (n - k) - n = i - k$$

for all $k \geq 2$. Let $C_\bullet^{tr}((X))$ denote the transverse locally finite PL chains.

As a precursor of intersection homology, McCrory [170, Theorem 5.2] proved the following result.

Theorem 2.3.2 (McCrory)

$$H_i(C_\bullet^{tr}((X))) \cong H^{n-i}(X; \mathbb{Z}).$$

Hence, if one could move every chain in X to make it homologous to a chain transverse to the stratification, then Poincaré duality would hold. However, as seen in Chapter 1, Poincaré duality fails for singular spaces, thus so does transversality. For example, the longitude generator of the first homology group H_1 of the pinched torus (see Example 1.2.2) passes through the singular point, so it cannot be made transversal to it.

In general, the cap product for an oriented PL pseudomanifold X is given by:

$$H^{n-i}(X; \mathbb{Z}) \xrightarrow{\cap [X]} H_i^{BM}(X; \mathbb{Z})$$

$$\downarrow \simeq \qquad\qquad\qquad \downarrow \simeq$$

$$H_i(C_\bullet^{tr}((X))) \qquad H_i(C_\bullet((X)))$$

with $[X] \in H_n^{BM}(X; \mathbb{Z})$ the fundamental class of X (see Definition 2.3.16 below).

In this section, we follow [81] to define intersection homology groups interpolating between $H_i(C_\bullet^{tr}((X)))$ and $H_i(C_\bullet((X)))$, i.e., factorizing the above map induced by the cap product. For this, one first needs to introduce a parameter $\overline{p}$ that measures the deviation of chains from transversality, which Goresky and MacPherson called a *perversity*. Associating to a perversity $\overline{p}$ a chain complex

$$IC_\bullet^{\overline{p}}((X)) \hookrightarrow C_\bullet((X)),$$

one gets a whole spectrum of groups $I^{BM}H_i^{\overline{p}}(X; \mathbb{Z})$ that, roughly speaking, range from $H^{n-i}(X; \mathbb{Z})$ (for $\overline{p} = \overline{0}$) to $H_i^{BM}(X; \mathbb{Z})$ (for $\overline{p} = \overline{t}$, the top-perversity).

First, it is clear that in order to make up for the above-mentioned lack of transversality in singular spaces, one needs to restrict how chains meet the singular locus $\Sigma = X_{n-2}$ of X. A first natural condition to impose on a PL i-chain $\xi \in C_i((X))$ with support $|\xi|$ is that

$$\dim(|\xi| \cap \Sigma) \leq \dim(|\xi|) - 2 = i - 2.$$

This condition says that every allowable i-chain must intersect the singular locus Σ transversally. For two-dimensional PL pseudomanifolds (e.g., the pinched torus), this suffices to define intersection homology. But for more complicated singularities, intersection homology also takes into account the singularities within the singular set and so forth, and hence additional intersection conditions are needed. These are all encoded in what is called a perversity.

Definition 2.3.3 ([81]) A *perversity* $\overline{p}$ is a function $\overline{p} : \mathbb{Z}_{(\geq 2)} \longrightarrow \mathbb{N}$ such that

(i) $\overline{p}(2) = 0$;
(ii) $\overline{p}(k) \leq \overline{p}(k+1) \leq \overline{p}(k) + 1$, for all integers $k \geq 2$.

For the purpose of defining intersection homology, the input of a perversity function is the (real) codimension of strata.

Example 2.3.4 The most common perversity functions are:

1. *Zero-perversity*: $\overline{0}(k) = 0$ for all $k \geq 2$.
2. *Top-perversity*: $\overline{t}(k) = k - 2$ for all $k \geq 2$.
3. *Lower-middle perversity*: $\overline{m}(k) = \lfloor \frac{k-2}{2} \rfloor$ for all $k > 2$.
4. *Upper-middle perversity*: $\overline{n}(k) = \lceil \frac{k-2}{2} \rceil$ for all $k > 2$.

Definition 2.3.5 Two perversities $\overline{p}, \overline{q}$ are called *complementary* if

$$\overline{p} + \overline{q} = \overline{t}.$$

Example 2.3.6 $\{\overline{0}, \overline{t}\}$ and $\{\overline{m}, \overline{n}\}$ are pairs of complementary perversities.

Definition 2.3.7 ([81]) Let X^n be an n-dimensional PL pseudomanifold with filtration

$$X = X_n \supseteq X_{n-1} = X_{n-2} \supseteq \cdots \supseteq X_0 \supseteq \emptyset.$$

A PL i-chain $\xi \in C_i((X))$ is called $\overline{p}$-allowable if the following conditions hold for all $k \geq 2$:

(a) $\dim(|\xi| \cap X_{n-k}) \leq i - k + \overline{p}(k)$,
(b) $\dim(|\partial \xi| \cap X_{n-k}) \leq i - k - 1 + \overline{p}(k)$.

Remark 2.3.8 The above definition uses $\mathbb{Z}$-coefficients. Similar considerations apply to arbitrary coefficients, e.g., a field.

Let $IC_i^{\overline{p}}((X))$ be the set of all $\overline{p}$-allowable PL i-chains on X. The boundary condition (b) is imposed so that one gets a chain subcomplex

$$\left(IC_\bullet^{\overline{p}}((X)), \partial \right) \subseteq C_\bullet((X)).$$

Similarly, one defines $IC_\bullet^{\overline{p}}(X) \subseteq C_\bullet(X)$, which are called $\overline{p}$-allowable finite chains.

Definition 2.3.9 ([81]) The perversity $\overline{p}$ intersection homology groups of X are defined as:

$$I^{BM} H_i^{\overline{p}}(X; \mathbb{Z}) := H_i \left(IC_\bullet^{\overline{p}}((X)) \right),$$

$$I H_i^{\overline{p}}(X; \mathbb{Z}) := H_i \left(IC_\bullet^{\overline{p}}(X) \right).$$

Remark 2.3.10 A locally finite version of intersection homology is convenient for sheafification, while a finite version of intersection homology is convenient for geometric intuition.

Remark 2.3.11 Here are some immediate observations based on the above definition of intersection homology:

1. If $\overline{p} \leq \overline{q}$, then $IC_\bullet^{\overline{p}} \subseteq IC_\bullet^{\overline{q}}$ so there is an induced map

$$I^{(BM)} H_i^{\overline{p}}(X; \mathbb{Z}) \to I^{(BM)} H_i^{\overline{q}}(X; \mathbb{Z})$$

   for all i. (Here we use the notation $I^{(BM)} H$ to indicate that the statement holds for both $I^{BM} H$ and $I H$.)
2. A PL i-chain ξ is dimensionally transverse to X_{n-k} if $\dim(|\xi| \cap X_{n-k}) \leq i - k$. So, $\overline{p}(k)$ measures the deviation from dimensional transversality of a $\overline{p}$-allowable i-chain.
3. Elements of $IC_\bullet^{\overline{0}}((X))$ are dimensionally transverse to all strata. On the other hand, $IC_\bullet^{\overline{t}}((X))$ consists of chains ξ satisfying only $\dim(|\xi| \cap X_{n-2}) \leq \dim(|\xi|) - 2$, and the same condition for $\partial \xi$.
4. A priori, the groups $I^{(BM)} H_i^{\overline{p}}(X; \mathbb{Z})$ depend on the choice of a PL structure and stratification. In fact, as we will see later on (Chapter 6), they are independent of such choices.

5. The definition of intersection homology groups does not use the cone-like structure of pseudomanifolds. Indeed, one can define intersection homology groups for any filtered PL space. The pseudomanifold structure is used to show that the intersection homology groups are topological invariants (not just invariants of the filtration).

6. If X is compact, then $I^{BM}H_*^{\overline{p}}(X; \mathbb{Z}) = IH_*^{\overline{p}}(X; \mathbb{Z})$ for every perversity function $\overline{p}$.

7. If X is manifold, then $I^{(BM)}H_*^{\overline{p}}(X; \mathbb{Z}) = H_*^{(BM)}(X; \mathbb{Z})$ for every perversity $\overline{p}$.

Remark 2.3.12 A *singular* version of intersection homology was developed by King [127] as follows (see also [75, Section 3.4] for a detailed account): a $\overline{p}$-allowable singular i-simplex on X is a singular i-simplex $\sigma : \Delta_i \longrightarrow X$ satisfying

$$\sigma^{-1}(X_{n-k} - X_{n-k-1}) \subseteq (i - k + \overline{p}(k))\text{-skeleton of } \Delta_i$$

for all $k \geq 2$. A singular i-chain is $\overline{p}$-allowable if it is a (locally finite) combination of $\overline{p}$-allowable singular i-simplices. In order to form a subcomplex of $\overline{p}$-allowable chains, one needs to also ask, just as in the simplicial context, that boundaries of $\overline{p}$-allowable singular chains are $\overline{p}$-allowable. King showed that the corresponding *singular intersection homology groups*, i.e., the homology groups defined by the complex of (locally finite) $\overline{p}$-allowable singular chains with $\overline{p}$-allowable boundaries, coincide with the ones defined by using simplices, provided X has a PL structure. So *one can define intersection homology for topological pseudomanifolds, independently of PL structures.* In this text, we will freely make use of both versions, and it should be clear from the context which version is used. However, as we will see in Corollary 6.3.10, intersection homology groups are independent of the underlying PL structure. So there is no harm in assuming tacitly for now that our pseudomanifolds have underlying PL-structures. Such an assumption will often simplify the exposition.

Remark 2.3.13 Unlike homology, intersection homology computed on a fixed triangulation is not preserved by simplicial subdivision. Nevertheless, intersection homology groups of a PL pseudomanifold X can still be computed simplicially, but with respect to a *full* triangulation T of X, i.e., there are group isomorphisms

$$I^{(BM)}H_i(X; \mathbb{Z}) \cong I^{(BM)}H_i^T(X; \mathbb{Z}),$$

with $I^{(BM)}H_i^T(X; \mathbb{Z})$ denoting the (locally finite) *simplicial* intersection homology groups corresponding to the full triangulation T, see [151, Appendix], or the more recent [75, Theorem 3.3.20]. Such full triangulations of a PL filtered space always exist, e.g., see [75, Lemma 3.3.19].

Let us now apply the definition of intersection homology on an easy example.

Example 2.3.14 Let $X := \Sigma(S^1 \sqcup S^1)$ be the suspension on a disjoint union of two circles, see Figure 2.3. Denote the two circles by A and B, with points $a \in A$ and

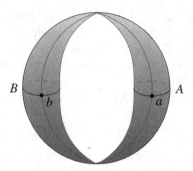

Fig. 2.3 $X = \Sigma(S^1 \sqcup S^1)$

$b \in B$. As in Figure 2.3, denote by cone(a) (resp., cone(b)) the path joining a (resp., b) to the top suspension point, and let susp(a) (resp., susp(b)) denote the geodesic path joining the two suspension points, which passes through a (resp., b). Denote by susp(A), susp(B) the two 2-spheres obtained by suspending the circles A and, resp., B. Then the homology groups of X are computed as:

1. $H_0(X; \mathbb{Z}) = \mathbb{Z} = \langle [a] \rangle = \langle [b] \rangle$, since $\partial(\text{cone}(a) - \text{cone}(b)) = b - a$.
2. $H_1(X; \mathbb{Z}) = \mathbb{Z} = \langle [\text{susp}(a) - \text{susp}(b)] \rangle$.
3. $H_2(X; \mathbb{Z}) = \mathbb{Z} \oplus \mathbb{Z} = \langle [\text{susp}(A)], [\text{susp}(B)] \rangle$.

Note that X can be considered as a pseudomanifold $X = X_2 \supset X_0$, where X_0 consists of two singular (suspension) points. To calculate intersection homology with, say middle-perversity $\overline{m}$, consider

$$IC_0^{\overline{m}}(X) = \left\{ \xi^0 \in C_0(X) : \dim(|\xi^0| \cap X_0) \leq 0 - 2 + \overline{m}(2) = -2 \right\},$$

so 0-chains are not allowed to intersect X_0. Similarly, one can see that 1-chains are not allowed to intersect X_0. Also,

$$IC_2^{\overline{m}}(X) = \left\{ \xi^2 \in C_2(X) : \begin{array}{l} \dim(|\xi^2| \cap X_0) \leq 2 - 2 + \overline{m}(2) = 0, \\ \dim(|\partial \xi^2| \cap X_0) \leq 2 - 2 - 1 + \overline{m}(2) = -1 \end{array} \right\}$$

so 2-chains can intersect X_0, but their boundaries cannot. Therefore,

1. $IH_0^{\overline{m}}(X; \mathbb{Z}) = \mathbb{Z} \oplus \mathbb{Z} = \langle [a], [b] \rangle$, since the 1-chain cone(a) − cone(b) (which passes through X_0) is not allowed.
2. $IH_1^{\overline{m}}(X; \mathbb{Z}) = 0$, since the 1-chain susp(a) − susp(b) is not allowed and cone(A), cone(B) are allowed 2-chains whose boundaries do not intersect X_0, so A and B are boundaries of intersection chains.
3. $IH_2^{\overline{m}}(X; \mathbb{Z}) = \mathbb{Z} \oplus \mathbb{Z} = \langle [\text{susp}(A)], [\text{susp}(B)] \rangle$.

Exercise 2.3.15 Calculate the middle-perversity intersection homology groups $IH_i^{\overline{m}}(\Sigma T^2; \mathbb{Z})$ of the suspension ΣT^2 of a torus.

Definition 2.3.16 A PL pseudomanifold X^n is said to be *orientable* if for some triangulation T of X, one can orient each n-simplex so that the n-chain with coefficient $1 \in \mathbb{Z}$ for each n-simplex with the chosen orientation is an n-cycle. A choice of such an n-cycle is called an *orientation* for X, and its homology class $[X] \in H_n^{BM}(X; \mathbb{Z})$ is the *fundamental class* of X. Equivalently, a PL pseudomanifold X^n is orientable (resp., oriented) if $X - X_{n-2}$ is.

Proposition 2.3.17 *Let X be an n-dimensional PL oriented pseudomanifold with orientation class $[X]$. Then, the cap product map*

$$\frown [X] : H^{n-k}(X; \mathbb{Z}) \longrightarrow H_k^{BM}(X; \mathbb{Z})$$

factors as

$$
\begin{array}{ccc}
H^{n-k}(X; \mathbb{Z}) & \xrightarrow{\quad \frown [X] \quad} & H_k^{BM}(X; \mathbb{Z}) \\
\downarrow & & \uparrow \\
I^{BM}H_k^{\overline{0}}(X; \mathbb{Z}) \rightarrow I^{BM}H_k^{\overline{p}}(X; \mathbb{Z}) & \rightarrow & I^{BM}H_k^{\overline{t}}(X; \mathbb{Z})
\end{array}
$$

for every perversity function $\overline{p}$.

Proof If $\overline{p} \leq \overline{q}$ (as integer-valued functions), one has an inclusion $IC_i^{\overline{p}} \subseteq IC_i^{\overline{q}}$, and hence an induced map $I^{BM}H_i^{\overline{p}} \rightarrow I^{BM}H_i^{\overline{q}}$. Clearly, $IC_i^{\overline{t}} \subset C_i$, which gives rise to $I^{BM}H_i^{\overline{t}} \longrightarrow H_i^{BM}$. It remains to look at

$$\frown [X] : H^{n-k}(X; \mathbb{Z}) \longrightarrow I^{BM}H_k^{\overline{0}}(X; \mathbb{Z}).$$

For this, we claim that

$$\text{Image}\big(\frown [X] : C^{n-k}(X) \rightarrow C_k((X)) \big) \subset IC_k^{\overline{0}}((X)).$$

Let T be a triangulation of X with barycentric subdivision T'. For every simplex σ in T, denote by $\hat{\sigma}$ its barycenter. For an $(n-k)$-simplex σ in T, define the cochain $1_\sigma \in C_T^{n-k}(X)$ by

$$
1_\sigma(\sigma') = \begin{cases} 0 & \text{if } \sigma' \neq \sigma, \\ 1 & \text{if } \sigma' = \sigma. \end{cases}
$$

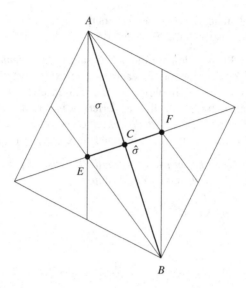

Fig. 2.4 Dual block of σ

Let $D_X(\sigma)$ be the *dual block* of σ, that is,

$$D_X(\sigma) = \{\tau \in T' \mid \tau \cap \sigma = \{\hat{\sigma}\}\}.$$

$D_X(\sigma)$ is a k-dimensional subcomplex of T'.[2]

The geometric interpretation of the cap product is that it maps the cochain 1_σ to the dual block $D_X(\sigma)$, i.e.,

$$1_\sigma \xmapsto{\cap [X]} D_X(\sigma).$$

A direct computation then shows that $D_X(\sigma) \in IC_k^{\bar{0}}((X)) \subseteq C_k^{T'}((X))$, thus finishing the proof. $\square$

Let X be a pseudomanifold of even (real) dimension $n = 2m$ (e.g., an m-dimensional complex algebraic variety) with only one isolated singular point x. Then X has a pseudomanifold stratification

$$X = X_{2m} \supset X_0 = \{x\}.$$

The (middle-perversity) intersection homology of such a space is computed as in the following result.

---

[2]In the notations of Figure 2.4, $\sigma = AB$, $\hat{\sigma} = C$, and $D_X(\sigma) = EC \cup CF$.

Proposition 2.3.18 *Let X be an $n = 2m$-dimensional pseudomanifold with only one isolated singular point x. Then,*

$$IH_i^{\bar{m}}(X; \mathbb{Z}) = \begin{cases} H_i(X; \mathbb{Z}), & i > m, \\ \text{Image}(H_m(X - x; \mathbb{Z}) \longrightarrow H_m(X; \mathbb{Z})), & i = m, \\ H_i(X - x; \mathbb{Z}), & i < m. \end{cases}$$

Before giving the proof of Proposition 2.3.18, let us apply it on some concrete examples.

Example 2.3.19 Let X be the pinched torus of Figure 2.5, with singular (pinch) point x. Then $X - x$ is homotopy equivalent to a circle S^1, and the generator of $H_1(X - x; \mathbb{Z})$ is a boundary in X, hence it maps to 0 in $H_1(X; \mathbb{Z})$. By the above proposition, it follows that

$$IH_i^{\bar{m}}(X; \mathbb{Z}) = \begin{cases} \mathbb{Z}, & i = 2, \\ 0, & i = 1, \\ \mathbb{Z}, & i = 0. \end{cases}$$

Example 2.3.20 Let $(M, \partial M)$ be a compact $n = 2m$-dimensional manifold with boundary. Let

$$X = M \cup_{\partial M} c(\partial M)$$

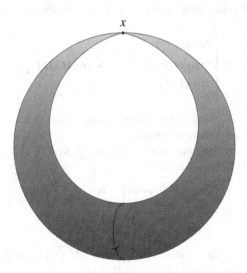

Fig. 2.5 Pinched torus

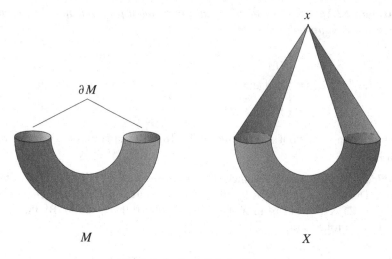

Fig. 2.6 Coning off the boundary

be obtained by coning off the boundary of M, as in Figure 2.6. This provides an example of a pseudomanifold as in Proposition 2.3.18. If x denotes the cone point, then $X - x$ is homotopy equivalent to M. Therefore,

$$H_i(X; \mathbb{Z}) \xrightarrow{\cong} H_i(X, c(\partial M); \mathbb{Z}) \xrightarrow{\cong} H_i(M, \partial M; \mathbb{Z})$$

for $i > 0$, where the first isomorphism follows from the long exact sequence of a pair, and the second is obtained by excision. Therefore, Proposition 2.3.18 yields:

$$IH_i^{\bar{m}}(X; \mathbb{Z}) = \begin{cases} H_i(M, \partial M; \mathbb{Z}), & i > m, \\ \text{Image}\big(H_i(M; \mathbb{Z}) \longrightarrow H_i(M, \partial M; \mathbb{Z})\big), & i = m, \\ H_i(M; \mathbb{Z}), & i < m. \end{cases}$$

Proof of Proposition 2.3.18 For simplicity, assume that X is compatibly triangulated. (The singular computation is similar.)

Since X has only one singular stratum X_0, elements of $IC_i^{\bar{m}}(X)$ consist of chains $\xi \in C_i(X)$ satisfying the conditions:

(a) $\dim(|\xi| \cap X_0) \le i - 2m + \bar{m}(2m) = i - m - 1$,
(b) $\dim(|\partial \xi| \cap X_0) \le i - m - 2$.

This means that:

1. If $i \le m$, then $\xi \cap X_0 = \emptyset$ and $\partial \xi \cap X_0 = \emptyset$, so allowable i-chains cannot intersect X_0. Therefore, $IC_i^{\bar{m}}(X) = C_i(X - x)$.
2. If $i \ge m + 2$, then ξ and $\partial \xi$ are allowed to pass through x, which yields that $IC_i^{\bar{m}}(X) = C_i(X)$.

3. If $i = m + 1$, then an i-chain ξ is allowed to go through x, but $\partial\xi$ is not.

So the intersection chain complex looks like

$$\cdots \longrightarrow C_{m+3}(X) \xrightarrow{\partial} C_{m+2}(X) \xrightarrow{\partial} IC_{m+1}^{\overline{m}}(X)$$

$$\xrightarrow{\partial} C_m(X-x) \xrightarrow{\partial} C_{m-1}(X-x) \longrightarrow \cdots$$

This implies that

$$IH_i^{\overline{m}}(X;\mathbb{Z}) = \begin{cases} H_i(X-x;\mathbb{Z}), & i \leq m-1, \\ H_i(X;\mathbb{Z}), & i \geq m+2. \end{cases}$$

Moreover,

$$\text{Ker}\,(\partial : IC_{m+1}^{\overline{m}}(X) \to IC_m^{\overline{m}}(X)) = \text{Ker}\,(\partial : C_{m+1}(X) \to C_m(X)),$$

so $IH_{m+1}^{\overline{m}}(X;\mathbb{Z}) \cong H_{m+1}(X;\mathbb{Z})$. Finally,

$$\partial(IC_{m+1}^{\overline{m}}(X)) = (\partial C_{m+1}(X)) \cap IC_m^{\overline{m}}(X)$$

and $IC_m^{\overline{m}}(X) = C_m(X-x)$, so

$$IH_m^{\overline{m}}(X;\mathbb{Z}) \cong \text{Image}\big(H_m(X-x;\mathbb{Z}) \longrightarrow H_m(X;\mathbb{Z})\big).$$

$\square$

Exercise 2.3.21 (Projective Cone Over a Complex Projective Manifold) Let Y be a nonsingular complex projective variety of complex dimension $m-1$, embedded in $\mathbb{C}P^{N-1}$ considered as a hyperplane in $\mathbb{C}P^N$. Let $X \subset \mathbb{C}P^N$ be the projective cone on Y, i.e., the union of all projective lines passing through a fixed point $x \notin \mathbb{C}P^{N-1}$ and a point on Y. So X is a pure m-dimensional complex projective variety with an isolated singularity at the cone point x. (Topologically, X is the *Thom space* of the line bundle over Y corresponding to a hyperplane section or, equivalently, the restriction to Y of the normal bundle of $\mathbb{C}P^{N-1}$ in $\mathbb{C}P^N$.) Use Proposition 2.3.18 to show that for the middle-perversity and rational coefficients, we have:

$$IH_i^{\overline{m}}(X;\mathbb{Q}) = \begin{cases} H_i(Y;\mathbb{Q}), & i \leq m, \\ H_{i-2}(Y;\mathbb{Q}), & i > m. \end{cases}$$

Since Y is nonsingular and projective, this calculation shows that the middle-perversity intersection homology groups of X have a pure Hodge structure. Moreover, it can be shown that they also satisfy the Hard Lefschetz theorem. (As we will see later on, these features persist for every complex projective variety.)

Remark 2.3.22 In the above situations, we considered only even-dimensional pseudomanifolds, since our main examples come from complex algebraic geometry. But similar considerations also apply to real pseudomanifolds, e.g., (open) cones on manifolds, etc. (see also Section 2.5 below).

2.4 Normalization

Definition 2.4.1 An n-dimensional pseudomanifold X^n is said to be (topologically) *normal* if it has connected links, that is, every point $x \in X$ has an open neighborhood U_x in X such that $U_x - X_{n-2}$ is connected.

Definition 2.4.2 A *normalization* of a pseudomanifold X^n is a normal pseudomanifold $\widetilde{X}^n$ together with a finite-to-one map $\pi : \widetilde{X} \to X$ such that for every $x \in X$, the induced map

$$\pi_* : \bigoplus_{y \in \pi^{-1}(x)} H_n(\widetilde{X}, \widetilde{X} - y; \mathbb{Z}) \longrightarrow H_n(X, X - x; \mathbb{Z})$$

is an isomorphism.

In order to get some intuition about the meaning of normalization, say X has only isolated singularities, i.e., X has a pseudomanifold stratification $X = X_n \supset X_0$. If $\pi : \widetilde{X} \to X$ denotes the normalization of X, then for every $x \in X_0$ one has by excision and the long exact sequence of a pair that:

$$\bigoplus_{y \in \pi^{-1}(x)} H_n(\widetilde{X}, \widetilde{X} - y; \mathbb{Z}) \cong \bigoplus_y H_n(cL_y, cL_y - y; \mathbb{Z}) \cong \bigoplus_y \widetilde{H}_{n-1}(L_y; \mathbb{Z}),$$

where L_y denotes the link of $y \in \pi^{-1}(x)$. So a normalization of X "separates" the connected components of links L_x of points $x \in X_0$, i.e.,

$$\bigoplus_{y \in \pi^{-1}(x)} \widetilde{H}_{n-1}(L_y; \mathbb{Z}) \cong \widetilde{H}_{n-1}(L_x; \mathbb{Z}).$$

Similar considerations apply to higher dimensional singular strata, by using the local conical structure of a pseudomanifold. This also shows that a normalization map $\pi : \widetilde{X} \to X$ is one-to-one on the open dense stratum of X, while in general the fiber $\pi^{-1}(x)$ has as many points as the number of connected components of $U_x - X_{n-2}$.

Example 2.4.3 As an example, consider the "pinched torus" (left) and its normalization (right), as pictured in Figure 2.7.

Remark 2.4.4 Suppose that X is a complex quasi-projective variety. If X is normal in the sense of algebraic geometry (i.e., all local rings $\mathcal{O}_{X,x}$ are integrally

Fig. 2.7 Normalization of pinched torus

closed), then using Zariski's Main Theorem one can prove that X is normal as a pseudomanifold. However, it should be noted here that the two notions of normal spaces are not equivalent. For example, the cuspidal cubic $x^3 = y^2$ in $\mathbb{C}^2$ is normal in the pseudomanifold sense (since the link of the singularity at the origin is the trefoil knot, hence connected), but it is not algebraically normal. Since the cuspidal cubic is topologically normal, it is homeomorphic to its algebraic normalization.

Exercise 2.4.5 Let X^n be an n-dimensional PL pseudomanifold. Fix a triangulation of X^n, and define a space $\widetilde{X}$ as the disjoint union of all n-simplices in X, with the identification of $(n-1)$-simplices that came from the same $(n-1)$-simplex in X. Show that the space $\widetilde{X}$ defined in this way is a normal pseudomanifold. More generally, it can be shown that every topological pseudomanifold has a normalization (e.g., see [128, Section 4.5]).

One can show by checking allowability conditions that, if $\widetilde{X}$ is a normalization of X, then for every perversity $\overline{p}$, there are isomorphisms

$$IC_i^{\overline{p}}((\widetilde{X})) \cong IC_i^{\overline{p}}((X)) \, , \quad IC_i^{\overline{p}}(\widetilde{X}) \cong IC_i^{\overline{p}}(X),$$

for all $i \geq 0$. This yields the following result of Goresky and MacPherson (cf. [81, Section 4.2]):

Theorem 2.4.6 *Suppose that X is a pseudomanifold and $\pi : \widetilde{X} \to X$ is a normalization of X. Then for every perversity $\overline{p}$ and for all integers $i \geq 0$,*

$$\pi_* : I^{(BM)}H_i^{\overline{p}}(\widetilde{X}; \mathbb{Z}) \to I^{(BM)}H_i^{\overline{p}}(X; \mathbb{Z})$$

is an isomorphism.

Remark 2.4.7 Intersection homology is not functorial in general, i.e., a continuous map of spaces does not necessarily induce a map of the corresponding intersection homology groups. But, as we will see later on, for certain maps (e.g., finite), such induced homomorphisms exist.

Example 2.4.8 As an application, we give an example that shows that intersection homology is *not* a homotopy invariant. Note that $S^4 \vee S^4$ and $S^4 \cup_{S^2} \mathbb{C}P^2$ are homotopy equivalent (with a homotopy defined by collapsing the S^2 inside S^4). However, their intersection homology groups are not isomorphic, and calculations can be done by using their normalizations. Indeed, first note that the normalization of $S^4 \vee S^4$ is a disjoint union of two copies of S^4, whereas the normalization of $S^4 \cup_{S^2} \mathbb{C}P^2$ is the disjoint union of S^4 and $\mathbb{C}P^2$ (as can be easily seen, e.g., from Exercise 2.4.5). Thus,

$$IH_i^{\overline{p}}(S^4 \vee S^4; \mathbb{Z}) \cong IH_i^{\overline{p}}(S^4 \sqcup S^4; \mathbb{Z}) \cong H_i(S^4; \mathbb{Z}) \oplus H_i(S^4; \mathbb{Z}),$$

while

$$IH_i^{\overline{p}}(S^4 \cup_{S^2} \mathbb{C}P^2; \mathbb{Z}) \cong IH_i^{\overline{p}}(S^4 \sqcup \mathbb{C}P^2; \mathbb{Z}) \cong H_i(S^4; \mathbb{Z}) \oplus H_i(\mathbb{C}P^2; \mathbb{Z}).$$

So these homotopy equivalent spaces have non-isomorphic intersection homology groups.

Normal pseudomanifolds are particularly important for understanding the "extreme" intersection homology groups, i.e., those corresponding to the top and, respectively, zero-perversity function. More precisely, one has the following result (see [81, Section 4.3]):

Theorem 2.4.9 *If X is an oriented, normal, compact n-dimensional pseudomanifold, then*

$$IH_i^{\overline{t}}(X; \mathbb{Z}) \cong H_i(X; \mathbb{Z})$$

and

$$IH_i^{\overline{0}}(X; \mathbb{Z}) \cong H^{n-i}(X; \mathbb{Z}).$$

Remark 2.4.10 The normality of X implies that $X - X_{n-3}$ is still a topological manifold. Indeed, the link of every $x \in X_{n-2} - X_{n-3}$ must be a connected compact real 1-dimensional manifold, hence homeomorphic to S^1. So any actual singularities occur in one codimension higher. Recall also that $\overline{t}$-allowability amounts to asking that $|\xi|$ and $|\partial\xi|$ are transversal to $\Sigma = X_{n-2}$. So the first isomorphism of Theorem 2.4.9 says that the normality condition gives enough flexibility to insure that every cycle is homologous to one transversal to the singular set Σ.

2.5 Intersection Homology of an Open Cone

In this section, we calculate the intersection homology of the open cone $\mathring{c}L$ of an $(n-1)$-dimensional pseudomanifold L. Notice that $\mathring{c}L$ is contractible, hence its ordinary (reduced) homology groups vanish. But while homology does not detect the singularities, i.e., the cone point, we will see below that intersection homology does so.

For simplicity of exposition, we perform the calculations below simplicially, assuming the spaces involved have an underlying PL structure. Singular versions of these calculations are similar (e.g., see [128]).

We start by computing the intersection homology of the product space $X \times \mathbb{R}$, for a PL pseudomanifold X^m of real dimension m. First stratify $X \times \mathbb{R}$ by defining

$$(X \times \mathbb{R})_{m+1-k} = X_{m-k} \times \mathbb{R}.$$

Then, for every $\xi \in IC_i^{\overline{p}}((X))$, consider the *suspension of* ξ, namely $\xi \times \mathbb{R} \in C_{i+1}((X \times \mathbb{R}))$. Note that

$$\dim(|\xi \times \mathbb{R}| \cap (X \times \mathbb{R})_{m+1-k}) = \dim(\mathbb{R} \times (|\xi| \cap X_{m-k}))$$
$$\leq 1 + i - k + \overline{p}(k)$$

and a similar statement holds for $\partial \xi$. Therefore, one gets maps

$$IC_i^{\overline{p}}((X)) \to IC_{i+1}^{\overline{p}}((X \times \mathbb{R}))$$

defined by $\xi \mapsto \xi \times \mathbb{R}$, hence a map of complexes

$$IC_\bullet^{\overline{p}}((X)) \to IC_{\bullet+1}^{\overline{p}}((X \times \mathbb{R})).$$

The later complex is commonly denoted by $IC_\bullet^{\overline{p}}((X \times \mathbb{R}))[1]$. Then the following holds (see [84, Section 1.6] or [15, Chapter II]):

Proposition 2.5.1 *The above map of complexes induces the following Künneth isomorphisms:*

$$IH_i^{\overline{p}}(X; \mathbb{Z}) \cong IH_i^{\overline{p}}(X \times \mathbb{R}; \mathbb{Z})$$

and

$$I^{BM}H_i^{\overline{p}}(X; \mathbb{Z}) \cong I^{BM}H_{i+1}^{\overline{p}}(X \times \mathbb{R}; \mathbb{Z}).$$

The Künneth isomorphisms play an important role in proving the following *cone formula* for intersection homology:

Theorem 2.5.2 *Suppose L is a compact PL pseudomanifold of dimension $m \geq 1$. Then, for every perversity $\overline{p}$,*

$$IH_i^{\overline{p}}(\mathring{c}L; \mathbb{Z}) \cong \begin{cases} IH_i^{\overline{p}}(L; \mathbb{Z}), & i < m - \overline{p}(m+1), \\ 0, & otherwise. \end{cases} \tag{2.1}$$

Similarly,

$$I^{BM}H_i^{\overline{p}}(\mathring{c}L; \mathbb{Z}) \cong \begin{cases} IH_{i-1}^{\overline{p}}(L; \mathbb{Z}), & i > m - \overline{p}(m+1), \\ 0, & otherwise. \end{cases} \tag{2.2}$$

Proof To prove formula (2.1), first note that the cone point $\{c\}$ has codimension $m + 1$ in $\mathring{c}L$. So for $\xi \in IC_i^{\overline{p}}(\mathring{c}L)$, one has that

$$\dim(|\xi| \cap \{c\}) \leq i - (m+1) + \overline{p}(m+1).$$

Hence, ξ cannot intersect $\{c\}$ for $i \leq m - \overline{p}(m + 1)$. Therefore, in this range, $IC_i^{\overline{p}}(\mathring{c}L) \cong IC_i^{\overline{p}}(\mathring{c}L - \{c\})$. Then for $i < m - \overline{p}(m + 1)$,

$$IH_i^{\overline{p}}(\mathring{c}L; \mathbb{Z}) \cong IH_i^{\overline{p}}(\mathring{c}L - \{c\}; \mathbb{Z}) \cong IH_i^{\overline{p}}(L \times \mathbb{R}; \mathbb{Z}) \cong IH_i^{\overline{p}}(L; \mathbb{Z}).$$

On the other hand, for $i \geq m - \overline{p}(m + 1)$, a $\overline{p}$-allowable i-chain ξ with $\partial\xi = 0$ satisfies $\xi = \partial(c\xi)$, and $c\xi$ is in fact a $\overline{p}$-allowable $(i + 1)$-chain. Hence, $[\xi] = 0 \in IH_i^{\overline{p}}(\mathring{c}L; \mathbb{Z})$.

For formula (2.2), note that the Borel–Moore intersection homology vanishes in low dimensions because chains that do not meet the vertex $\{c\}$ can be *coned off to infinity*. On the other hand, in high dimensions, Borel–Moore intersection homology classes arise as the open cone on classes in the intersection homology of $\mathring{c}L - \{c\}$, hence the dimension shift (cf. Proposition 2.5.1). $\square$

Remark 2.5.3 By taking $L = S^{n-1}$, we have $\mathring{c}L = \mathbb{R}^n$. The above theorem reduces to the calculation for the local model of a topological manifold:

$$H_i(\mathbb{R}^n; \mathbb{Z}) \cong \begin{cases} 0, & i \neq 0, \\ \mathbb{Z}, & i = 0. \end{cases}$$

$$H_i^{BM}(\mathbb{R}^n; \mathbb{Z}) \cong \begin{cases} \mathbb{Z}, & i = n, \\ 0, & i \neq n. \end{cases}$$

Exercise 2.5.4 Let X be a $(2m - 1)$-dimensional pseudomanifold. Compute the middle-perversity intersection homology groups $IH_i^{\overline{m}}(\Sigma X; \mathbb{Z})$ of the suspension of X in terms of the middle-perversity intersection homology groups of X.

2.6 Poincaré Duality for Pseudomanifolds

In this section, we explain the geometric idea behind Poincaré duality for intersection homology. This topic will be revisited later on (in Section 6.4), through a sheaf-theoretic approach via Verdier duality. When working with chains or (intersection) homology, $\mathbb{Q}$-*coefficients* will be used in this section, even though not explicitly mentioned.

Let us start by recalling the following consequence of Poincaré duality for manifolds. Suppose M^n is an oriented, closed, connected, n-dimensional manifold. Then there exists a non-degenerate (intersection) pairing:

$$H_i(M) \otimes H_{n-i}(M) \xrightarrow{\frown} \mathbb{Q}.$$

Geometrically, if $a \in H_i(M)$ and $b \in H_{n-i}(M)$, then a and b have chain representatives $\alpha \in C_i(M)$ and $\beta \in C_{n-i}(M)$ such that $|\alpha|$ and $|\beta|$ are subspaces of M of complementary dimension. Moreover, by transversality, one can choose them in such a way that $|\alpha| \cap |\beta|$ is a finite set. The number of these points counted with multiplicities does not depend on the choice of representatives and is exactly $a \frown b \in \mathbb{Q}$. To prove the duality statement, one first proves it for $\mathbb{R}^n$, then covers M with finitely many open sets homeomorphic to $\mathbb{R}^n$, and patches the local dualities by the Mayer–Vietoris sequence.

The duality statement in the singular context is formulated in terms of intersection homology as follows (see [81, Section 3.3]):

Theorem 2.6.1 (Poincaré Duality for Pseudomanifolds, Chain Version) *If X^n is an oriented n-dimensional topological pseudomanifold, and $\overline{p}$ and $\overline{q}$ are complementary perversities, then there is a non-degenerate bilinear pairing*

$$I H_i^{\overline{p}}(X) \times I^{BM} H_{n-i}^{\overline{q}}(X) \xrightarrow{\frown} \mathbb{Q}.$$

Before discussing the proof, let us explain the geometric intuition behind Theorem 2.6.1. Fix a stratification $X^n = X_n \supseteq X_{n-2} \supseteq \cdots \supseteq X_0 \supseteq \emptyset$, and assume, for simplicity, that X has a compatible triangulation. For $a \in I H_i^{\overline{p}}(X)$ and $b \in I^{BM} H_{n-i}^{\overline{q}}(X)$ one can choose simplicial intersection chains $\xi \in I C_i^{\overline{p}}(X)$ and $\eta \in I^{BM} C_{n-i}^{\overline{q}}(X)$ so that $|\xi| \cap |\eta| \subset X - X_{n-2}$ and $|\xi| \cap |\eta|$ is a finite number of points. The number of these points counted with multiplicities (depending on coefficients of ξ, η, and on the orientation) does not depend on the representatives ξ, η for a and b. This number is $a \frown b$.

A proof of Poincaré duality for pseudomanifolds, similar to the one for manifolds, would consist of the following steps:

(a) Induction for proving (local) Poincaré duality for open cones $\mathring{c}L$.
(b) Show that Poincaré duality holds for conical neighborhoods of the form $\mathring{c}L \times \mathbb{R}^k$.

(c) Cover X by conical neighborhoods and patch local Poincaré dualities for such neighborhoods by a Mayer–Vietoris argument.[3]

We only deal here with the first step (the theorem will be proved later on by using sheaves, which are designed to relate local and global information; see Corollary 6.4.2). We prove that, if L is a compact m-dimensional pseudomanifold, and if Poincaré duality holds for L, then Poincaré duality holds also for $\mathring{c}L$. Recall the calculation of intersection homology of cones from the previous section:

$$IH_i^{\overline{p}}(\mathring{c}L; \mathbb{Z}) \cong \begin{cases} IH_i^{\overline{p}}(L; \mathbb{Z}), & i < m - \overline{p}(m+1), \\ 0, & \text{otherwise.} \end{cases}$$

$$I^{BM}H_i^{\overline{p}}(\mathring{c}L; \mathbb{Z}) \cong \begin{cases} 0, & i \leq m - \overline{p}(m+1), \\ IH_{i-1}^{\overline{p}}(L; \mathbb{Z}), & \text{otherwise.} \end{cases}$$

Assume now that L satisfies Poincaré duality, i.e.,

$$IH_i^{\overline{p}}(L; \mathbb{Q}) \cong IH_{m-i}^{\overline{q}}(L; \mathbb{Q})^{\vee}$$

for $\overline{p}$ and $\overline{q}$ complementary perversities. Then, if $i < m - \overline{p}(m+1)$, one has:

$$IH_i^{\overline{p}}(\mathring{c}L; \mathbb{Q}) \cong IH_i^{\overline{p}}(L; \mathbb{Q}) \cong IH_{m-i}^{\overline{q}}(L; \mathbb{Q})^{\vee} \cong I^{BM}H_{m+1-i}^{\overline{q}}(\mathring{c}L; \mathbb{Q})^{\vee},$$

while if $i \geq m - \overline{p}(m+1)$ one has

$$IH_i^{\overline{p}}(\mathring{c}L) = 0 = I^{BM}H_{m+1-i}^{\overline{q}}(\mathring{c}L),$$

since $m - 1 = \overline{p}(m+1) + \overline{q}(m+1)$. This proves the claim. $\qquad\square$

2.7 Signature of Pseudomanifolds

Intersection homology groups are also useful for the study of manifolds. In this section, we give an application of intersection homology to an important question in manifold theory, namely we use the signature invariant associated to the middle-perversity intersection homology of a Witt space to prove the Novikov additivity property for the signature of manifolds with boundary.

---

[3]We leave it as an exercise for the reader to formulate and prove the corresponding Mayer–Vietoris result for intersection homology groups; it is a simple adaptation of the analogous result in simplicial/singular homology.

In general, there is no perversity $\bar{p}$ that is self-complementary, such that the groups $IH_i^{\bar{p}}(X; \mathbb{Q})$ are dual to each other. However, by restricting X to the class of Witt spaces, Siegel [218] noted that the lower-middle perversity intersection homology groups have this property. This class already includes the pseudomanifolds with only even codimension strata, for which a signature invariant was defined earlier in [81].

Definition 2.7.1 (Witt Space) A pseudomanifold X is a *Witt space* if for every stratum S of odd codimension $2r + 1$ with link L_S, one has $IH_r^{\bar{m}}(L_S; \mathbb{Q}) = 0$.

Example 2.7.2 Manifolds are Witt spaces. Complex algebraic varieties are Witt spaces since the above condition is void (all strata have even codimension). But the suspension of the 2-torus, ΣT^2, is not a Witt space, since the link of a suspension point is T^2 and $H_1(T^2; \mathbb{Q}) \neq 0$.

The following result was proved in [218, Theorem 3.4]:

Theorem 2.7.3 (P. Siegel) *If X is a Witt space, then the homomorphism*

$$I^{(BM)}H_*^{\bar{m}}(X; \mathbb{Q}) \longrightarrow I^{(BM)}H_*^{\bar{n}}(X; \mathbb{Q})$$

induced by inclusion of chain complexes is an isomorphism.

It follows that, if X is a compact, oriented, Witt pseudomanifold of $\dim_{\mathbb{R}}(X) = 4r$ (for example a complex projective variety X of complex dimension $2r$), then there is a non-degenerate and symmetric rational *intersection pairing*

$$IH_{2r}^{\bar{m}}(X; \mathbb{Q}) \otimes IH_{2r}^{\bar{m}}(X; \mathbb{Q}) \longrightarrow \mathbb{Q}$$

on the middle middle-perversity intersection homology group of X. In particular, this pairing can be diagonalized with only real eigenvalues.

Definition 2.7.4 The *signature* $\sigma(X)$ of a $4r$-dimensional compact, oriented, Witt pseudomanifold X^{4r} is defined as the difference between the number of positive and negative eigenvalues of the intersection form

$$IH_{2r}^{\bar{m}}(X; \mathbb{Q}) \otimes IH_{2r}^{\bar{m}}(X; \mathbb{Q}) \longrightarrow \mathbb{Q}.$$

For the following statement see, e.g., [75, Theorem 9.3.16]:

Proposition 2.7.5 *The following properties hold:*

(a) $\sigma(X)$ is a topological invariant.
(b) If X is a manifold, then $\sigma(X)$ is the usual signature.

Recall that the signature of closed oriented manifolds is a *bordism invariant*. This means that if M and M' are closed, oriented, $4r$-dimensional manifolds such that there is a compact oriented $(4r + 1)$-dimensional manifold with boundary $\partial W \cong M \sqcup (-M')$ (that is, M and M' are *bordant*), then $\sigma(M) = \sigma(M')$. (Here, $-M'$

denotes the manifold M' with the opposite orientation, and $\sqcup$ is disjoint union.) In particular, if $M^{4r} = \partial W^{4r+1}$, then $\sigma(M^{4r}) = 0$. This property still holds for signatures of Witt spaces. Let us introduce the necessary definitions.

Definition 2.7.6

1. $(X^n, \partial X)$ is a *stratified Witt space with boundary* if the following conditions are satisfied:

   (i) $X, \partial X$ are stratified spaces, with a stratified inclusion map $\partial X \hookrightarrow X$,
   (ii) $X - \partial X, \partial X$ are Witt spaces of dimension n and $n - 1$, respectively,
   (iii) ∂X has a collared neighborhood, i.e., a neighborhood homeomorphic to $\partial X \times [0, 1)$ with $\partial X \cong \partial X \times \{0\}$.

2. $(X, \partial X)$ is an *oriented Witt space with boundary* if $X - \partial X$ is oriented and the collar neighborhood induces an orientation on ∂X.

Definition 2.7.7 Closed oriented Witt spaces X and X' are *Witt bordant* if there exists a compact oriented Witt space with boundary $(Y, \partial Y)$ and an orientation preserving homeomorphism $\partial Y \cong X \sqcup (-X')$.

Theorem 2.7.8 (P. Siegel) *If X and X' are Witt bordant, then $\sigma(X) = \sigma(X')$.*

This theorem is a consequence of the following result of Siegel [218, Chapter 2, Section 2], see also [75, Theorem 9.3.17] (in the case when X has only even-codimension strata, the result is proved in [81] for PL spaces, and a sheaf-theoretic proof that works for topological pseudomanifold is indicated in [6, Theorem 6.1.4]):

Theorem 2.7.9 *If $(X^{4r+1}, \partial X)$ is a compact oriented Witt space with boundary, then $\sigma(\partial X) = 0$.*

An important consequence of Theorem 2.7.8 is a simple and geometric proof of the *Novikov additivity property* for the signature of manifolds with boundary, see [218, Chapter II, Section 3]. Recall first that if $(M, \partial M)$ is a manifold with boundary of dimension $4r$, its signature $\sigma(M, \partial M)$ is the signature of the Lefschetz–Poincaré pairing defined as follows: let j_* be the usual map $H_{2r}(M; \mathbb{Q}) \to H_{2r}(M, \partial M; \mathbb{Q})$ induced by inclusion of pairs, and consider the Poincaré duality map:

$$PD : H_{2r}(M, \partial M; \mathbb{Q}) \to H^{2r}(M; \mathbb{Q}).$$

The *Lefschetz–Poincaré pairing*

$$H_{2r}(M; \mathbb{Q}) \otimes H_{2r}(M; \mathbb{Q}) \longrightarrow \mathbb{Q}$$

of $(M, \partial M)$ is then defined by:

$$(a, b) \mapsto \langle PD(j_*a), b \rangle.$$

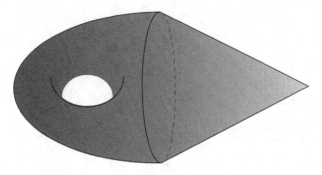

Fig. 2.8 $\widehat{M}$

(Here, $\langle -, - \rangle$ denotes the usual Kronecker pairing between cohomology and homology.) Note that this pairing is not necessarily non-degenerate, so in order to define its signature $\sigma(M, \partial M)$ one has to discard the kernel of j_*, i.e., $\sigma(M, \partial M)$ is the signature of the induced pairing on $H_{2r}(M; \mathbb{Q}) / \mathrm{Ker}\,(j_*)$.

Corollary 2.7.10 (Novikov Additivity) *Let* $(M^{4r}, \partial M)$, $(M'^{4r}, \partial M')$ *be 4r-dimensional compact oriented manifolds with boundary, with an orientation reversing homeomorphism* $\partial M \cong \partial M'$. *Let* $M \cup_\partial M'$ *be the space obtained from the disjoint union* $M \sqcup M'$ *by identifying* ∂M *with* $\partial M'$ *by the above homeomorphism. Then the following additivity property holds:*

$$\sigma(M \cup_\partial M') = \sigma(M, \partial M) + \sigma(M', \partial M'). \tag{2.3}$$

Exercise 2.7.11 Let $(M, \partial M)$ be an even-dimensional compact oriented manifold with boundary. Show that the space $\widehat{M} = M \cup_{\partial M} \mathrm{cone}(\partial M)$ obtained by coning off the boundary of M is a Witt space (Fig. 2.8). Furthermore, show that:

$$\sigma(M, \partial M) = \sigma(\widehat{M}). \tag{2.4}$$

Hint: Recall that for the middle-perversity $\overline{m}$, one has by Proposition 2.3.18 that

$$IH_{2r}^{\overline{m}}(\widehat{M}) = \mathrm{Image}\big(j_* : H_{2r}(M) \to H_{2r}(M, \partial M)\big).$$

Sketch of Proof of Corollary 2.7.10 Let $X := M \cup_\partial M'$. In view of equation (2.4), one has to show that

$$\sigma(X) = \sigma(\widehat{M}) + \sigma(\widehat{M}').$$

It is sufficient to create a Witt bordism between X and the one point union $\widehat{M} \vee \widehat{M}'$ of $\widehat{M}$ and $\widehat{M}'$, since it can be shown easily (e.g., by using a normalization) that the middle intersection homology group of the one point union and for the disjoint union are isomorphic.

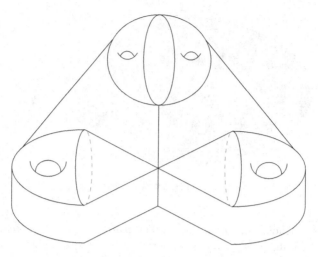

Fig. 2.9 P

Let $Z = \partial M = \partial M'$ and let $X' = X/Z$. So

$$X' = (M \cup_\partial \text{cone}(Z)) \cup_{\text{cone point}} (M' \cup_\partial \text{cone}(Z)) = \widehat{M} \vee \widehat{M}'.$$

Let $\pi : X \to X'$ be the quotient map and denote by c the common cone (wedge) point. Let P be the mapping cylinder of π with a collar $X' \times [0,1]$ attached to X', see Figure 2.9 for a sketch.

The claim is that P is a Witt space. In order to see this, one needs to check that

$$I H_{2r}^{\overline{m}}(L_P(\{c\} \times \{0\}); \mathbb{Q}) = 0,$$

where $L_P(\{c\} \times \{0\})$ denotes the link of $\{c\} \times \{0\}$ in P. It can be seen that

$$L_P(\{c\} \times \{0\}) = Z \times [-1,1] \cup \text{cone}\partial(Z \times [-1,1]) \cong \Sigma(Z)/(c_+ = c_-),$$

where $c_\pm$ denote the cone points in the suspension $\Sigma(Z)$. Since intersection homology does not change under normalization, one gets that

$$I H_{2r}^{\overline{m}}(L_P(\{c\} \times \{0\}); \mathbb{Q}) \cong I H_{2r}^{\overline{m}}(\Sigma(Z); \mathbb{Q}),$$

so it remains to show that the latter group vanishes. If $\xi \in I C_{2r}^{\overline{m}}(\Sigma(Z))$, then by allowability one has that $\xi \cap \{c_\pm\} = \emptyset$. Hence one can find a representative $[\xi'] = [\xi]$ so that $|\xi'| \subset Z$. But $\xi' = \partial(c_+\xi')$, and $c_+\xi'$ is $\overline{m}$-allowable. Thus $[\xi'] = 0$.

Since P is a Witt bordism between X and X', Theorem 2.7.8 yields the desired signature identification. $\square$

Chapter 3
L-Classes of Stratified Spaces

The signature invariant and characteristic classes (e.g., *L*-classes) play a fundamental role in classification schemes for smooth manifolds, e.g., in various *bordism theories* or *surgery theory*. For example, if M is a closed oriented manifold, the signature $\sigma(M)$ is a homotopy invariant of M. On the other hand, its *L*-classes $L^i(M) \in H^{4i}(M; \mathbb{Q})$ are not homotopy invariants, and are therefore much more refined. In particular, *L*-classes can help to distinguish diffeomorphism types of manifolds in a given homotopy type. As a sample (but hopefully convincing) result, we mention the following (cf. [27]):

Theorem 3.0.1 (Browder–Novikov) *The homotopy type and L-classes determine the diffeomorphism class of a smooth simply connected manifold of dimension ≥ 5 up to a finite number of possibilities.*

Given the importance that the signature and *L*-classes play in manifold theory, it is therefore desirable to extend such concepts to stratified spaces and to develop similar classification schemes in the singular context. The signature of a compact, oriented, Witt pseudomanifold was defined in Section 2.7 by using the duality pairing on the middle-perversity intersection homology. In Section 3.4, this signature invariant will be used to define *L*-classes of such singular spaces.

In Section 3.1, we review the basics of (multiplicative) characteristic classes of vector bundles, as explained e.g., in [102] and [180] (see also [216] and [6, Chapter 5] for nice accounts of this story). The main examples to be discussed here are the Chern classes and *L*-classes. Characteristic classes of manifolds, e.g., the above-mentioned *L*-classes, are defined via their tangent bundles; we discuss this and the associated *index formulae* in Section 3.2. *L*-classes of manifolds can be also defined by the Pontrjagin–Thom construction, by using maps to spheres and signatures of submanifolds with trivial normal bundles. The latter construction is adapted to Witt spaces in Section 3.4, leading to the extension of *L*-classes to such singular spaces. As expected, the *L*-classes of a singular space X coincide with the classical definition of *L*-classes when X is actually a manifold.

© Springer Nature Switzerland AG 2019
L. G. Maxim, *Intersection Homology & Perverse Sheaves*, Graduate Texts in Mathematics 281, https://doi.org/10.1007/978-3-030-27644-7_3

3.1 Multiplicative Characteristic Classes of Vector Bundles: Examples

A *characteristic class* of vector bundles over a topological space X is an assignment

$$c\ell : \mathrm{Vect}(X) \longrightarrow H^*(X; \Lambda)$$

from the set $\mathrm{Vect}(X)$ of isomorphism classes of vector bundles over X to the cohomology ring $H^*(X; \Lambda)$ with a coefficient ring Λ, which is required to be compatible with pullbacks, i.e., for a continuous map $f : Y \to X$ the following diagram commutes:

$$
\begin{array}{ccc}
\mathrm{Vect}(X) & \xrightarrow{\;c\ell\;} & H^*(X; \Lambda) \\[2pt]
{\scriptstyle f^*}\big\downarrow & & \big\downarrow{\scriptstyle f^*} \\[2pt]
\mathrm{Vect}(Y) & \xrightarrow[\;c\ell\;]{} & H^*(Y; \Lambda).
\end{array}
$$

The most important characteristic classes of a real vector bundle $\mathcal{E}$ over X are the *Stiefel–Whitney classes* $w^i(\mathcal{E}) \in H^i(X; \mathbb{Z}/2)$ [225, 240, 241], *Pontrjagin classes* $p^i(\mathcal{E}) \in H^{4i}(X; \mathbb{Z}[1/2])$ [198], and for a complex vector bundle $\mathcal{E}$ the *Chern classes* $c^i(\mathcal{E}) \in H^{2i}(X; \mathbb{Z})$ [41, 42]. These characteristic classes $c\ell^i(\mathcal{E}) \in H^*(X; \Lambda)$ obey the same formalism, and as a consequence they can be described axiomatically in a unified way as follows:

Definition 3.1.1 The Stiefel–Whitney, resp., Pontrjagin classes of real vector bundles, resp., Chern classes of complex vector bundles, are defined by the operator assigning to each real (resp., complex) vector bundle $\mathcal{E} \to X$ cohomology classes

$$
c\ell^i(\mathcal{E}) := \begin{cases}
w^i(\mathcal{E}) & \in H^i(X; \mathbb{Z}/2) \\
p^i(\mathcal{E}) & \in H^{4i}(X; \mathbb{Z}[1/2]) \\
c^i(\mathcal{E}) & \in H^{2i}(X; \mathbb{Z})
\end{cases}
$$

of the base space X such that the following axioms are satisfied:

Axiom 1 (*Finiteness*)

For each vector bundle $\mathcal{E}$ one has $c\ell^0(\mathcal{E}) := 1$ and $c\ell^i(\mathcal{E}) = 0$ for $i > \mathrm{rank}\,\mathcal{E}$ (moreover, $p^i(\mathcal{E}) = 0$ for $i > [\mathrm{rank}\,\mathcal{E}/2]$). The sum

$$c\ell^*(\mathcal{E}) := \sum_{i \geq 0} c\ell^i(\mathcal{E})$$

is called the *total characteristic class* of $\mathcal{E}$. In particular, $c\ell^*(0_X) = 1$ for the zero vector bundle 0_X of rank zero on X.

Axiom 2 (*Naturality*)

For a continuous map $f : Y \to X$ and a vector bundle $\mathcal{E} \to X$, one has:

$$c\ell^*(f^*\mathcal{E}) = f^*c\ell^*(\mathcal{E}).$$

Axiom 3 (*Multiplicativity*)

The following *Whitney sum formula* holds in $H^*(X; \Lambda)$:

$$c\ell^*(\mathcal{E} \oplus \mathcal{F}) = c\ell^*(\mathcal{E}) \smile c\ell^*(\mathcal{F}),$$

or, more generally,

$$c\ell^*(\mathcal{E}) = c\ell^*(\mathcal{E}') \smile c\ell^*(\mathcal{E}'')$$

for every short exact sequence $0 \to \mathcal{E}' \to \mathcal{E} \to \mathcal{E}'' \to 0$ of vector bundles on X. (Here, multiplication on the right-hand side is with respect to the cohomology cup product.)

Axiom 4 (*Normalization*)

For the canonical (i.e., the dual of the tautological) line bundle $\gamma_n^1(\mathbb{K}) := \mathcal{O}_{\mathbb{K}P^n}(1)$ over the projective space $\mathbb{K}P^n$ (with $\mathbb{K} = \mathbb{R}, \mathbb{C}$) one has:

(w^1): $w^1(\gamma_n^1(\mathbb{R})) \neq 0$.
(p^1): $p^1(\gamma_n^1(\mathbb{C})) = c^1(\gamma_n^1(\mathbb{C}))^2$.
(c^1): $c^1(\gamma_n^1(\mathbb{C})) = c := [\mathbb{C}P^{n-1}] \in H^2(\mathbb{C}P^n; \mathbb{Z})$ is the cohomology class represented (under Poincaré duality) by the hyperplane $\mathbb{C}P^{n-1}$.

The existence of such characteristic classes for vector bundles of rank n can be shown, for example, with the help of a *classifying space*, i.e., the infinite dimensional Grassmann manifolds $\mathbf{G}_n(\mathbb{K}^\infty)$ (with $\mathbb{K} = \mathbb{R}, \mathbb{C}$), and the fact that the cohomology ring of this Grassmann manifold is a polynomial ring. More precisely,

$$H^*(\mathbf{G}_n(\mathbb{K}^\infty); \Lambda) = \begin{cases} \mathbb{Z}/2[w^1, w^2, \cdots, w^n], & \mathbb{K} = \mathbb{R}, \Lambda = \mathbb{Z}/2, \\ \mathbb{Z}[1/2][p^1, p^2, \cdots, p^{[n/2]}], & \mathbb{K} = \mathbb{R}, \Lambda = \mathbb{Z}[1/2], \\ \mathbb{Z}[c^1, c^2, \cdots, c^n], & \mathbb{K} = \mathbb{C}, \Lambda = \mathbb{Z}. \end{cases}$$

A rank n $\mathbb{K}$-vector bundle $\mathcal{E}$ on X is "classified" by a map $f_\mathcal{E} : X \to \mathbf{G}_n(\mathbb{K}^\infty)$, and characteristic classes of $\mathcal{E}$ are obtained by pulling back under $f_\mathcal{E}$ the generators of $H^*(\mathbf{G}_n(\mathbb{K}^\infty); \Lambda)$, that is, $w^i(\mathcal{E}) = f_\mathcal{E}^*(w^i) \in H^i(X; \mathbb{Z}/2)$, etc.

Remark 3.1.2 By what is usually referred to as the "splitting principle," one can assume (after pulling back to a suitable bundle, whose pullback on the cohomology level is injective) that a given non-zero vector bundle $\mathcal{E}$ splits into a sum of line (or 2-plane) bundles. Then Axiom 3 reduces the calculation of characteristic classes to the case of (real, resp., complex) line bundles (for $c\ell = w, c$) or real 2-plane bundles (for $c\ell = p$). By naturality, these are then uniquely characterized by Axiom 4.

The Stiefel–Whitney, Pontrjagin, and Chern classes are the building blocks of characteristic class theory, in the sense that every *multiplicative* characteristic class $c\ell^*$ (i.e., satisfying Axiom 3) over a finite dimensional base space (i.e., with $H^i(X; \Lambda) = 0$ for i large enough) is expressed uniquely as a power series in these classes. More precisely, the "splitting principle" implies the following:

Theorem 3.1.3 *Let Λ be a $\mathbb{Z}/2$-algebra (resp., a $\mathbb{Z}[1/2]$-algebra) for the case of real vector bundles, or a $\mathbb{Z}$-algebra for the case of complex vector bundles. There is a one-to-one correspondence between*

(a) *multiplicative characteristic classes $c\ell^*$ over finite dimensional base spaces, and*

(b) *formal power series $\hbar \in \Lambda[[z]]$*

such that $c\ell^*(\mathcal{L}) = \hbar(w^1(\mathcal{L}))$ or $c\ell^*(\mathcal{L}) = \hbar(c^1(\mathcal{L}))$ for every real or complex line bundle $\mathcal{L}$, respectively, $c\ell^*(\mathcal{L}) = \hbar(p^1(\mathcal{L}))$ for every real 2-plane bundle $\mathcal{L}$.

In the above correspondence, $\hbar$ is called the *characteristic power series* of the corresponding multiplicative characteristic class $c\ell_{\hbar}^*$. Moreover $c\ell_{\hbar}^*$ is invertible with inverse $c\ell_{1/\hbar}^*$ if $\hbar \in \Lambda[[z]]$ is invertible, i.e., if $\hbar(0) \in \Lambda$ is a unit (e.g., $\hbar$ is a normalized power series with $\hbar(0) = 1$). So the corresponding multiplicative characteristic class $c\ell_{\hbar}^*$ extends over a finite dimensional base space X to a natural transformation of groups

$$c\ell_{\hbar}^* : (\mathbf{K}(X), \oplus) \longrightarrow (H^*(X; \Lambda), \smile)$$

on the Grothendieck group $\mathbf{K}(X)$ of real or complex vector bundles over X.

3.2 Characteristic Classes of Manifolds: Tangential Approach

In this section, assume that M is a smooth (stably almost complex) manifold (i.e., $TM \oplus \mathbb{R}_M^a$ has the structure of a complex vector bundle for some a). If $c\ell^*$ is a multiplicative characteristic class as in the previous section, the characteristic class

$$c\ell^*(M) := c\ell^*(TM)$$

of the real (or complex) tangent bundle TM is called a *characteristic cohomology class* of the manifold M. Let

$$c\ell_*(M) := c\ell^*(TM) \frown [M] \in H_*^{BM}(M; \Lambda)$$

be the corresponding *characteristic homology class* of the manifold M, with $[M] \in H_*^{BM}(M; \Lambda)$ the fundamental class in the Borel–Moore homology of the (oriented) manifold M.

Example 3.2.1 With c as in Axiom 4, the Chern and Pontrjagin classes of $\mathbb{C}P^n$ are computed by (e.g., see [180, Theorem 14.10, Example 15.6]):

$$c^*(\mathbb{C}P^n) = (1+c)^{n+1}, \quad p^*(\mathbb{C}P^n) = (1+c^2)^{n+1}. \tag{3.1}$$

In particular, $c^i(\mathbb{C}P^n) = \binom{n+1}{i}c^i$ for $i \leq n$, and $p^i(\mathbb{C}P^n) = \binom{n+1}{i}c^{2i}$ for $i \leq n/2$.

Thom [229] proved that the Stiefel–Whitney classes $w^*(M)$ of a smooth manifold M are *topological* invariants. Subsequently, in [231] Thom introduced *rational Pontrjagin and L-classes* for compact rational PL-homology manifolds so that the *rational* Pontrjagin classes $p^*(M) \in H^*(M; \mathbb{Q})$ of a smooth closed manifold M are *combinatorial (or piecewise linear) invariants*. Finally, Novikov [190] proved the *topological* invariance of these *rational* Pontrjagin classes $p^*(M) \in H^*(M; \mathbb{Q})$ of a smooth manifold M.

For a closed complex manifold M, the following *Gauss–Bonnet–Chern formula* holds (see [40]):

$$deg(c_*(M)) := \langle c^*(TM), [M] \rangle = \chi(M).$$

(Here, $\langle -, - \rangle$ denotes the usual Kronecker pairing between cohomology and homology. So the total Chern class can be viewed as a higher cohomology class version of the Euler–Poincaré characteristic. Similarly, for a closed manifold M one has:

$$deg(w_*(M)) = \langle w^*(TM), [M] \rangle = \chi(M) \; mod \; 2.$$

More generally, let Iso_n^G be the set of isomorphism classes of smooth closed (and oriented) pure n-dimensional manifolds M for $G = O$ (resp., $G = SO$), or of pure n-dimensional stably almost complex manifolds M for $G = U$, (i.e., $TM \oplus \mathbb{R}_M^a$ is a complex vector bundle for suitable a, with $\mathbb{R}_M$ the trivial real line bundle over M). Then

$$\text{Iso}_*^G := \bigoplus_n \text{Iso}_n^G$$

is a commutative graded semiring with addition and multiplication given by disjoint union and cartesian product, and with 0 and 1 defined by the classes of the empty set and the one-point space, respectively. A multiplicative characteristic class $c\ell_\hbar$ defined by a power series $\hbar$ in the variable $z = w^1, p^1,$ or c^1 induces by the assignment

$$M \mapsto deg(c\ell_{\hbar *}(M)) := \langle c\ell_\hbar^*(TM), [M] \rangle$$

a semiring homomorphism

$$\Phi_\hbar : \mathrm{Iso}_*^G \to \Lambda = \begin{cases} \text{a } \mathbb{Z}/2\text{-algebra for } G = O \text{ and } z = w^1, \\ \text{a } \mathbb{Z}[1/2]\text{-algebra for } G = SO \text{ and } z = p^1, \\ \text{a } \mathbb{Z}\text{-algebra for } G = U \text{ and } z = c^1. \end{cases}$$

Let

$$\Omega_*^G := \mathrm{Iso}_*^G / \sim$$

be the *bordism ring* of closed ($G = O$) and oriented ($G = SO$) or stably almost complex manifolds ($G = U$), with $M \sim 0$ for a closed pure n-dimensional G-manifold M if and only if there is a compact pure $(n + 1)$-dimensional G-manifold W with boundary $\partial W \cong M$. Note that Ω_*^G is indeed a ring with $-[M] = [M]$ for $G = O$ or $-[M] = [-M]$ for $G = SO$ or U, where $-M$ has the opposite orientation of M. Moreover, for W as above with $\partial W \cong M$ one has

$$TW|_{\partial W} \simeq TM \oplus \mathbb{R}_M.$$

Hence, by naturality,

$$c\ell_\hbar^*(TM) = i^* c\ell_\hbar^*(TW)$$

for $i : M \cong \partial W \hookrightarrow W$ the closed inclusion of the boundary. (This also explains the use of the stable tangent bundle for the definition of a stably almost complex manifold.) Then

$$M \sim 0 \implies deg(c\ell_{\hbar*}(M)) = \langle c\ell_\hbar^*(TM), [M] \rangle = 0.$$

Indeed, in the above notations,

$$\begin{aligned} \langle c\ell_\hbar^*(TM), [M] \rangle &= \langle i^* c\ell_\hbar^*(TW), [M] \rangle \\ &= \langle c\ell_\hbar^*(TW), i_*[M] \rangle \\ &= \langle c\ell_\hbar^*(TW), i_* \partial[W] \rangle \\ &= 0, \end{aligned}$$

since $i_* \circ \partial = 0$ in the homology long exact sequence of the pair (W, M). It follows that a multiplicative characteristic class $c\ell_\hbar^*$ defined by a power series $\hbar$ in the variable $z = w^1, p^1,$ or c^1 induces a ring homomorphism (that is, a *genus*)

$$\Phi_\hbar : \Omega_*^G \to \Lambda = \begin{cases} \text{a } \mathbb{Z}/2\text{-algebra for } G = O \text{ and } z = w^1, \\ \text{a } \mathbb{Z}[1/2]\text{-algebra for } G = SO \text{ and } z = p^1, \\ \text{a } \mathbb{Z}\text{-algebra for } G = U \text{ and } z = c^1. \end{cases} \qquad (3.2)$$

Moreover, for Λ a $\mathbb{Q}$-algebra, this induces a one-to-one correspondence between

(i) normalized power series $\hbar$ in the variable $z = p^1$ (or c^1),
(ii) normalized and multiplicative characteristic classes $c\ell^*_{\hbar}$ over finite dimensional base spaces, and
(iii) genera $\Phi_{\hbar} : \Omega^G_* \to \Lambda$ for $G = SO$ (or $G = U$).

Definition 3.2.2 Given $\hbar$ a normalized (i.e., $\hbar(0) = 1$) power series as above, with corresponding class $c\ell^*_{\hbar}$, the associated genus $\Phi_{\hbar}$ is defined by:

$$\Phi_{\hbar}(M) = deg(c\ell^*_{\hbar}(M)) := \langle c\ell^*_{\hbar}(TM), [M] \rangle.$$

Every genus is completely determined by its values on all (even-dimensional) complex projective spaces, by the following structure theorem (see [230] and [177]):

Theorem 3.2.3

(a) *(Thom)* $\Omega^{SO}_* \otimes \mathbb{Q} = \mathbb{Q}[[\mathbb{C}P^{2n}] \mid n \in \mathbb{N}]$ *is a polynomial algebra in the classes of the complex even-dimensional projective spaces.*
(b) *(Milnor)* $\Omega^U_* \otimes \mathbb{Q} = \mathbb{Q}[[\mathbb{C}P^n] \mid n \in \mathbb{N}]$ *is a polynomial algebra in the classes of the complex projective spaces.*

Moreover, a genus $\Phi_{\hbar} : \Omega^U_* \otimes \mathbb{Q} \to \Lambda$ factorizes over the canonical map

$$\Omega^U_* \otimes \mathbb{Q} \to \Omega^{SO}_* \otimes \mathbb{Q}$$

if and only if $\hbar(z)$ is an even power series in $z = c^1$, i.e., $\hbar(z) = g(z^2)$ with $z^2 = (c^1)^2 = p^1$.

Example 3.2.4 (Signature of Manifolds) Let $\sigma(M)$ be the *signature* of a closed oriented manifold M of real dimension $4n$, with $\sigma(M) := 0$ in all other dimensions. Thom showed that the signature defines a genus

$$\sigma : \Omega^{SO}_* \otimes \mathbb{Q} \to \mathbb{Q}$$

with $\sigma(\mathbb{C}P^{2n}) = 1$, for all n. The signature σ comes from the power series $g(z) = \sqrt{z}/\tanh(\sqrt{z})$ in the variable $z = p^1$ (or $\hbar(z) = z/\tanh(z)$ in the variable $z = c^1$), whose corresponding characteristic class $c\ell^* = L^*$ is by definition the *Hirzebruch L-class*. The above correspondence between the signature and L-class is given by the *Hirzebruch signature theorem* [100, 103]:

$$\sigma(M) = \langle L^*(TM), [M] \rangle.$$

Example 3.2.5 (Hirzebruch Genus) The *Hirzebruch polynomial* of a compact complex manifold M is defined as

$$\chi_y(M) := \sum_j \chi(M, \Omega_M^j) y^j.$$

This defines a genus

$$\chi_y : \Omega_*^U \to \mathbb{Q}[y],$$

with $\chi_y(\mathbb{C}P^n) = \sum_{i=0}^n (-y)^i$. The corresponding normalized power series in $z = c^1$ is given by

$$\hbar_y(z) = \frac{z(1+y)}{1 - e^{-z(1+y)}} - zy \in \mathbb{Q}[y][[z]],$$

and the associated characteristic class is the *Hirzebruch class* T_y^*. The correspondence between the χ_y-genus and the Hirzebruch class is given by the *generalized Hirzebruch–Riemann–Roch theorem* [101, 102]:

$$\chi_y(X) = \langle T_y^*(TM), [M] \rangle.$$

Definition 3.2.6 The value $\Phi(M)$ of a genus Φ on the closed manifold M is called a *characteristic number* of M.

Characteristic numbers can be used to classify closed manifolds up to bordism. For example, one has the following result (see [177, 189, 180, 237]):

Theorem 3.2.7

(a) *(Pontrjagin–Thom) Two closed smooth manifolds are bordant (i.e., represent the same element in Ω_*^O) if and only if all their Stiefel–Whitney numbers are the same.*

(b) *(Thom–Wall) Two closed oriented smooth manifolds are bordant up to 2-torsion (i.e., represent the same element in $\Omega_*^{SO} \otimes \mathbb{Z}[1/2]$) if and only if all their Pontrjagin numbers are the same.*

(c) *(Milnor–Novikov) Two closed stably almost complex manifolds are bordant (i.e., represent the same element in Ω_*^U) if and only if all their Chern numbers are the same.*

3.3   L-Classes of Manifolds: Pontrjagin–Thom Construction

In this section, we describe an alternative way (due to Thom [231], see also [180, Section 20]) of building L-classes that applies to triangulated manifolds, for which the tangent bundle is not directly available. As shown by Thom, these L-classes are combinatorial (i.e., piecewise linear) invariants. Moreover, in the case of a smooth manifold, suitably triangulated, they coincide with the (Poincaré duals of the) Hirzebruch *L*-classes defined via the tangent bundle. In 1965, Novikov [190] proved that the rational Pontrjagin classes, hence all the *L*-classes are topological invariants.

The main tools used in the construction of Thom's combinatorial *L*-classes are the Hirzebruch signature theorem and the transversality principle. In more detail, Thom noticed that, given the Hirzebruch signature theorem, one can build an *L*-class provided one has the signature defined on *enough* submanifolds. Such submanifolds are obtained by considering "generic" fibers of maps to spheres.

Construction

For simplicity, consider here the case when M^n is a smooth closed oriented manifold, but see Remark 3.3.1.

If $f : M \to S^{n-4i}$ is a smooth map, it follows by Sard's theorem that the set of regular values of f is dense in S^{n-4i}. For every regular value $p \in S^{n-4i}$, the fiber $f^{-1}(p)$ of f over p is a smooth closed oriented submanifold of M of dimension $4i$, with trivial normal bundle (since it is induced from the normal bundle of the point $\{p\}$ in S^{n-4i}). Consider the signature $\sigma(f^{-1}(p)) \in \mathbb{Z}$ of $f^{-1}(p)$, for p such a regular value of f. It is easy to see that $\sigma(f^{-1}(p))$ depends only on the homotopy class of f. Indeed, if $f \simeq g$, then one can construct a smooth homotopy $H : M \times I \to S^{n-4i}$ having p as a regular value. The compact manifold with boundary $H^{-1}(p)$ is a bordism from $f^{-1}(p)$ to $g^{-1}(p)$, hence by the bordism invariance of the signature one gets that $\sigma(f^{-1}(p)) = \sigma(g^{-1}(p))$. Similarly, if q is another regular value of f, then changing from p to q yields a bordism between $f^{-1}(p)$ and $f^{-1}(q)$, hence $\sigma(f^{-1}(p)) = \sigma(f^{-1}(q))$. Denote by $\sigma(f)$ the common value of $\sigma(f^{-1}(p))$, as p runs through the regular values of f. Since, by the smooth approximation theorem (e.g., see [19, Proposition 17.8]), every continuous map is homotopic to a smooth one, the above considerations yield a map

$$\sigma : [M, S^{n-4i}] \longrightarrow \mathbb{Z}, \quad [f] \mapsto \sigma(f), \tag{3.3}$$

where $[M, S^{n-4i}]$ is the set of homotopy classes of maps $M \to S^{n-4i}$. For obvious reasons, $[M, S^{n-4i}]$ is also called the *cohomotopy set* $\pi^{n-4i}(M)$, and it is in fact a group when $4i < \frac{n-1}{2}$. Moreover, in this range, the map from (3.3) is a group homomorphism. In what follows, it will be more convenient to rationalize (3.3), i.e., to work with

$$\sigma \otimes \mathbb{Q} : \pi^{n-4i}(M) \otimes \mathbb{Q} \to \mathbb{Q}. \tag{3.4}$$

Next, consider the Hurewicz map

$$\pi^k(M^n) \longrightarrow H^k(M; \mathbb{Z}),$$

which is defined by the assignment

$$[f : M^n \to S^k] \mapsto f^*(u),$$

with $u \in H^k(S^k; \mathbb{Z})$ a generator chosen so that $\langle u, [S^k] \rangle = +1$. Serre proved that the rational Hurewicz map

$$\pi^k(M^n) \otimes \mathbb{Q} \longrightarrow H^k(M; \mathbb{Q}) \tag{3.5}$$

is an isomorphism for $n < 2k - 1$ (and in fact this statement holds for every compact CW complex). Note that for $k = n - 4i$, this range translates to $4i < \frac{n-1}{2}$, i.e., it coincides with the range for which the cohomotopy sets are groups.

Assume now that i is chosen so that $4i < \frac{n-1}{2}$. By combining (3.4) and (3.5), one gets a homomorphism

$$\sigma \otimes \mathbb{Q} : H^{n-4i}(M; \mathbb{Q}) \longrightarrow \mathbb{Q}, \tag{3.6}$$

which, by the Universal Coefficient Theorem (cf. [98, Section 3.1]), defines a homology class

$$\ell_{n-4i}(M) := \sigma \otimes \mathbb{Q} \in H_{n-4i}(M; \mathbb{Q}). \tag{3.7}$$

For i varying in the above range, these are called the *Thom homology L-classes* of the smooth manifold M.

Remark 3.3.1 The above arguments can be adapted, with minor modifications, to the case of a closed oriented PL rational homology n-manifold M^n. A fundamental class $[M] \in H_n(M; \mathbb{Z})$ can still be defined by noting that each $(n - 1)$-simplex in a triangulation of M is incident to exactly two n-simplices, and by the orientation assumption it is possible to assign an orientation to each n-simplex so that the sum of all n-simplices forms an n-cycle. By definition, this n-cycle represents the fundamental class $[M]$. The key fact needed in the PL setting (which replaces the argument based on Sard's theorem) is the following:

If $f : M^n \to S^{n-4i}$ is a PL map, then for almost all $p \in S^{n-4i}$, the fiber $f^{-1}(p)$ is a compact PL rational homology $4i$-manifold, with an orientation induced from those of M^n and S^{n-4i}. Moreover, the signature $\sigma(f^{-1}(p))$ of this oriented homology manifold is independent of p for almost all p.

Here, "almost all p" means "except for p belonging to some lower dimensional subcomplex." Furthermore, every map $f : M^n \to S^{n-4i}$ that is not necessarily PL, is homotopic by the PL approximation theorem (e.g., see [108, Lemma 4.2]) to a PL map, and every continuous homotopy of PL maps is homotopic to a PL homotopy of the same PL maps. Hence $[M, S^k]_{PL} = [M, S^k]$, and the above construction applies.

As we will see in Section 3.4, such arguments and construction can be further extended for defining the Goresky–MacPherson L-classes of PL Witt spaces, by making use of the Goresky–MacPherson signature of fibers of maps to spheres. In particular, the L-class of a singular space will in general only be a homology class, which need not lift to cohomology under cap product with the fundamental class.

Coincidence with Hirzebruch L-Classes

In the above discussion, it was assumed that the space M is smooth, so that the Hirzebruch L-classes $L^i(M) \in H^{4i}(M; \mathbb{Q})$ are also available. Consider their Poincaré duals

$$L_{n-4i}(M) := L^i(M) \frown [M] \in H_{n-4i}(M; \mathbb{Q}).$$

The next result asserts that the two notions of homology L-classes of smooth manifolds coincide, namely:

Proposition 3.3.2

$$L_{n-4i}(M) = \ell_{n-4i}(M) \in H_{n-4i}(M; \mathbb{Q}). \tag{3.8}$$

In the proof of Proposition 3.3.2, the following auxiliary result, left here as an exercise, is needed (see, e.g., [6, Lemma 5.7.1]):

Exercise 3.3.3 Let $f : M^n \to S^{n-4i}$ be a smooth map with regular value $p \in S^{n-4i}$. Let $N := f^{-1}(p)$ and denote by $j : N \hookrightarrow M$ the inclusion map. Then the following identity holds in $H_{4i}(M)$:

$$f^*(u) \frown [M] = j_*[N]. \tag{3.9}$$

(Here $u \in H^k(S^k; \mathbb{Z})$ denotes as before a generator chosen so that $\langle u, [S^k] \rangle = +1$.)

Proof of Proposition 3.3.2 By the non-degeneracy of the Kronecker pairing, it suffices to show that

$$\langle x, L_{n-4i}(M) \rangle = \langle x, \ell_{n-4i}(M) \rangle \tag{3.10}$$

for all $x \in H^{n-4i}(M; \mathbb{Q})$.

Recall that i was chosen so that $4i < \frac{n-1}{2}$, so by Serre's theorem we can represent such a cohomology class x as $x = f^*(u)$, for some $f : M^n \to S^{n-4i}$, which moreover can be assumed to be smooth. Then, by definition, for a regular value p of f one has that:

$$\langle f^*(u), \ell_{n-4i}(M) \rangle = \sigma(f^{-1}(p)).$$

So it suffices to show that

$$\langle f^*(u), L_{n-4i}(M) \rangle = \sigma(f^{-1}(p)). \tag{3.11}$$

By the Hirzebruch signature theorem, the right-hand side of (3.11) is computed as

$$\sigma(f^{-1}(p)) = \langle L^i(f^{-1}(p)), [f^{-1}(p)] \rangle.$$

Set $N := f^{-1}(p)$ and denote by $j : N \hookrightarrow M$ the inclusion map. Using the fact that the normal bundle ν_N of N in M is trivial, together with the naturality and multiplicativity of L-classes, the i-th Hirzebruch L-class of N can be computed as follows:

$$\begin{aligned}
j^* L^i(M) &= j^* L^i(TM) \\
&= L^i(j^* TM) \\
&= L^i(TN \oplus \nu_N) \tag{3.12} \\
&= L^i(TN) \\
&= L^i(N).
\end{aligned}$$

Therefore, by using (3.9), one gets:

$$\begin{aligned}
\langle f^*(u), L_{n-4i}(M) \rangle &= \langle f^*(u), L^i(M) \frown [M] \rangle \\
&= \langle L^i(M), f^*(u) \frown [M] \rangle \\
&= \langle L^i(M), j_*[N] \rangle \\
&= \langle j^* L^i(M), [N] \rangle \\
&= \langle L^i(N), [N] \rangle.
\end{aligned}$$

$\square$

Removing the Dimension Restriction

To complete the definition of the Thom L-class $\ell_*(M)$, it remains to define $\ell_{n-4i}(M)$ when $4i \geq \frac{n-1}{2}$. This is done by crossing with a high-dimensional sphere. Let

$$\tilde{M} = M^n \times S^m,$$

with m large, and write

$$\tilde{n} := \dim \tilde{M} = n + m.$$

For m large enough, one has that $4i < \frac{n+m-1}{2} = \frac{\tilde{n}-1}{2}$, and so $\ell_{\tilde{n}-4i}(\widetilde{M})$ is defined as in Section 3.3. Furthermore, it follows by the Künneth isomorphism that

$$\ell_{\tilde{n}-4i}(\widetilde{M}) \in H_{n+m-4i}(M^n \times S^m; \mathbb{Q})$$

$$\cong H_{n+m-4i}(M^n; \mathbb{Q}) \otimes H_0(S^m; \mathbb{Q})$$

$$\oplus H_{n-4i}(M^n; \mathbb{Q}) \otimes H_m(S^m; \mathbb{Q})$$

$$\cong H_{n-4i}(M^n; \mathbb{Q}).$$

Let

$$\eta : H_{n+m-4i}(M^n \times S^m; \mathbb{Q}) \xrightarrow{\cong} H_{n-4i}(M^n; \mathbb{Q})$$

denote this isomorphism, and define

$$\ell_{n-4i}(M) := \eta(\ell_{\tilde{n}-4i}(\widetilde{M})). \tag{3.13}$$

Exercise 3.3.4 Show that the definition in (3.13) is independent of the choice of large m, i.e., $\ell_{n-4i}(M)$ is well defined for all i.

3.4 Goresky–MacPherson L-Classes

In this section, we sketch the construction of the Goresky–MacPherson L-classes $\ell_k(X) \in H_k(X; \mathbb{Q})$ for a closed oriented n-dimensional PL Witt space X. Full details can be found in [81, 83, 218], or in Friedman's more recent treatment [75] on intersection homology. The construction is a direct extension of the Thom–Pontrjagin approach of Section 3.3 to the PL Witt setting, by making use of the Goresky–MacPherson intersection homology signature of "generic" fibers of maps $X \to S^k$. The basic idea is that such generic fibers admit stratifications with respect to which they are PL Witt spaces, so a signature invariant of such fibers can be defined via intersection homology.

Let $\ell_0(X) \in H_0(X; \mathbb{Q})$ be defined as the class represented by a point in each component carrying the coefficient corresponding to the (Witt) signature of the component. When $k > 0$, the following key result implies that, in suitable situations, one can assign Witt signature invariants to maps $f : X \to S^k$.

Proposition 3.4.1 *Let X be a closed oriented n-dimensional PL Witt space, and suppose that S^k, $k > 0$, has been given an orientation so that $f : X \to S^k$ is a PL map. Then for almost all $p \in S^k$, the inverse image $f^{-1}(p)$ can be stratified as a closed oriented $(n-k)$-dimensional PL Witt space embedded in X. Furthermore, for almost all $p, q \in S^k$, the Witt spaces $f^{-1}(p)$ and $f^{-1}(q)$ have the same signature; this common signature, denoted by $\sigma(f)$, depends only on the PL homotopy class of f in $[X, S^k]_{PL}$.*

Here, $[-, -]_{PL}$ denotes the set of PL homotopy classes of PL maps, and "almost all p" means as before that the statement applies to all p not belonging to the simplicial $(k-1)$-skeleton of some appropriate triangulation of S^k, in particular one with respect to which $f : X \to S^k$ is simplicial. The key idea for the proof of the above statement is that for almost all p, the fibers $f^{-1}(p)$ have stratifications with respect to which they are PL Witt spaces, and changing from p to q or changing from f to a homotopic map will yield PL Witt space bordisms between these fiber Witt spaces. Since signature is independent of stratification and Witt bordism class, almost all fibers $f^{-1}(p)$ have the same Witt signature.

Proposition 3.4.1 shows how to assign a number, $\sigma(f)$, to each element of $[X, S^k]_{PL}$, the PL homotopy set of PL maps from X to S^k. If $g : X \to S^k$ is a map that is not necessarily PL, then by the PL approximation theorem (e.g., see [108, Lemma 4.2]), g is homotopic to a PL map. Furthermore, by the same theorem, every continuous homotopy of PL maps is homotopic to a PL homotopy of the same PL maps. Therefore,

$$[X, S^k]_{PL} = [X, S^k] = \pi^k(X),$$

the full set of topological homotopy classes of maps $X \to S^k$, i.e., the k-th cohomotopy set of X. So one gets by Proposition 3.4.1 a well-defined function

$$\sigma : \pi^k(X) \longrightarrow \mathbb{Z}.$$

Moreover, one has the following:

Lemma 3.4.2 *If $k > \frac{n+1}{2}$, $\sigma : \pi^k(X) \longrightarrow \mathbb{Z}$ is a homomorphism of abelian groups.*

At this point, one can formally repeat all arguments of Section 3.3 (i.e., Serre's theorem for the Hurewicz map, and resp., crossing with a large sphere for removing the dimensionality constraint in Serre's theorem), to obtain homology classes $\ell_k(X) \in H_k(X; \mathbb{Q})$, called the *Goresky–MacPherson*[1] *L-classes of X*. Furthermore, if X is a rational homology (or smooth) manifold, these classes coincide with those constructed in Section 3.3.

Remark 3.4.3 If $n - k$ is not a multiple of 4, then the signature of every fiber of a map $X \to S^k$ must be 0, by definition. In this case the L-class is trivial. Therefore the L-classes are typically only defined in dimensions $k = n - 4i, i \geq 0$.

Remark 3.4.4 In general, the L-classes $\ell_k(X) \in H_k(X; \mathbb{Q})$ are not in the image of the Poincaré map $H^*(X; \mathbb{Q}) \xrightarrow{\cap [X]} H_*(X; \mathbb{Q})$. Thus the Goresky–MacPherson L-classes of (singular) PL Witt spaces are typically only homology characteristic classes.

---

[1]In fact, Goresky and MacPherson defined L-classes only for pseudomanifolds with only even codimension strata, e.g., irreducible complex algebraic varieties.

We conclude this section with a single characterizing property of *L*-classes in terms of signatures of subspaces.

Definition 3.4.5 A subspace $Z \subset X$ is called a *trivial normally nonsingular subspace* if the inclusion $i : Z \hookrightarrow X$ extends to a stratified homeomorphism from $\mathbb{R}^k \times Z$ onto some neighborhood U of Z for some k.

If X and Z are PL pseudomanifolds, a trivial normally nonsingular PL subspace is defined in the obvious way, and if X is a Witt space then so is Z. Given a trivial normally nonsingular subspace Z of X, one can define a map $f_Z : X \to S^k$ that "projects to the nonsingular direction." Specifically, identifying S^k with the one-point compactification of $\mathbb{R}^k$ with p_0 being the point at infinity, the map $f_Z : X \to S^k$ takes each image fiber in U of the form $i(\mathbb{R}^k \times \{z\})$ for $z \in Z$ identically to $\mathbb{R}^k = S^k - p_0$, and it maps $X - U$ to p_0. If $k = 0$, the only trivial normally nonsingular subspaces are the components of X, in which case each component in Z is sent to $1 \in S^0$ and the other components are sent to $0 \in S^0$. With this notation, the following result holds:

Theorem 3.4.6 *Let X^n be a closed oriented PL Witt space. Then, for every k, there exists a unique homology L-class $\ell_k = \ell_k(X) \in H_k(X; \mathbb{Q})$ such that for every $(n - k)$-dimensional trivial normally nonsingular PL subspace Z one has*

$$\langle f_Z^*(u), \ell_k \rangle = \sigma(Z), \tag{3.14}$$

where $u \in H^k(S^k; \mathbb{Q})$ is the element such that $\langle u, [S^k] \rangle = 1$.

Remark 3.4.7 Further generalizations of *L*-classes to similar invariants of self-dual sheaf complexes have been carried out by Cappell–Shaneson [31], Banagl [4, 5], and Woolf [246]. Chern classes and Whitney classes for singular varieties have been developed in the early 1970s by MacPherson [148] and, respectively, Sullivan [226]. In 1975, Baum–Fulton–MacPherson [9] defined (homology) Todd classes for singular varieties, generalizing the Riemann–Roch theorem in the singular context. More recently, Brasselet–Schürmann–Yokura [21] defined (homology) Hirzebruch classes for singular varieties, which provide a functorial unification of the above-mentioned Chern-, Baum–Fulton–MacPherson Todd-, and *L*-classes, respectively (a feature already existent in the nonsingular case, as described in Hirzebruch's seminal book [102]). For a nice survey on characteristic classes of singular spaces, the reader may consult [216].

Chapter 4
Brief Introduction to Sheaf Theory

In this chapter, we introduce the prerequisites needed later on (in Chapter 6) for the sheaf-theoretic description of intersection homology groups. For more detailed accounts, see [113, 23, 122], or [15].

We fix a commutative noetherian ring A of finite cohomological dimension and (unless otherwise specified) we work with sheaves of A-modules.

4.1 Sheaves

Definition 4.1.1 A *presheaf* $\mathcal{F}$ (of A-modules) on a topological space X is a contravariant functor from the category of open sets and inclusions to the category of A-modules. Equivalently, a presheaf $\mathcal{F}$ is given by the following data:

(a) to every open subset $U \subseteq X$ associate an A-module $\mathcal{F}(U)$, with $\mathcal{F}(\emptyset) = 0$.
(b) to every inclusion $i : U \hookrightarrow V$ of open subsets in X associate an A-module homomorphism $\rho_{VU} : \mathcal{F}(V) \to \mathcal{F}(U)$ (called *restriction*), so that $\rho_{UU} = id$ and $\rho_{VU} \circ \rho_{WV} = \rho_{WU}$, for open subsets $U \subset V \subset W$ in X.

The A-module $\mathcal{F}(U)$ is referred to as the *sections of $\mathcal{F}$ over U*, and it will be often denoted by $\Gamma(U, \mathcal{F})$. If $U \subset V$ and $s \in \mathcal{F}(V)$, then $\rho_{VU}(s)$ will be usually denoted by $s|_U$.

Definition 4.1.2 A presheaf $\mathcal{F}$ is called a *sheaf* (of A-modules) if, in addition, it satisfies the following *gluing property*:

(c) Let $\{V_i \mid i \in I\}$ be a collection of opens in X and suppose that sections $s_i \in \Gamma(V_i, \mathcal{F})$ are given such that $s_i|_{V_i \cap V_j} = s_j|_{V_i \cap V_j}$ for all $i, j \in I$. Then there exists a unique section $s \in \Gamma(\bigcup_{i \in I} V_i, \mathcal{F})$ so that $s|_{V_i} = s_i$.

Remark 4.1.3 An essential point in the gluing property is that the index set I may have infinite cardinality.

© Springer Nature Switzerland AG 2019
L. G. Maxim, *Intersection Homology & Perverse Sheaves*, Graduate Texts in Mathematics 281, https://doi.org/10.1007/978-3-030-27644-7_4

Remark 4.1.4 One may, of course, consider sheaves taking values in the category of sets, without any additional algebraic structure.

A sheaf can be regarded as a device for dealing with properties that are local in nature and measuring the transition from local to global.

Example 4.1.5

(i) if X is a point, a sheaf on X is an A-module.

(ii) The *constant sheaf* $\underline{M}_X$ on X for an A-module M is defined by $\underline{M}_X(\emptyset) = 0$, $\underline{M}_X(U) = \{$continuous functions $f : U \to M\}$, where M is equipped with the discrete topology. If X is locally connected, then $\underline{M}_X(U) = M$ for every connected open set U in X.

(iii) The continuous $\mathbb{R}$-valued functions on a topological space X form a sheaf C_X^0 (with $A = \mathbb{R}$), where $C_X^0(U)$ is the real vector space of continuous maps $U \to \mathbb{R}$ and, for $U \subset V$, ρ_{VU} is the usual restriction map. Similarly, if X is a smooth manifold, one defines the sheaf C_X^∞ of smooth functions on X. If X is an analytic manifold, then one also has a sheaf of analytic functions on X. For an algebraic variety one can define similarly a sheaf of regular (algebraic) functions.

(iv) If $\pi : Y \to X$ is a continuous map, define for every open $U \subseteq X$:

$$\mathcal{F}(U) = \{s : U \to Y \mid s \text{ is continuous, } \pi \circ s = id_U\}.$$

This assignment defines a sheaf of sets, called the *sheaf of sections of π*. In particular, a vector bundle whose fiber is an A-vector space gives rise to a sheaf of A-vector spaces. Important examples are the sheaf of vector fields coming from the tangent bundle, or the sheaf of 1-forms associated to the cotangent bundle.

(v) Let $\mathcal{L}^1$ be the presheaf on $\mathbb{R}$ whose sections on an open $U \subseteq \mathbb{R}$ are given by $\Gamma(U, \mathcal{L}^1) = L^1(U, dx)$, the Lebesgue integrable functions over U. Cover $\mathbb{R}$ by open bounded intervals U_α, and consider the constant function $1|_{U_\alpha}$ (as section) on U_α. Then $1|_{U_\alpha} \in \mathcal{L}^1(U_\alpha)$. Moreover, on overlaps $U_\alpha \cap U_\beta$, these sections agree, but one cannot glue them together since the global function 1 on $\mathbb{R}$ is not integrable. So $\mathcal{L}^1$ is *not* a sheaf.

Definition 4.1.6 The *stalk* of a presheaf $\mathcal{F}$ at a point $x \in X$ is defined as the limit

$$\mathcal{F}_x := \lim_{x \in U} \mathcal{F}(U)$$

over the directed set of open neighborhoods of x.

An element of the stalk $\mathcal{F}_x$ is represented by a pair (U, s), where $U \subseteq X$ is an open neighborhood of x and $s \in \mathcal{F}(U)$. Moreover, two such representatives (U, s) and (V, t) are equivalent, and one writes $(U, s) \sim (V, t)$, if there exists an open set $W \subseteq U \cap V$ such that $s|_W = t|_W$. Write $s_x \in \mathcal{F}_x$ for such a representative (U, s).

As seen in Example 4.1.5(v), not every presheaf is a sheaf. However, there is a canonical way to assign to a presheaf $\mathcal{F}$ a sheaf

$$\mathcal{F}^+ := \mathit{Sheaf}(\mathcal{F}),$$

called the *sheafification of* $\mathcal{F}$, so that stalk information is preserved. (Details of this construction will not be given here; the reader may consult instead any of the standard references on sheaves.) For this reason, from now on we work only with sheaves.

Definition 4.1.7 A homomorphism of (pre)sheaves $\mathcal{F} \to \mathcal{G}$ is a collection of A-module homomorphisms $\mathcal{F}(U) \to \mathcal{G}(U)$ associated to open subsets $U \subseteq X$, which are compatible with the restrictions.

With the obvious notions of subsheaf, quotient sheaf, kernel, cokernel, etc., it can be shown that:

Lemma 4.1.8 *Sheaves of A-modules on a topological space X form an abelian category, denoted by $Sh(X)$.*

Moreover, the following holds:

Proposition 4.1.9 *A homomorphism of sheaves $\mathcal{F} \to \mathcal{G}$ on X is injective (resp., surjective) if and only if the stalk homomorphism $\mathcal{F}_x \to \mathcal{G}_x$ is injective (resp., surjective) for all $x \in X$.*

We continue with the definition of some important functors.

Definition 4.1.10 *Pushforward and pullback*

- If $f : X \to Y$ is a continuous map and $\mathcal{F} \in Sh(X)$, the *pushforward* $f_* \mathcal{F} \in Sh(Y)$ is defined by the assignment

$$f_* \mathcal{F}(U) := \mathcal{F}(f^{-1}(U))$$

for every open set $U \subseteq Y$.
- If $f : X \to Y$ is a continuous map and $\mathcal{G} \in Sh(Y)$, the *pullback* $f^* \mathcal{G} \in Sh(X)$ is defined by assigning to every open $V \subseteq X$ the A-module

$$f^* \mathcal{G}(V) := \lim_{f(V) \subseteq U} \mathcal{G}(U)$$

where the limit is over opens $U \subseteq Y$ containing $f(V)$. This is only a presheaf, and we sheafify.

Example 4.1.11 Let $f : X \to pt$ be the constant map to a point space. Then $f_* \mathcal{F} = \Gamma(X, \mathcal{F})$ is the *global section functor*, and if M is an A-module then $f^* M = \underline{M}_X$ is the constant sheaf.

Example 4.1.12 If $i : \{x\} \hookrightarrow X$ is the inclusion of a point, and M is an A-module, then i_*M is a *skyscraper sheaf* at x (so-called because its stalks are 0, except for the stalk at x). Moreover, if $\mathcal{F} \in Sh(X)$, then $i^*\mathcal{F} = \mathcal{F}_x$.

Exercise 4.1.13 Show that for a continuous map $f : X \to Y$ and $\mathcal{G} \in Sh(Y)$ one has

$$(f^*\mathcal{G})_x \cong \mathcal{G}_{f(x)}.$$

Proposition 4.1.14 $g_* \circ f_* = (g \circ f)_*$ *and* $f^* \circ g^* = (g \circ f)^*$.

If $i : X \hookrightarrow Y$ is an inclusion and $\mathcal{G} \in Sh(Y)$, denote $i^*\mathcal{G}$ by $\mathcal{G}|_X$, and call it the *restriction* of $\mathcal{G}$ to the subset X.

Definition 4.1.15 (Sections of a Sheaf Over a Closed Subset) Let $Z \hookrightarrow X$ be a closed subset in X and let $\mathcal{F} \in Sh(X)$. The group of sections of $\mathcal{F}$ over Z is defined as:

$$\Gamma(Z, \mathcal{F}) := \Gamma(Z, \mathcal{F}|_Z).$$

The next result states that f^* and f_* are adjoint functors.

Proposition 4.1.16 *Let* $f : X \to Y$ *be a continuous map, and let* $\mathcal{F} \in Sh(X)$, $\mathcal{G} \in Sh(Y)$. *There are canonical adjunction morphisms* $f^*f_*\mathcal{F} \longrightarrow \mathcal{F}$ *and* $\mathcal{G} \longrightarrow f_*f^*\mathcal{G}$, *which yield an isomorphism*

$$\mathrm{Hom}_{Sh(X)}(f^*\mathcal{G}, \mathcal{F}) \cong \mathrm{Hom}_{Sh(Y)}(\mathcal{G}, f_*\mathcal{F}).$$

Definition 4.1.17 (Hom Sheaf) If $\mathcal{F}, \mathcal{G} \in Sh(X)$, the sheaf $\mathcal{H}om(\mathcal{F}, \mathcal{G})$ is defined by the assignment:

$$U \mapsto \mathrm{Hom}_{Sh(U)}(\mathcal{F}|_U, \mathcal{G}|_U).$$

Exercise 4.1.18 Show that the group of global sections of $\mathcal{H}om(\mathcal{F}, \mathcal{G})$ is the A-module of all sheaf homomorphisms from $\mathcal{F}$ to $\mathcal{G}$, i.e.,

$$\Gamma(X, \mathcal{H}om(\mathcal{F}, \mathcal{G})) = \mathrm{Hom}_{Sh(X)}(\mathcal{F}, \mathcal{G}).$$

Exercise 4.1.19 Show that for any $\mathcal{G} \in Sh(X)$, one has the identifications: $\mathrm{Hom}_{Sh(X)}(\underline{A}_X, \mathcal{G}) = \Gamma(X, \mathcal{G})$ and $\mathcal{H}om(\underline{A}_X, \mathcal{G}) = \mathcal{G}$.

Remark 4.1.20 Note that in general,

$$\mathcal{H}om(\mathcal{F}, \mathcal{G})_x \neq \mathrm{Hom}(\mathcal{F}_x, \mathcal{G}_x).$$

(See Exercise 4.1.21 below.)

Exercise 4.1.21 Let $X = \mathbb{C}$ and let $\mathcal{F} = \mathbb{Q}_0$ be the skyscraper sheaf with stalks $\mathcal{F}_x = 0$ for $x \neq 0$ and $\mathcal{F}_0 = \mathbb{Q}$. Let $\mathcal{G} = \mathbb{Q}_X$ and let $u : \mathcal{F} \to \mathcal{G}$ be the natural injective homomorphism. Let $\mathcal{K} = \operatorname{Coker} u$. Show that $\operatorname{Hom}(\mathcal{G}_0, \mathcal{K}_0) \neq \mathcal{Hom}(\mathcal{G}, \mathcal{K})_0$.

Definition 4.1.22 (Direct Sum, Tensor Product) If $\mathcal{F}, \mathcal{G} \in Sh(X)$, the *direct sum* $\mathcal{F} \oplus \mathcal{G}$ is the sheaf defined by:

$$U \mapsto \mathcal{F}(U) \oplus \mathcal{G}(U).$$

The *tensor product* $\mathcal{F} \otimes \mathcal{G} \in Sh(X)$ is the sheaf associated to the presheaf:

$$U \mapsto \mathcal{F}(U) \otimes_A \mathcal{G}(U).$$

Remark 4.1.23 Since direct sums and tensor products commute with direct limits, one has the stalk identities:

$$(\mathcal{F} \oplus \mathcal{G})_x = \mathcal{F}_x \oplus \mathcal{G}_x, \quad (\mathcal{F} \otimes \mathcal{G})_x = \mathcal{F}_x \otimes \mathcal{G}_x.$$

For completeness, we include the following definition, which will be expanded on in the next chapter:

Definition 4.1.24 (Direct Image with Proper Support) If $s \in \mathcal{F}(U)$, then the *support of s* is defined as

$$\operatorname{supp}(s) = \overline{\{x \in U \mid s_x \neq 0\}}.$$

If $f : X \to Y$ is a continuous map of locally compact topological spaces and $\mathcal{F} \in Sh(X)$, define $f_! \mathcal{F} \in Sh(Y)$ by the assignment

$$U \mapsto \{s \in \mathcal{F}(f^{-1}(U)) \mid f_{|\operatorname{supp}(s)} : \operatorname{supp}(s) \to Y \text{ is proper}\}.$$

(Recall that a map between locally compact topological spaces is called proper if the inverse image of a compact set is compact.)

Example 4.1.25 In the notations and assumptions of Definition 4.1.24, let us consider some of its special cases:

(a) If $f : X \to pt$ is the constant map to a point space, then $f_! \mathcal{F} = \Gamma_c(X, \mathcal{F})$ is the group (A-module) of *global sections* of $\mathcal{F}$ with *compact support*.
(b) If $f : X \hookrightarrow Y$ is the inclusion of a subspace, then for every open $U \subseteq Y$, $f_! \mathcal{F}(U)$ consists of sections $s \in \mathcal{F}(U \cap X)$ whose support is compact. In particular, $f_! \mathcal{F}$ vanishes outside X.
(c) If $i : X \hookrightarrow Y$ is a closed inclusion, then $i_! = i_*$. In general, if $f : X \to Y$ is proper, then $f_! = f_*$.

Definition 4.1.26 (Sections with Support in a Closed Subset) If $U \subseteq X$ is open and K is a closed subset of U, set

$$\Gamma_K(U, \mathcal{F}) := \mathrm{Ker}\left(\mathcal{F}(U) \to \mathcal{F}(U - K)\right)$$

for the subgroup (submodule) of $\Gamma(U, \mathcal{F})$ consisting of sections whose support is contained in K.

4.2 Local Systems

Definition 4.2.1 An A-*local system* on a topological space X is a *locally constant sheaf* $\mathcal{L}$ of A-modules on X, i.e., there is an open covering $\{U_i\}_i$ of X and a family of A-modules $\{M_i\}_i$ so that $\mathcal{L}|_{U_i} \simeq \underline{M_i}$, the constant sheaf on U_i associated to the A-module M_i. If X is connected, one can replace the family $\{M_i\}_i$ by a single A-module M. If M_i are all free of finite rank r (and A is a field or a principal ideal domain), then $\mathcal{L}$ is said to have rank r. If the covering $\{U_i\}_i$ can be chosen to be the trivial cover $\{X\}$, then $\mathcal{L}$ is a trivial (constant) local system.

Example 4.2.2 For an n-dimensional manifold X, define its *orientation sheaf* by

$$Or_X := Sheaf\left(U \mapsto H_n(X, X - U; A)\right),$$

that is, the sheafification of the presheaf defined by the assignment $U \mapsto H_n(X, X - U; A)$. If $\partial X = \emptyset$, then Or_X is a locally constant sheaf with stalks isomorphic to A. It is constant if X is orientable. If $\partial X \neq \emptyset$, then $Or_X|_{\partial X} = 0$.

From now on, we assume that X is paracompact, Hausdorff, path-connected and locally simply connected (e.g., X is a connected complex algebraic variety endowed with the complex topology). In particular, X has a universal cover. Under these assumptions, there are several classical descriptions of local systems, e.g., see [113, Page 252] and [219, Page 58, Page 360]:

Proposition 4.2.3 *The following categories are equivalent:*

(i) *A-local systems on X;*
(ii) *covariant functors from the fundamental groupoid of X to the category of A-modules;*
(iii) *representations $\rho : \pi_1(X, x_0) \to Aut(M)$, where x_0 is a basepoint in X, and M is an A-module.*

Given an A-local system $\mathcal{L}$ on X, a representation $\rho : \pi_1(X, x_0) \to Aut(\mathcal{L}_{x_0})$ can be obtained as follows: pick a loop $\gamma : [0, 1] \to X$ at x_0 and use local trivializations of $\mathcal{L}$ along the loop γ to move elements in $\mathcal{L}_{x_0}$ along $\mathcal{L}_{\gamma(t)}$, back to $\mathcal{L}_{x_0}$. Conversely, given a representation $\rho : \pi_1(X, x_0) \to Aut(M)$, a local system associated to ρ can be defined as the sheaf of local sections of the natural projection map $\widetilde{X} \times_G M \to X$, where $(\widetilde{X}, \widetilde{x}) \to (X, x_0)$ is the universal cover of X, $G = \pi_1(X, x_0)$ acts on $\widetilde{X}$ by deck transformations, and M is endowed with the discrete topology.

Exercise 4.2.4 Use Proposition 4.2.3 to show that a local system on a contractible space is isomorphic to a constant sheaf.

Exercise 4.2.5 Use Proposition 4.2.3 to show that if $\mathcal{L}$ and $\mathcal{M}$ are A-local systems on X, then $\mathcal{L} \oplus \mathcal{M}$, $\mathcal{L} \otimes \mathcal{M}$, and $\mathcal{H}om(\mathcal{L}, \mathcal{M})$ are also local systems.

Remark 4.2.6 In view of the correspondence of Proposition 4.2.3, terminology from representation theory can be carried over to local systems. For example, an irreducible (or simple) local system is defined by an irreducible representation, while a semi-simple local system corresponds to a completely reducible representation of the fundamental group.

Definition 4.2.7 If $\mathcal{L}$ is an A-local system on X, its *dual* is defined as

$$\mathcal{L}^\vee := \mathcal{H}om(\mathcal{L}, \underline{A}_X).$$

Exercise 4.2.8 Given an A-local system $\mathcal{L}$ on X, show that its dual $\mathcal{L}^\vee$ is also an A-local system. Find its stalk $\mathcal{L}_x^\vee$ and the corresponding representation $\rho^\vee : \pi_1(X, x) \to Aut(\mathcal{L}_x^\vee)$ as given by Proposition 4.2.3.

It is easy to see that:

Proposition 4.2.9 *The category $Loc(X)$ of A-local systems on X is a full abelian subcategory of the category $Sh(X)$ of sheaves of A-modules on X.*

Using Proposition 4.2.3(ii), a local system $\mathcal{L}$ on X corresponds to an A-module $\mathcal{L}_x$ for each $x \in X$, and an isomorphism $\alpha^* : \mathcal{L}_{\alpha(0)} \to \mathcal{L}_{\alpha(1)}$ for every continuous path $\alpha : [0, 1] \to X$, so that:

(a) $\alpha^* = \beta^*$ if α and β are homotopic paths with the same endpoints;
(b) $(\alpha * \beta)^* = \beta^* \circ \alpha^*$ if $\alpha(1) = \beta(0)$, where $\alpha * \beta$ denotes the concatenation of paths α and β.

Denote the local system corresponding to a representation as in Proposition 4.2.3(iii) by (M, ρ). Then one easily gets the following:

Proposition 4.2.10 *If $f : X \to Y$ is a continuous map, and $\mathcal{L}$ is a local system on Y given by a representation (M, ρ), then $f^*\mathcal{L}$ is a local system on X given by the representation $(M, \rho \circ f_*)$, with $f_* : \pi_1(X, x_0) \to \pi_1(Y, f(x_0))$ induced by f.*

Remark 4.2.11 If $f : X \to Y$ is a continuous map, and $\mathcal{L}$ is a local system on X, then $f_*\mathcal{L}$ is *not* necessarily a local system on Y. For example, if $\mathcal{L}$ is a non-trivial local system on $\mathbb{C}^*$ and $f : \mathbb{C}^* \to \mathbb{C}$ is the inclusion map, then $f_*\mathcal{L}$ is not a local system on $\mathbb{C}$ (as it is not the constant sheaf).

Exercise 4.2.12 Show that if $f : X \to Y$ is a finite (unramified) covering map and $\mathcal{L}$ is a sheaf of A-modules on X, then $\mathcal{L}$ is a local system on X if and only if

$f_*\mathcal{L}$ is a local system on Y. If $\mathcal{L}$ is a local system on X, describe the representation corresponding to $f_*\mathcal{L}$.

Homology with Local Coefficients

If $\sigma : \Delta_i \to X$ is a singular i-simplex of X, and $\mathcal{L}$ is a local system on X, then $\sigma^*\mathcal{L}$ is a local system on Δ_i. Since $\pi_1(\Delta_i) = 0$, it follows that $\sigma^*\mathcal{L}$ is a constant sheaf on Δ_i. Let

$$\mathcal{L}_\sigma := \sigma^*\mathcal{L}(\Delta_i).$$

Then the restriction maps

$$\rho_p^\sigma : \mathcal{L}_\sigma \longrightarrow (\sigma^*\mathcal{L})_p = \mathcal{L}_{\sigma(p)}$$

are isomorphisms for all $p \in \Delta_i$. If τ is a face of σ, and $p \in \tau$, the composition

$$\rho_\tau^\sigma := (\rho_p^\tau)^{-1} \circ \rho_p^\sigma : \mathcal{L}_\sigma \longrightarrow \mathcal{L}_\tau$$

is independent of the chosen point p.

Let $S_i(X, \mathcal{L})$ be the A-module consisting of all formal finite linear combinations

$$\xi = \sum_\sigma \ell_\sigma \sigma$$

with σ running through the singular i-simplices of X and $\ell_\sigma \in \mathcal{L}_\sigma$. Define a boundary operator

$$\partial_i : S_i(X, \mathcal{L}) \longrightarrow S_{i-1}(X, \mathcal{L})$$

$$\partial_i(\xi) = \sum_\sigma \sum_{\tau \text{ face of } \sigma} \pm \rho_\tau^\sigma(\ell_\sigma) \cdot \tau,$$

with the sign depending on orientation as in the classical case. It then follows that

$$\partial_i \circ \partial_{i+1} = 0.$$

The *homology of X with local coefficients* $\mathcal{L}$ is defined as:

$$H_i(X; \mathcal{L}) := \operatorname{Ker} \partial_i / \operatorname{Image} \partial_{i+1}.$$

Similarly, cohomology groups $H^i(X; \mathcal{L})$ can be defined by dualizing the above construction. In particular, when A is a field and $\mathcal{L}$ is an A-local system, one gets

$$H^i(X; \mathcal{L}^\vee) \cong H_i(X; \mathcal{L})^\vee \tag{4.1}$$

for all integers i, where $-^\vee$ on the right-hand side denotes the dual A-vector space.

Exercise 4.2.13 Let $X = \mathbb{C}^*$ and assume that A is a field. Let $\mathcal{L}_T$ be the local system on X corresponding to the representation

$$\rho : \pi_1(X, 1) \cong \mathbb{Z} \longrightarrow GL_r(A) = Aut(A^r)$$

$$1 \longmapsto T.$$

Show that

$$H_i(X; \mathcal{L}_T) = \begin{cases} \text{Coker } (T - id), & i = 0, \\ \text{Ker } (T - id), & i = 1, \\ 0, & i > 1. \end{cases}$$

Remark 4.2.14 An alternative definition of homology with local coefficients goes as follows (see, e.g., [61, page 50] and the references therein). Let $G = \pi_1(X, x_0)$ be the fundamental group of X, and let $\rho : G \to Aut(M)$ be the representation associated to the local system $\mathcal{L}$. Let $H \subseteq G$ be a normal subgroup so that $H \subseteq \text{Ker } \rho$. Let $X_H \to X$ be the (unramified) covering corresponding to H, with group of covering transformations $G' = G/H$. Let $\rho' : G' \to Aut(M)$ be the induced representation. As above, let $S_*(X_H)$ be the singular chain complex of X_H with A-coefficients. Consider the equivariant tensor product

$$S_*(X, \mathcal{L}) = S_*(X_H) \otimes_{G'} M$$

and the equivariant Hom

$$S^*(X, \mathcal{L}) = \text{Hom}^*_{G'}(S_*(X_H), M).$$

The homology groups of these chain complexes of A-modules yield the (co)homology groups $H_*(X; \mathcal{L})$ and $H^*(X; \mathcal{L})$ of X with local coefficients $\mathcal{L}$.

Exercise 4.2.15 Let A be a field and let $\mathcal{L}$ be an A-local system of rank r on a finite CW complex X. Show that the A-vector space $H_i(X; \mathcal{L})$ is finite dimensional for every integer i, and the corresponding Euler characteristic is computed by the formula:

$$\chi(X, \mathcal{L}) := \sum_i (-1)^i \dim_A H_i(X; \mathcal{L}) = r \cdot \chi(X).$$

Exercise 4.2.16 Let X be a connected space, with fundamental group $G = \pi_1(X)$, and let $\rho : G \to Aut(M)$ be the representation associated to the local system $\mathcal{L}$ on X. Show that

$$H^0(X; \mathcal{L}) \cong M^{\pi_1(X)},$$

where $M^{\pi_1(X)}$ denotes the fixed part of M under the $\pi_1(X)$-action.

Remark 4.2.17 When $A = \mathbb{C}$ and X is a complex analytic manifold, there is an equivalence of categories between the category of finite rank $\mathbb{C}$-local systems $\mathcal{L}$ on X and the category of holomorphic vector bundles $\mathcal{V}$ on X with a flat connection

$$\nabla : \mathcal{V} \to \mathcal{V} \otimes \Omega_X^1,$$

with Ω_X^1 denoting the (locally free) sheaf of holomorphic 1-forms on X. Under this equivalence, the vector bundle associated to a local system $\mathcal{L}$ is

$$\mathcal{V} := \mathcal{L} \otimes_{\mathbb{C}_X} \mathcal{O}_X,$$

and the connection is the unique integrable connection whose sheaf of horizontal sections $\operatorname{Ker} \nabla$ is exactly $\mathcal{L}$. (As customary, $\mathcal{O}_X$ denotes the sheaf of germs of holomorphic functions on X.) A similar equivalence holds between $\mathbb{R}$-local systems on a real smooth manifold and real smooth vector bundles $\mathcal{V}$ with a flat integrable connection, with $\mathcal{O}_X$ and Ω_X^1 the sheaf of smooth real-valued functions and, respectively, one-forms on X.

Intersection Homology with Local Coefficients

Suppose now that X is a topological pseudomanifold with a fixed stratification

$$X = X_n \supseteq X_{n-2} \supseteq \cdots \supseteq X_0.$$

To make the above construction work in intersection homology, one only needs the local system $\mathcal{L}$ to be defined on the dense open part $X - X_{n-2}$ of X. Indeed, the allowability conditions on intersection i-chains ξ (e.g., using King's singular version of the theory) guarantee that if the coefficient ℓ_σ of ξ is non-zero, then $\sigma^{-1}(X - X_{n-2}) \neq \emptyset$, and similarly $\tau^{-1}(X - X_{n-2}) \neq \emptyset$ for every face τ of σ. So one can define as above the intersection homology groups $IH_i^{\overline{p}}(X; \mathcal{L})$ for a local system $\mathcal{L}$ defined on $X - X_{n-2}$. For more details, see e.g., [75, Section 6.3.3].

4.3 Sheaf Cohomology

In this section, we extend the notion of cohomology of locally constant sheaves to arbitrary sheaves.

Definition 4.3.1 A *resolution* $\mathcal{K}^\bullet$ of a sheaf $\mathcal{F}$ is a collection of sheaves $\{\mathcal{K}^i\}_{i \geq 0}$ fitting into an exact sequence:

$$0 \longrightarrow \mathcal{F} \xrightarrow{d^{-1}} \mathcal{K}^0 \xrightarrow{d^0} \mathcal{K}^1 \xrightarrow{d^1} \cdots.$$

Example 4.3.2 (de Rham Complex) Let X be a smooth manifold and let $A = \mathbb{Z}$. Associate to an open set $U \subseteq X$ the group $\mathbf{A}^i(U)$ of differential i-forms on U. The presheaf $U \mapsto \mathbf{A}^i(U)$ is a sheaf on X, denoted $\mathbf{A}^i$. Exterior derivation gives a map $d : \mathbf{A}^i(U) \to \mathbf{A}^{i+1}(U)$, which induces sheaf maps

$$d : \mathbf{A}^i \to \mathbf{A}^{i+1},$$

and one can form the *de Rham complex* $\mathbf{A}^\bullet_X$ by:

$$\mathbf{A}^\bullet_X : \quad \mathbf{A}^0 \xrightarrow{d} \mathbf{A}^1 \xrightarrow{d} \mathbf{A}^2 \xrightarrow{d} \cdots$$

Let $\mathbb{R}_X$ be the constant sheaf with stalk $\mathbb{R}$ on X. There is a monomorphism

$$\mathbb{R}_X \hookrightarrow \mathbf{A}^0$$

defined by assigning to a real number $r \in (\mathbb{R}_X)_x$ the germ of the constant function r in $(\mathbf{A}^0)_x$. The Poincaré lemma states that every closed form on the Euclidean space is exact, therefore the sequence:

$$0 \longrightarrow \mathbb{R}_X \longrightarrow \mathbf{A}^0 \xrightarrow{d} \mathbf{A}^1 \xrightarrow{d} \mathbf{A}^2 \xrightarrow{d} \cdots$$

is exact on Euclidean balls, hence exact. Thus the de Rham complex $\mathbf{A}^\bullet_X$ is a resolution of the constant sheaf $\mathbb{R}_X$.

Definition 4.3.3 (Injective Sheaf) A sheaf $\mathcal{I} \in Sh(X)$ is *injective* if, for every sheaf monomorphism $\mathcal{F} \hookrightarrow \mathcal{G}$ and a sheaf map $\mathcal{F} \to \mathcal{I}$, there exists an extension $\mathcal{G} \to \mathcal{I}$. In other words, $\mathcal{I}$ is injective if and only if $\mathrm{Hom}_{Sh(X)}(-, \mathcal{I})$ is an exact functor.[1]

---

[1] A functor $F : \mathcal{A} \to \mathcal{B}$ of abelian categories is *left exact* if a short exact sequence $0 \to A' \to A \to A'' \to 0$ in $\mathcal{A}$ is sent by F to an exact sequence $0 \to F(A') \to F(A) \to F(A'')$. F is *right exact* if $F(A') \to F(A) \to F(A'') \to 0$ is exact. F is *exact* if F preserves exact sequences, i.e., $0 \to F(A') \to F(A) \to F(A'') \to 0$ is exact.

The category of sheaves $Sh(X)$ has *enough injectives* in the following sense:

Lemma 4.3.4 *Every sheaf $\mathcal{F} \in Sh(X)$ is a subsheaf of an injective sheaf.*

(Here, one uses in an essential way the fact that the category of A-modules has enough injectives.)

This fact can be used to show the following:

Proposition 4.3.5 *Every sheaf $\mathcal{F} \in Sh(X)$ has a canonical injective resolution.*

Proof This can be constructed inductively as follows. Construct $\mathcal{J}^0$ and d^{-1} : $\mathcal{F} \to \mathcal{J}^0$ as in Lemma 4.3.4. Assuming $\mathcal{J}^i$ has been constructed, embed the cokernel $\mathcal{J}^i / \text{Image } d^{i-1}$ into an injective $\mathcal{J}^{i+1}$ using Lemma 4.3.4, and let d^i be the composition

$$\mathcal{J}^i \twoheadrightarrow \mathcal{J}^i / \text{Image } d^{i-1} \hookrightarrow \mathcal{J}^{i+1}.$$

□

Definition 4.3.6 (Sheaf Cohomology) The i-th *sheaf cohomology of X with coefficients in $\mathcal{F} \in Sh(X)$* is the A-module defined by:

$$H^i(X; \mathcal{F}) := H^i \Gamma(X, \mathcal{J}^\bullet),$$

where $\mathcal{F} \to \mathcal{J}^\bullet$ is the canonical injective resolution of $\mathcal{F}$.

Remark 4.3.7 Since $\Gamma(X, -)$ is a left exact functor, there is an exact sequence

$$0 \longrightarrow \Gamma(X, \mathcal{F}) \longrightarrow \Gamma(X, \mathcal{J}^0) \longrightarrow \Gamma(X, \mathcal{J}^1).$$

In particular,

$$H^0(X; \mathcal{F}) = \Gamma(X, \mathcal{F}).$$

Exercise 4.3.8 Show that a homomorphism $\mathcal{F} \to \mathcal{G}$ of sheaves induces corresponding sheaf cohomology homomorphisms:

$$H^i(X; \mathcal{F}) \longrightarrow H^i(X; \mathcal{G}).$$

Exercise 4.3.9 Show that $H^i(X; \mathcal{F})$ is independent of the choice of injective resolution of $\mathcal{F}$.

Proposition 4.3.10 *Given a short exact sequence*

$$0 \longrightarrow \mathcal{E} \longrightarrow \mathcal{F} \longrightarrow \mathcal{G} \longrightarrow 0$$

of sheaves on X, there is an associated long exact sequence on sheaf cohomology:

$$\cdots \to H^i(X; \mathcal{E}) \to H^i(X; \mathcal{F}) \to H^i(X; \mathcal{G}) \to H^{i+1}(X; \mathcal{E}) \to \cdots$$

Definition 4.3.11 (Acyclic Sheaf) A sheaf $\mathcal{F} \in Sh(X)$ is called *acyclic* if $H^i(X; \mathcal{F}) = 0$, for all $i > 0$.

Example 4.3.12 Injective sheaves are acyclic.

Exercise 4.3.13 Show that sheaf cohomology $H^i(X; \mathcal{F})$ can be computed by using a resolution of $\mathcal{F}$ by acyclic sheaves (e.g., see [23, Theorem II.4.1]).

Definition 4.3.14 (Soft Sheaf) A sheaf $\mathcal{F} \in Sh(X)$ is *soft* if the restriction map $\Gamma(X, \mathcal{F}) \longrightarrow \Gamma(K, \mathcal{F})$ is surjective for all closed subsets $K \subset X$.

Proposition 4.3.15 ([23, Theorem II.9.11]) *If X is a paracompact space, then soft sheaves are acyclic.*

Example 4.3.16 Let X be a (paracompact) smooth manifold, and let $\mathbb{R}_X \longrightarrow \mathbf{A}_X^\bullet$ be the de Rham resolution of Example 4.3.2. It can be shown (by using a partition of unity) that every $\mathbf{A}^i$ is a soft sheaf, so for computing the sheaf cohomology $H^i(X; \mathbb{R}_X)$ of X with constant sheaf coefficients $\mathbb{R}_X$ one can use the de Rham resolution, to get what is usually called the *smooth de Rham theorem*:

$$H^i(X; \mathbb{R}_X) \cong H^i\Gamma(X, \mathbf{A}_X^\bullet) = H^i(\mathbf{A}_X^\bullet(X)) =: H_{DR}^i(X),$$

with $H_{DR}^i(X)$ the *de Rham cohomology* of X.

Example 4.3.17 If X is a complex analytic manifold, and Ω_X^i denotes the sheaf of holomorphic i-forms on X, then one obtains a resolution

$$\mathbb{C}_X \longrightarrow \Omega_X^\bullet$$

of the constant sheaf $\mathbb{C}_X$ on X. However, the sheaves Ω_X^i are no longer acyclic in general (since there is no holomorphic partition of unity). The acyclicity property holds though when X is a Stein manifold (see [125, Section 52]).[2] So in this case one has the *complex analytic de Rham theorem*:

$$H^i(X; \mathbb{C}_X) \cong H^i\Gamma(X, \Omega_X^\bullet) = H^i(\Omega_X^\bullet(X)).$$

More generally, using Remark 4.2.17, one has the following:

Example 4.3.18 If X is a real smooth or complex analytic manifold, one can compute the cohomology groups with coefficients in a $\mathbb{C}$-local system $\mathcal{L}$ by using the *twisted de Rham complex*

---

[2] A *Stein manifold* is a complex manifold X so that every coherent $\mathcal{O}_X$-sheaf $\mathcal{F}$ is acyclic. (Recall that $\mathcal{F}$ is coherent if it has locally a finite presentation $\mathcal{O}_X^n \to \mathcal{O}_X^m \to \mathcal{F} \to 0$, e.g., $\mathcal{F}$ is the locally free sheaf of sections of a holomorphic vector bundle.) For example, a closed complex submanifold of some $\mathbb{C}^N$ (e.g., a nonsingular affine complex variety) is a Stein manifold.

$$(\Omega_X^\bullet(\mathcal{V}), \nabla): \quad 0 \longrightarrow \mathcal{V} \xrightarrow{\nabla} \Omega_X^1(\mathcal{V}) \xrightarrow{\nabla} \Omega_X^2(\mathcal{V}) \xrightarrow{\nabla} \cdots,$$

where $(\mathcal{V}, \nabla)$ is the integrable (flat) connection corresponding to the local system $\mathcal{L}$ and

$$\Omega_X^i(\mathcal{V}) := \Omega_X^i \otimes_{\mathcal{O}_X} \mathcal{V}.$$

In the real case, the twisted de Rham complex provides a soft resolution of $\mathcal{L}$, while in the complex case it is a resolution of $\mathcal{L}$ that moreover is acyclic if X is a Stein manifold. Therefore, if X is a Stein manifold, one gets (e.g., see [61, Theorem 2.5.11]):

$$H^i(X; \mathcal{L}) \cong H^i(\Omega_X^\bullet(\mathcal{V})(X)).$$

In particular, $H^i(X; \mathcal{L}) = 0$ for X Stein and $i > n = \dim_{\mathbb{C}} X$.

Let us conclude this section with the following important observation. If X is a paracompact Hausdorff space that is locally contractible, then the singular cohomology groups $H^*(X; A)$ of X with A-coefficients are isomorphic to the sheaf cohomology of the constant sheaf $\underline{A}_X$, that is,

$$H^*(X; A) \cong H^*(X; \underline{A}_X). \tag{4.2}$$

(See, e.g., [23, Chapter III] and also [61, Remark 2.5.12].) For example, the isomorphism (4.2) holds when X is a topological manifold, a CW complex, or a complex algebraic variety. So, in these cases, results about singular cohomology can be deduced from sheaf cohomology considerations. However, for arbitrary topological spaces, singular cohomology and sheaf cohomology with constant coefficients may be different. In fact, this can already be seen at the level of H^0. Indeed, the 0-th singular cohomology $H^0(X; \mathbb{Z})$ is the group of all functions from the set of path components of X to the integers $\mathbb{Z}$, whereas the sheaf cohomology $H^0(X; \underline{\mathbb{Z}}_X)$ is the group of locally constant functions from X to $\mathbb{Z}$. These are different, for example, when X is the *Cantor set*. We leave the details of the calculation in this case as an exercise for the interested reader.

4.4 Complexes of Sheaves

A resolution of a sheaf is what is called a complex of sheaves.

Definition 4.4.1 A *complex of sheaves* (or *differential graded sheaf*, DGS) $\mathcal{F}^\bullet$ on X is a collection of sheaves $\mathcal{F}^i$, $i \in \mathbb{Z}$, and homomorphisms (called differentials) $d^i: \mathcal{F}^i \to \mathcal{F}^{i+1}$ so that $d^i \circ d^{i-1} = 0$.

Definition 4.4.2 The *i-th cohomology sheaf* of a DGS $\mathcal{F}^\bullet$ is

$$\mathcal{H}^i(\mathcal{F}^\bullet) := \operatorname{Ker} d^i \,/\, \operatorname{Image} d^{i-1},$$

that is, the sheaf associated to the presheaf defined by $U \to H^i\Gamma(U, \mathcal{F}^\bullet)$.

Remark 4.4.3 As taking stalks is an exact functor, one gets

$$\mathcal{H}^i(\mathcal{F}^\bullet)_x \cong H^i(\mathcal{F}^\bullet_x).$$

Example 4.4.4 A sheaf $\mathcal{F}$ can be regarded as a complex of sheaves $\mathcal{F}^\bullet$ with $\mathcal{F}^0 = \mathcal{F}$, $\mathcal{F}^i = 0$ for $i \neq 0$, and $d^i = 0$ for all i. In this case, the complex $\mathcal{F}^\bullet$ is said to be concentrated in degree 0.

Definition 4.4.5 A *morphism of complexes* $f^\bullet : \mathcal{F}^\bullet \to \mathcal{G}^\bullet$ is a collection of sheaf homomorphisms $f^i : \mathcal{F}^i \to \mathcal{G}^i$, $i \in \mathbb{Z}$, so that each square

$$
\begin{array}{ccc}
\mathcal{F}^i & \xrightarrow{\ d^i\ } & \mathcal{F}^{i+1} \\
\downarrow{\scriptstyle f^i} & & \downarrow{\scriptstyle f^{i+1}} \\
\mathcal{G}^i & \xrightarrow{\ d^i\ } & \mathcal{G}^{i+1}
\end{array}
$$

commutes.

Proposition 4.4.6 *A morphism of complexes* $f^\bullet : \mathcal{F}^\bullet \to \mathcal{G}^\bullet$ *induces sheaf maps*

$$\mathcal{H}^i(f^\bullet) : \mathcal{H}^i(\mathcal{F}^\bullet) \longrightarrow \mathcal{H}^i(\mathcal{G}^\bullet)$$

for all $i \in \mathbb{Z}$.

Definition 4.4.7 A morphism of complexes $f^\bullet : \mathcal{F}^\bullet \to \mathcal{G}^\bullet$ is called a *quasi-isomorphism* if $\mathcal{H}^i(f^\bullet)$ is a sheaf isomorphism for all $i \in \mathbb{Z}$.

Example 4.4.8 A resolution $\mathcal{K}^\bullet$ of a sheaf $\mathcal{F}$ is a quasi-isomorphism $\mathcal{F} \longrightarrow \mathcal{K}^\bullet$.

Definition 4.4.9 A *resolution* $\mathcal{K}^\bullet$ of a complex $\mathcal{F}^\bullet$ is a quasi-isomorphism $\mathcal{F}^\bullet \longrightarrow \mathcal{K}^\bullet$.

Denote by $C(X)$ the category of complexes of sheaves with the above morphisms. With the obvious definition of kernel and cokernel complexes, it can be seen that $C(X)$ is an abelian category.

To a bounded (from below) complex of sheaves one can associate global cohomological objects called hypercohomology groups (or A-modules). These agree with the sheaf cohomology for a single sheaf, when that sheaf is regarded as a complex concentrated in degree zero.

Definition 4.4.10 A complex $\mathcal{F}^\bullet \in C(X)$ is *bounded from below* (in which case we write $\mathcal{F}^\bullet \in C^+(X)$) if there is $n \in \mathbb{Z}$ so that $\mathcal{F}^i = 0$ for all $i < n$. Notions of a *bounded from above* complex $\mathcal{F}^\bullet \in C^-(X)$, and, resp., *bounded complex* $\mathcal{F}^\bullet \in C^b(X)$ are defined similarly.

The following result holds (e.g., see [15, Corollary V.1.18]):

Proposition 4.4.11 *Every bounded (from below) complex $\mathcal{F}^\bullet$ has a canonical injective resolution $\mathcal{F}^\bullet \to \mathcal{J}^\bullet$ that is bounded (from below).*

Definition 4.4.12 (Hypercohomology) The *hypercohomology groups* (or A-modules) $\mathbb{H}^i(X; \mathcal{F}^\bullet)$ of a space X with coefficients in a bounded from below complex of A-sheaves $\mathcal{F}^\bullet$ are defined by

$$\mathbb{H}^i(X; \mathcal{F}^\bullet) := H^i \Gamma(X, \mathcal{J}^\bullet),$$

where $\mathcal{F}^\bullet \to \mathcal{J}^\bullet$ is the canonical injective resolution of $\mathcal{F}^\bullet$.

The following result is very useful in computations:

Proposition 4.4.13 (Hypercohomology Spectral Sequence) *If $\mathcal{F}^\bullet \in C^+(X)$ is a bounded from below complex of sheaves, there is a convergent spectral sequence with*

$$E_2^{p,q} = H^p(X; \mathcal{H}^q(\mathcal{F}^\bullet)) \implies \mathbb{H}^{p+q}(X; \mathcal{F}^\bullet). \tag{4.3}$$

As an application of the hypercohomology spectral sequence, one has the following:

Corollary 4.4.14 *A quasi-isomorphism $f^\bullet : \mathcal{F}^\bullet \to \mathcal{G}^\bullet$ of bounded from below complexes induces isomorphisms*

$$\mathbb{H}^i(X; \mathcal{F}^\bullet) \cong \mathbb{H}^i(X; \mathcal{G}^\bullet)$$

on hypercohomology groups.

Typically, one is more interested in the (hyper)cohomology of a complex than in the complex itself. So it would be convenient to work in a category of complexes of sheaves where two quasi-isomorphic complexes are interchangeable. Moreover, such a category should extend the category of sheaves in the sense that the sheaf functors should extend to functors between complexes. However, there are obvious technical problems with such demands, as can already be seen from Example 4.3.16: the quasi-isomorphism $\mathbb{R}_X \longrightarrow \mathbf{A}_X^\bullet$ does not induce in general a cohomology isomorphism upon applying the global section functor $\Gamma(X, -)$, as the de Rham cohomology of X does not necessarily vanish in positive degrees. These issues will be remedied at once, with the introduction of derived categories and derived functors.

Let us first define some further operations on complexes that will be needed later on.

Definition 4.4.15 (Shift Functor) The *shift functor* $[n] : C(X) \to C(X)$ is defined by $(\mathcal{F}^\bullet[n])^i = \mathcal{F}^{i+n}$ and $d^i_{\mathcal{F}^\bullet[n]} = (-1)^n d^{i+n}_{\mathcal{F}^\bullet}$. Given a morphism of complexes $f^\bullet : \mathcal{F}^\bullet \to \mathcal{G}^\bullet$, there is a shifted morphism $f^\bullet[n] : \mathcal{F}^\bullet[n] \to \mathcal{G}^\bullet[n]$ given by $(f^\bullet[n])^i = f^{i+n}$.

Definition 4.4.16 (Truncation Functors) The *truncation functors* $\tau_{\leq n}, \tau_{\geq n} : C(X) \to C(X)$ are defined by:

$$\tau_{\leq n}\mathcal{F}^\bullet = \cdots \to \mathcal{F}^{n-2} \to \mathcal{F}^{n-1} \to \mathrm{Ker}\,(d^n) \to 0 \to 0 \to \cdots$$

$$\tau_{\geq n}\mathcal{F}^\bullet = \cdots \to 0 \to 0 \to \mathrm{Coker}\,(d^{n-1}) \to \mathcal{F}^{n+1} \to \mathcal{F}^{n+2} \to \cdots$$

Induced morphisms $\tau_{\leq n}f, \tau_{\geq n}f$ are obtained in the obvious way.

The truncation functors are designed to satisfy the following:

Proposition 4.4.17

$$\mathcal{H}^i(\tau_{\leq n}\mathcal{F}^\bullet) \cong \begin{cases} \mathcal{H}^i(\mathcal{F}^\bullet), & i \leq n, \\ 0, & i > n. \end{cases}$$

$$\mathcal{H}^i(\tau_{\geq n}\mathcal{F}^\bullet) \cong \begin{cases} 0, & i < n, \\ \mathcal{H}^i(\mathcal{F}^\bullet) & i \geq n. \end{cases}$$

These truncation functors will play a crucial role in Chapter 6, where the Deligne construction of the *intersection cohomology complex* will be presented.

Definition 4.4.18 (Complex of Homomorphisms) For $\mathcal{F}^\bullet, \mathcal{G}^\bullet \in C(X)$, we define $\mathcal{H}om^\bullet(\mathcal{F}^\bullet, \mathcal{G}^\bullet) \in C(X)$ by

$$\mathcal{H}om^n(\mathcal{F}^\bullet, \mathcal{G}^\bullet) = \prod_{p \in \mathbb{Z}} \mathcal{H}om(\mathcal{F}^p, \mathcal{G}^{p+n})$$

with

$$d^n = \prod_p d^n_p : \mathcal{H}om^n(\mathcal{F}^\bullet, \mathcal{G}^\bullet) \to \mathcal{H}om^{n+1}(\mathcal{F}^\bullet, \mathcal{G}^\bullet)$$

defined as follows: on an open subset U of X, we let

$$d^n_p : \mathrm{Hom}(\mathcal{F}^p|_U, \mathcal{G}^{p+n}|_U) \to \mathrm{Hom}(\mathcal{F}^p|_U, \mathcal{G}^{p+n+1}|_U)$$

be given by

$$d_p^n(f_p) = d_{\mathcal{G}^\bullet}^{p+n} \circ f_p + (-1)^{n+1} f_{p+1} \circ d_{\mathcal{F}^\bullet}^p, \tag{4.4}$$

for $\{f_p\}_{p \in \mathbb{Z}} \in \prod_p \mathrm{Hom}(\mathcal{F}^p|_U, \mathcal{G}^{p+n}|_U)$.

Definition 4.4.19 (Direct Image and Pullback Complexes) If $f : X \to Y$ is a continuous map, and $\mathcal{F}^\bullet \in C(X)$ and $\mathcal{G}^\bullet \in C(Y)$ are complexes of sheaves, define the *direct image complex* $f_*\mathcal{F}^\bullet \in C(Y)$ by

$$(f_*\mathcal{F}^\bullet)^i = f_*(\mathcal{F}^i), \quad d_{f_*\mathcal{F}^\bullet}^i = f_*(d_{\mathcal{F}^\bullet}^i),$$

and the *pullback complex* $f^*\mathcal{G}^\bullet \in C(X)$ by

$$(f^*\mathcal{G}^\bullet)^i = f^*(\mathcal{G}^i), \quad d_{f^*\mathcal{G}^\bullet}^i = f^*(d_{\mathcal{G}^\bullet}^i).$$

Remark 4.4.20 Since f^* is an exact functor, it follows that

$$\mathcal{H}^i(f^*\mathcal{G}^\bullet) \cong f^*\mathcal{H}^i(\mathcal{G}^\bullet).$$

Moreover, pullback commutes with truncation.

Definition 4.4.21 (Tensor Product of Complexes) Given two bounded from above complexes $\mathcal{F}^\bullet, \mathcal{G}^\bullet \in C^-(X)$, define their *tensor product* $\mathcal{F}^\bullet \otimes \mathcal{G}^\bullet \in C^-(X)$ by:

$$(\mathcal{F}^\bullet \otimes \mathcal{G}^\bullet)^i = \bigoplus_{p+q=i} \mathcal{F}^p \otimes \mathcal{G}^q,$$

with differential

$$d(x^p \otimes y^q) = dx^p \otimes y^q + (-1)^p x^p \otimes dy^q$$

for $x^p \in \mathcal{F}^p, y^q \in \mathcal{G}^q$.

Exercise 4.4.22 Show that a pair of morphisms $u_i : \mathcal{F}_i^\bullet \to \mathcal{G}_i^\bullet, i = 1, 2$, induces a morphism $u_1 \otimes u_2 : \mathcal{F}_1^\bullet \otimes \mathcal{F}_2^\bullet \to \mathcal{G}_1^\bullet \otimes \mathcal{G}_2^\bullet$.

Definition 4.4.23 (External Tensor Product of Complexes) Let X_1 and X_2 be topological spaces, with projections $p_i : X_1 \times X_2 \to X_i, i = 1, 2$. The *external tensor product* of bounded (above) complexes $\mathcal{F}_1^\bullet \in C^-(X_1)$ and $\mathcal{F}_2^\bullet \in C^-(X_2)$ is defined as:

$$\mathcal{F}_1^\bullet \boxtimes \mathcal{F}_2^\bullet := p_1^*\mathcal{F}_1^\bullet \otimes p_2^*\mathcal{F}_2^\bullet.$$

4.5 Homotopy Category

Definition 4.5.1 Two morphisms of complexes $f^\bullet, g^\bullet : \mathcal{F}^\bullet \to \mathcal{G}^\bullet$ in $C(X)$ are *homotopic* if there exists a collection $\{s^i\}_{i \in \mathbb{Z}}$ of sheaf maps $s^i : \mathcal{F}^i \to \mathcal{G}^{i-1}$ (called a *homotopy*), so that:

$$d_{\mathcal{G}^\bullet}^{i-1} \circ s^i + s^{i+1} \circ d_{\mathcal{F}^\bullet}^i = f^i - g^i$$

for all $i \in \mathbb{Z}$. A morphism is called *null-homotopic* if it is homotopic to the zero morphism.

When working with complexes, many diagrams commute only up to homotopy. In order to have these diagrams actually commute, one needs to work in a category where homotopy equivalences become isomorphisms. This is the *homotopy category* $K(X)$ defined as follows:

(i) Objects:

$$Ob(K(X)) = Ob(C(X)),$$

(ii) Morphisms:

$$\mathrm{Hom}_{K(X)}(\mathcal{F}^\bullet, \mathcal{G}^\bullet) = [\mathcal{F}^\bullet, \mathcal{G}^\bullet] := \{[f^\bullet] \mid f^\bullet \in \mathrm{Hom}_{C(X)}(\mathcal{F}^\bullet, \mathcal{G}^\bullet)\},$$

where $[f^\bullet]$ denotes the homotopy class of $f^\bullet$.

Composition of morphisms $[f^\bullet] \circ [g^\bullet] = [f^\bullet \circ g^\bullet]$ is well defined since compositions of homotopic maps are homotopic.

Remark 4.5.2 One has the following identification:

$$\Gamma(X, \mathcal{H}om^\bullet(\mathcal{F}^\bullet, \mathcal{G}^\bullet)) = [\mathcal{F}^\bullet, \mathcal{G}^\bullet].$$

In fact, by (4.4), the n-cycles of $\Gamma(X, \mathcal{H}om^\bullet(\mathcal{F}^\bullet, \mathcal{G}^\bullet))$ are in one-to-one correspondence with morphisms of complexes $\mathcal{F}^\bullet \to \mathcal{G}^\bullet[n]$, and the n-boundaries correspond to morphisms that are homotopic to zero. Thus,

$$H^n \Gamma(X, \mathcal{H}om^\bullet(\mathcal{F}^\bullet, \mathcal{G}^\bullet)) = [\mathcal{F}^\bullet, \mathcal{G}^\bullet[n]] . \tag{4.5}$$

It is easy to see that $K(X)$ is an additive category: there is a zero element for the obvious addition of homotopy classes, and $\mathrm{Hom}(-,-)$ is an abelian group.

Note that if $f^\bullet$ is a homotopy equivalence, then $[f^\bullet]$ is invertible in $K(X)$, with inverse $[g^\bullet]$, where $g^\bullet$ is a homotopy inverse of $f^\bullet$. However, $K(X)$ is *not* an abelian category, as the following example shows:

Example 4.5.3 Assume X is a point and work over $\mathbb{Z}$, so $Sh(X) = Ab$, the category of abelian groups. Under this assumption, let $F : C(X) \to K(X)$ be the quotient functor, i.e., $F(\mathcal{F}^\bullet) = \mathcal{F}^\bullet$ and $F(f^\bullet) = [f^\bullet]$. This example shows that $K(X)$ is not an abelian category such that F is an exact functor.

Consider the following short exact sequence in $C(X)$:

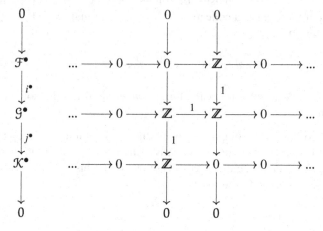

The map $\mathcal{G}^\bullet \to 0^\bullet$ is a homotopy equivalence (with homotopy inverse $0^\bullet \to \mathcal{G}^\bullet$), since $\mathcal{G}^\bullet \xrightarrow{1} \mathcal{G}^\bullet$ and $\mathcal{G}^\bullet \xrightarrow{0} \mathcal{G}^\bullet$ are homotopic. So, $F(\mathcal{G}^\bullet) \cong F(0^\bullet) = 0^\bullet$ in $K(X)$. If $K(X)$ were an abelian category such that F is an exact functor, then the short exact sequence

$$0 \longrightarrow F(\mathcal{F}^\bullet) \xrightarrow{F(i^\bullet)} F(\mathcal{G}^\bullet) \xrightarrow{F(j^\bullet)} F(\mathcal{K}^\bullet) \longrightarrow 0$$

would imply that $F(\mathcal{F}^\bullet) \cong \text{Image } F(i^\bullet) \cong \text{Ker } F(j^\bullet) \cong 0^\bullet$. But cohomology is a homotopy invariant, and $H^*(F(\mathcal{F}^\bullet)) = H^*(\mathcal{F}^\bullet) \neq 0$, which gives a contradiction.

Therefore, one has to look for a substitute for short exact sequences in $K(X)$. Such a substitute should still have the property that it induces long exact sequences in (hyper)cohomology. In fact, $K(X)$ has a structure of a *triangulated category*, in which the role of exact sequences is played by *triangles*, i.e., objects of the form:

$$\mathcal{A}^\bullet \longrightarrow \mathcal{B}^\bullet \longrightarrow \mathcal{C}^\bullet \xrightarrow{[1]} \mathcal{A}^\bullet[1].$$

We shall not get here into details about triangulated categories, but list instead the basic notions needed in the sequel. For a very instructive overview, the reader may consult [15, V.5], or see [6, Chapter II] for a more detailed account.

Given complexes $\mathcal{A}^\bullet, \mathcal{B}^\bullet \in C(X)$ and a morphism $u^\bullet : \mathcal{A}^\bullet \to \mathcal{B}^\bullet$, *the mapping cone of* $u^\bullet$ is the complex

$$\mathcal{C}_u^\bullet := \mathcal{A}^\bullet[1] \oplus \mathcal{B}^\bullet, \tag{4.6}$$

i.e., $\mathcal{C}_u^i = \mathcal{A}^{i+1} \oplus \mathcal{B}^i$, with differential $d^i : \mathcal{C}_u^i \to \mathcal{C}_u^{i+1}$ defined by

$$\begin{pmatrix} -d_A^{i+1} & 0 \\ u^{i+1} & d_B^i \end{pmatrix}.$$

There is a natural inclusion $i^\bullet : \mathcal{B}^\bullet \to \mathcal{C}_u^\bullet$ and the projection $p^\bullet : \mathcal{C}_u^\bullet \to \mathcal{A}^\bullet[1]$. The triangle

$$\mathcal{A}^\bullet \xrightarrow{u^\bullet} \mathcal{B}^\bullet \xrightarrow{i^\bullet} \mathcal{C}_u^\bullet \xrightarrow{p^\bullet} \mathcal{A}^\bullet[1] \qquad (4.7)$$

is called a *standard (or distinguished) triangle*. Every triangle in $K(X)$ is "isomorphic" to a standard triangle. Every morphism in $K(X)$ can be "embedded" in a distinguished triangle.

A triangle $\mathcal{A}^\bullet \to \mathcal{B}^\bullet \to \mathcal{C}^\bullet \to \mathcal{A}^\bullet[1]$ in $K(X)$ induces a long exact sequence in sheaf cohomology:

$$\cdots \to \mathcal{H}^i(\mathcal{A}^\bullet) \to \mathcal{H}^i(\mathcal{B}^\bullet) \to \mathcal{H}^i(\mathcal{C}^\bullet) \to \mathcal{H}^{i+1}(\mathcal{A}^\bullet) \to \cdots.$$

Moreover, if all complexes in question are bounded from below (i.e., they are elements of $K^+(X)$), then the above triangle also induces a long exact sequence of hypercohomology groups

$$\cdots \to \mathbb{H}^i(X; \mathcal{A}^\bullet) \to \mathbb{H}^i(X; \mathcal{B}^\bullet) \to \mathbb{H}^i(X; \mathcal{C}^\bullet) \to \mathbb{H}^{i+1}(X; \mathcal{A}^\bullet) \to \cdots.$$

This fact is usually referred to by saying that the functors $\mathcal{H}^0 : K(X) \to Sh(X)$ and $\mathbb{H}^0(X; -) : K^+(X) \to A-\text{mod}$ are *cohomological functors*.

4.6 Derived Category

Next, one would like to be able to associate a triangle to every short exact sequence of complexes in $C(X)$. More precisely, given such a short exact sequence

$$0 \to \mathcal{A}^\bullet \xrightarrow{u^\bullet} \mathcal{B}^\bullet \xrightarrow{v^\bullet} \mathcal{C}^\bullet \to 0 \qquad (4.8)$$

in $C(X)$, one would like to make sense of a triangle

$$\mathcal{A}^\bullet \xrightarrow{u^\bullet} \mathcal{B}^\bullet \xrightarrow{v^\bullet} \mathcal{C}^\bullet \xrightarrow{[1]} \mathcal{A}^\bullet[1]$$

in $K(X)$. Let $\mathcal{C}_u^\bullet$ be the mapping cone of $u^\bullet$ as in (4.6) and define $f^i : \mathcal{C}_u^i \to \mathcal{C}^i$ by $f^i = (0, v^i)$. This gives a morphism

$$f^\bullet : \mathcal{C}_u^\bullet \to \mathcal{C}^\bullet,$$

which is a quasi-isomorphism (as can be seen easily by comparing the associated long exact sequences induced by (4.8) and (4.7) in hypercohomology). Moreover, it can be shown that if the given short exact sequence (4.8) splits, then $f^\bullet$ is a homotopy equivalence.

Thus, if the sequence (4.8) splits in $C(X)$, then one can replace $\mathcal{C}_u^\bullet$ by $\mathcal{C}^\bullet$ in (4.7) and get an associated triangle

$$\mathcal{A}^\bullet \to \mathcal{B}^\bullet \to \mathcal{C}^\bullet \to \mathcal{A}^\bullet[1]$$

in $K(X)$ for the given short exact sequence in $C(X)$.

However, this is not the case in general, so one needs to devise a new (triangulated) category $D(X)$, the *derived category*, where quasi-isomorphisms in $C(X)$ become isomorphisms in $D(X)$, so there will be a distinguished triangle in $D(X)$ associated to every short exact sequence in $C(X)$, even if the given short exact sequence does not split.

The derived category was introduced by J.L. Verdier in his doctoral dissertation [236] (a summary of which had earlier appeared in SGA $4\frac{1}{2}$ [234]), and the construction is based on localization of a category, a generalization of localization of a ring.

More concretely, the derived category $D(X)$ is constructed by *localizing $K(X)$* at the multiplicative system of quasi-isomorphisms. The objects of $D(X)$ are

$$Ob(D(X)) = Ob(K(X)) = Ob(C(X)).$$

But a morphism $\mathcal{A}^\bullet \longrightarrow \mathcal{B}^\bullet$ in $D(X)$ is defined as an equivalence class of *roofs (or fractions)*:

$$\{\mathcal{A}^\bullet \xleftarrow{q.i.} \mathcal{C}^\bullet \to \mathcal{B}^\bullet\}/\sim,$$

with "q.i" labelling a quasi-isomorphism, where two roofs $\mathcal{A}^\bullet \xleftarrow{q.i.} \mathcal{C}_1^\bullet \to \mathcal{B}^\bullet$ and $\mathcal{A}^\bullet \xleftarrow{q.i.} \mathcal{C}_2^\bullet \to \mathcal{B}^\bullet$ are equivalent if there is a third roof $\mathcal{A}^\bullet \xleftarrow{q.i.} \mathcal{C}_3^\bullet \to \mathcal{B}^\bullet$ such that the following diagram commutes:

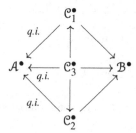

Remark 4.6.1 For concreteness and simplicity of exposition, all considerations above were applied to the abelian category $Sh(X)$ of sheaves of A-modules on X. But one can associate an abelian category of complexes $C(\mathcal{A})$, a homotopy category $K(\mathcal{A})$ and a derived category $D(\mathcal{A})$ to every abelian category $\mathcal{A}$. Moreover, the definition of truncation functors $\tau_{\leq n}, \tau_{\geq n}$ for complexes in $C(X)$ clearly also applies to complexes in $C(\mathcal{A})$, for every abelian category $\mathcal{A}$. Since

$$\tau_{\leq n}, \tau_{\geq n} : C(\mathcal{A}) \longrightarrow C(\mathcal{A})$$

take null-homotopic morphisms to null-homotopic morphisms, truncations are well defined on the homotopy category:

$$\tau_{\leq n}, \tau_{\geq n} : K(\mathcal{A}) \longrightarrow K(\mathcal{A}).$$

Moreover, since $\tau_{\leq n}, \tau_{\geq n}$ preserve quasi-isomorphisms, truncation functors are also well defined on the derived category:

$$\tau_{\leq n}, \tau_{\geq n} : D(\mathcal{A}) \longrightarrow D(\mathcal{A}).$$

Exercise 4.6.2 In the notations of Definition 4.4.16, show that if $\mathcal{F}^\bullet, \mathcal{G}^\bullet \in C(X)$ are complexes of sheaves on X so that $\mathcal{F}^\bullet \simeq \tau_{\leq s}(\mathcal{F}^\bullet)$ and $\mathcal{G}^\bullet \simeq \tau_{\geq s}(\mathcal{G}^\bullet)$ for some integer s, then the natural map

$$\mathrm{Hom}_{D(X)}(\mathcal{F}^\bullet, \mathcal{G}^\bullet) \longrightarrow \mathrm{Hom}_{Sh(X)}(\mathcal{H}^s(\mathcal{F}^\bullet), \mathcal{H}^s(\mathcal{G}^\bullet))$$

is an isomorphism of abelian groups.

The following elementary splitting criterion plays an important role in the de Cataldo–Migliorini proof of the BBDG decomposition theorem (see Section 9.3):

Proposition 4.6.3 *Let* $\mathcal{C}^\bullet \overset{u}{\to} \mathcal{A}^\bullet \overset{v}{\to} \mathcal{B}^\bullet \overset{[1]}{\to} \mathcal{C}^\bullet[1]$ *be a distinguished triangle in* $D(X)$, *and let s be an integer so that* $\mathcal{A}^\bullet \simeq \tau_{\leq s}(\mathcal{A}^\bullet)$ *and* $\mathcal{C}^\bullet \simeq \tau_{\geq s}(\mathcal{C}^\bullet)$. *Then* $\mathcal{H}^s(u) : \mathcal{H}^s(\mathcal{C}^\bullet) \to \mathcal{H}^s(\mathcal{A}^\bullet)$ *is an isomorphism if, and only if,*

$$\mathcal{A}^\bullet \simeq \tau_{\leq s-1}\mathcal{B}^\bullet \oplus \mathcal{H}^s(\mathcal{A}^\bullet)[-s]$$

and the map v is the direct sum of the natural map $\tau_{\leq s-1}\mathcal{B}^\bullet \to \mathcal{B}^\bullet$ *and the zero map.*

Proof See [48, Proposition 3.1.2]. $\square$

4.7 Derived Functors

Let $\mathcal{A}$ and $\mathcal{B}$ be abelian categories. An additive functor $F : \mathcal{A} \to \mathcal{B}$ induces in an obvious way a functor $F : C(\mathcal{A}) \to C(\mathcal{B})$ between the associated categories of complexes. If $f^{\bullet} \in \mathrm{Hom}_{C(\mathcal{A})}(\mathcal{A}^{\bullet}, \mathcal{B}^{\bullet})$ is null-homotopic with homotopy $\{s^i\}$, then $\{F(s^i)\}$ is a null-homotopy for $F(f^{\bullet})$. Thus F induces a functor

$$F : K(\mathcal{A}) \longrightarrow K(\mathcal{B})$$

between the homotopy categories. As an example, if $f : X \to Y$ is a continuous map of topological spaces, the direct image $f_* : K(X) \to K(Y)$ and inverse image $f^* : K(Y) \to K(X)$ are well-defined functors on the homotopy category of complexes of sheaves.

The next natural question to ask is the following: *does an additive functor $F : \mathcal{A} \to \mathcal{B}$ of abelian categories furthermore induce a functor $F : D(\mathcal{A}) \to D(\mathcal{B})$ on derived categories?*

The answer in general is *no*, as $F : K(\mathcal{A}) \longrightarrow K(\mathcal{B})$ need not take quasi-isomorphisms to quasi-isomorphisms, as the following example shows.

Example 4.7.1 Let $F : \mathcal{A} \to \mathcal{B}$ be an additive functor that is not exact. Let $\mathcal{A}^{\bullet}$ be an acyclic complex in $C(\mathcal{A})$, so that $F(\mathcal{A}^{\bullet})$ is not acyclic. Then $\mathcal{A}^{\bullet} \xrightarrow{0} 0^{\bullet}$ is a quasi-isomorphism. On the other hand, $F(0) = 0$ as F is additive, and $F(\mathcal{A}^{\bullet}) \xrightarrow{F(0)=0} F(0^{\bullet}) = 0^{\bullet}$ is not a quasi-isomorphism since $F(\mathcal{A}^{\bullet})$ is not acyclic.

An *exact* functor $F : \mathcal{A} \to \mathcal{B}$ indeed gives rise to a functor $F : D(\mathcal{A}) \to D(\mathcal{B})$ on derived categories, since the homotopy category functor $F : K(\mathcal{A}) \longrightarrow K(\mathcal{B})$ transforms quasi-isomorphisms to quasi-isomorphisms. The induced functor $F : D(\mathcal{A}) \to D(\mathcal{B})$ will then transform distinguished triangles into distinguished triangles. However, most important functors on abelian categories, such as Hom, $\otimes$, $\Gamma(X, -)$, direct image, are *not* exact. This shows the need of a mechanism to extend additive functors $F : \mathcal{A} \to \mathcal{B}$ of abelian categories to corresponding (right or left) *derived functors* $DF : D(\mathcal{A}) \to D(\mathcal{B})$ on derived categories.

We shall begin our discussion with right derived functors RF and $\mathcal{A}$ an abelian category with enough injectives, such as $Sh(X)$. Let $D^+(\mathcal{A}), D^-(\mathcal{A}), D^b(\mathcal{A})$ denote the corresponding derived categories of complexes in $C(\mathcal{A})$ that are bounded from below, bounded from above, resp., bounded.

Definition 4.7.2 Let $F : \mathcal{A} \to \mathcal{B}$ be a *left exact* functor on an abelian category $\mathcal{A}$ *with enough injectives*. Then the *right derived functor*

$$RF : D^+(\mathcal{A}) \longrightarrow D(\mathcal{B})$$

of F is defined as

$$RF(\mathcal{A}^{\bullet}) = F(\mathcal{I}^{\bullet}),$$

where $\mathcal{J}^\bullet$ is an injective resolution of $\mathcal{A}^\bullet$. The *i-th derived functor of* F is defined as:

$$R^i F(\mathcal{A}^\bullet) := \mathcal{H}^i F(\mathcal{J}^\bullet).$$

The same definition also applies to every functor $F : K^+(\mathcal{A}) \rightarrow K(\mathcal{B})$ of triangulated categories.

Remark 4.7.3 The relevance of the bounded below complexes in the above definition comes from the existence of injective resolution of complexes via Proposition 4.4.11.

Example 4.7.4 The i-th hypercohomology group $\mathbb{H}^i(X; \mathcal{A}^\bullet)$ of $\mathcal{A}^\bullet \in D^+(X)$ is the i-th derived functor of the global section functor $\Gamma(X, -)$.

Example 4.7.5 Given a continuous map $f : X \rightarrow Y$, the direct image functor $f_* : Sh(X) \rightarrow Sh(Y)$ is left exact, so it has a right derived functor

$$Rf_* : D^+(X) \longrightarrow D(Y).$$

On the other hand, $f^* : Sh(Y) \rightarrow Sh(X)$ is an exact functor, so it induces directly a functor

$$f^* : D(Y) \longrightarrow D(X)$$

on the corresponding derived categories of sheaves.

One has the following (e.g., see [15, Theorem V.10.6]):

Proposition 4.7.6 $R(g \circ f)_* = Rg_* \circ Rf_*$. *More generally,* $R(G \circ F) = RG \circ RF$ *if F maps injectives into G-acyclic objects.*

Example 4.7.7 For $\mathcal{F} \in \mathcal{A}$, the functor

$$\mathrm{Hom}(\mathcal{F}, -) : \mathcal{A} \longrightarrow Ab$$

is left exact, so it has a right derived functor

$$R\mathrm{Hom}^\bullet(\mathcal{F}, -) : D^+(\mathcal{A}) \longrightarrow D(Ab)$$

with i-th derived functors

$$\mathrm{Ext}^i(\mathcal{F}, -) := R^i\mathrm{Hom}(\mathcal{F}, -) = H^i(R\mathrm{Hom}^\bullet(\mathcal{F}, -))$$

and

$$\mathrm{Ext}^0(\mathcal{F}, -) = \mathrm{Hom}(\mathcal{F}, -).$$

More generally, if $F : \mathcal{A} \longrightarrow Ab$ is a left exact functor and $\mathcal{F} \in \mathcal{A}$, then

$$F(\mathcal{F}) \cong R^0 F(\mathcal{F}),$$

where $\mathcal{F}$ on the right-hand side is regarded as a complex concentrated in degree zero.

Example 4.7.8 Let $\mathcal{F}^\bullet$, $\mathcal{G}^\bullet$ be (bounded below) complexes of sheaves on X. To define $R\mathcal{H}om^\bullet(\mathcal{F}^\bullet, \mathcal{G}^\bullet)$, we consider $\mathcal{H}om^\bullet(\mathcal{F}^\bullet, \mathcal{G}^\bullet)$ as a functor of $\mathcal{G}^\bullet$ (with $\mathcal{F}^\bullet$ fixed) and we take its right derived functor as in Definition 4.7.2. Then it can be shown by using Remark 4.5.2 that (see [15, V.5.17(3)])

$$\mathbb{H}^n(X; R\mathcal{H}om^\bullet(\mathcal{F}^\bullet, \mathcal{G}^\bullet)) = \mathrm{Hom}_{D^+(X)}(\mathcal{F}^\bullet, \mathcal{G}^\bullet[n]). \tag{4.9}$$

Exercise 4.7.9 Show that if $\mathcal{F}^\bullet = \underline{A}_X$ is the constant sheaf on X and $\mathcal{G}^\bullet \in C^+(X)$, the identification in (4.9) yields the following interpretation of hypercohomology groups:

$$\mathbb{H}^n(X; \mathcal{G}^\bullet) = \mathrm{Hom}_{D^+(X)}(\underline{A}_X, \mathcal{G}^\bullet[n]). \tag{4.10}$$

In particular, if $\mathcal{G}^\bullet = \underline{A}_X$, then (4.10) yields that:

$$H^n(X; A) = \mathrm{Hom}_{D^+(X)}(\underline{A}_X, \underline{A}_X[n]), \tag{4.11}$$

i.e., an n-cohomology class $u \in H^n(X; A)$ can be viewed as a map $\underline{A}_X \to \underline{A}_X[n]$ in the derived category; the induced map in cohomology is of course the cup product with u.

To this end, let us say a few words about left derived functors.

If $F : \mathcal{A} \to \mathcal{B}$ is a *right exact* functor, and $\mathcal{A}$ has enough *projective* objects, then one can use projective resolutions of complexes to define a *left derived functor*

$$LF : D^-(\mathcal{A}) \to D(\mathcal{B})$$

of F. The main example to consider here is the left derived functor $\overset{L}{\otimes}$ of the tensor product. However, the category $Sh(X)$ of sheaves on a topological space X does not have enough projectives. (For example, the constant sheaf with stalk $\mathbb{Z}$ on the unit interval is not the quotient of a projective sheaf.) Thus, for the right exact functor

$$\mathcal{F} \otimes - : Sh(X) \longrightarrow Sh(X),$$

where $\mathcal{F} \in Sh(X)$, a left derived functor cannot be defined via projective resolutions. Instead, flat resolutions are used.

Let us assume till the end of this chapter that spaces are locally compact of finite cohomological dimension over A.

Definition 4.7.10 A sheaf $\mathcal{K} \in Sh(X)$ is *flat* if, for every sheaf monomorphism $\mathcal{F} \hookrightarrow \mathcal{G}$, the morphism $\mathcal{F} \otimes \mathcal{K} \to \mathcal{G} \otimes \mathcal{K}$ is injective as well. A complex $\mathcal{K}^\bullet$ is flat if every $\mathcal{K}^i$ is flat.

Exercise 4.7.11 Show that if A is a field, then every sheaf of A-vector spaces is flat.

Concerning the existence of flat resolutions for sheaf complexes, one has the following result (e.g., see [15, V.6.1]):

Proposition 4.7.12 *Every sheaf is a quotient of a flat sheaf. Therefore, every sheaf has a left flat resolution. Moreover, for every complex $\mathcal{G}^\bullet \in K^-(X)$, there exists a flat resolution $\mathcal{B}^\bullet \longrightarrow \mathcal{G}^\bullet$ with $\mathcal{B}^\bullet \in K^-(X)$.*

Definition 4.7.13 (Left Derived Tensor Product) For $\mathcal{F}^\bullet \in D^-(X)$, a left derived functor

$$\mathcal{F}^\bullet \overset{L}{\otimes} - : D^-(X) \to D^-(X)$$

is defined as:

$$\mathcal{F}^\bullet \overset{L}{\otimes} \mathcal{G}^\bullet = \mathcal{F}^\bullet \otimes \mathcal{B}^\bullet,$$

where $\mathcal{B}^\bullet \to \mathcal{G}^\bullet$ is a flat resolution of $\mathcal{G}^\bullet$.

In fact, for $\mathcal{F}^\bullet, \mathcal{G}^\bullet \in D^-(X)$ with flat resolutions $\mathcal{A}^\bullet \to \mathcal{F}^\bullet$ and $\mathcal{B}^\bullet \to \mathcal{G}^\bullet$, respectively, one has:

$$\mathcal{F}^\bullet \overset{L}{\otimes} \mathcal{G}^\bullet \simeq \mathcal{A}^\bullet \otimes \mathcal{G}^\bullet \simeq \mathcal{F}^\bullet \otimes \mathcal{B}^\bullet \simeq \mathcal{A}^\bullet \otimes \mathcal{B}^\bullet.$$

The corresponding i-th derived functor is the i-th *hypertor sheaf*:

$$\mathcal{T}or_i(\mathcal{F}^\bullet, \mathcal{G}^\bullet) := \mathcal{H}^{-i}(\mathcal{F}^\bullet \overset{L}{\otimes} \mathcal{G}^\bullet),$$

and note that $\mathcal{T}or_0 = \otimes$.

Exercise 4.7.14 Let $f : X \to Y$ be a continuous map. Show that

$$f^*(\mathcal{F}^\bullet \overset{L}{\otimes} \mathcal{G}^\bullet) \simeq f^*\mathcal{F}^\bullet \overset{L}{\otimes} f^*\mathcal{G}^\bullet.$$

We conclude with the following:

Definition 4.7.15 (External Derived Tensor Product) Let X_1 and X_2 be topological spaces, with projections $p_i : X_1 \times X_2 \to X_i$, $i = 1, 2$. The *external derived tensor product* of bounded (above) complexes $\mathcal{F}_1^\bullet \in C(X_1)$ and $\mathcal{F}_2^\bullet \in C(X_2)$ is defined as:

$$\mathcal{F}_1^\bullet \overset{L}{\boxtimes} \mathcal{F}_2^\bullet := p_1^* \mathcal{F}_1^\bullet \overset{L}{\otimes} p_2^* \mathcal{F}_2^\bullet.$$

Chapter 5
Poincaré–Verdier Duality

Starting with this chapter, we assume for simplicity that all spaces are locally compact of finite (cohomological) dimension, with coefficient ring A noetherian commutative of finite cohomological dimension, so that one has the "six functor formalism" (that is, Grothendieck's six operations Rf_*, $Rf_!$, f^*, $f^!$, $R\mathcal{H}om^\bullet$, and $\overset{L}{\otimes}$) for bounded derived categories of sheaves.

Recall from Proposition 4.1.16 that, given a continuous map $f : X \to Y$, the functors f_* and f^* are adjoint functors on sheaves. The derived version of this statement yields the isomorphism:

$$\mathrm{Hom}_{D^b(Y)}(\mathcal{G}^\bullet, Rf_*\mathcal{F}^\bullet) = \mathrm{Hom}_{D^b(X)}(f^*\mathcal{G}^\bullet, \mathcal{F}^\bullet), \tag{5.1}$$

where $D^b(-) = D^b(Sh(-))$ denotes the derived category of *bounded* complexes of sheaves. In this chapter, we recall the definition of the direct image with proper support functor $f_!$, and sketch the construction of a right adjoint $f^!$ for $Rf_!$, originally introduced by Verdier [235]. We also define the *dualizing functor* and *dualizing complex*, and show how these can be used to deduce Poincaré and Alexander duality statements for manifolds. For comprehensive references, see [15, V.7] or [122, Chapter III].

5.1 Direct Image with Proper Support

Recall from Definition 4.1.24 that if X is a topological space and $\mathcal{F} \in Sh(X)$, the *support of a section* $s \in \Gamma(X; \mathcal{F})$ is the closed set:

$$\mathrm{supp}(s) = \overline{\{x \in X \mid s(x) \neq 0\}}.$$

© Springer Nature Switzerland AG 2019
L. G. Maxim, *Intersection Homology & Perverse Sheaves*, Graduate Texts in Mathematics 281, https://doi.org/10.1007/978-3-030-27644-7_5

Let $\Gamma_c(X, \mathcal{F})$ be the subgroup of $\Gamma(X, \mathcal{F})$ consisting of those sections that have compact support. The functor $\Gamma_c(X, -)$ is left exact, and has a right derived functor $R\Gamma_c(X, -)$. The i-th derived functor of Γ_c is called the *hypercohomology in degree i with compact support*, denoted by:

$$\mathbb{H}_c^i(X; \mathcal{F}^\bullet) := H^i(R\Gamma_c(X, \mathcal{F}^\bullet)).$$

These groups can be computed by a corresponding hypercohomology spectral sequence, with E_2-term given by:

$$E_2^{p,q} = H_c^p(X; \mathcal{H}^q(\mathcal{F}^\bullet)) \Longrightarrow \mathbb{H}_c^{p+q}(X; \mathcal{F}^\bullet).$$

Let us also recall here the following:

Definition 5.1.1 (Direct Image with Proper Support) For $\mathcal{F} \in Sh(X)$ and $f : X \to Y$ a continuous map, define a sheaf $f_!\mathcal{F}$ on Y by assigning to every open subset $U \subseteq Y$ the A-module:

$$\Gamma(U, f_!\mathcal{F}) = \{s \in \Gamma(f^{-1}(U), \mathcal{F}) \mid f|_{\mathrm{supp}(s)} : \mathrm{supp}(s) \to U \text{ is proper}\}.$$

The sheaf $f_!\mathcal{F} \in Sh(Y)$ is called the *direct image with proper support of $\mathcal{F}$*.

Example 5.1.2 If Y is a point, then $f_!(\mathcal{F}) = \Gamma_c(X, \mathcal{F})$.

It can be shown (e.g., see [15, VI, Proposition 2.6]) that for $y \in Y$,

$$(f_!\mathcal{F})_y \cong \Gamma_c(f^{-1}(y), \mathcal{F}). \tag{5.2}$$

Note that there is a sheaf monomorphism $f_!\mathcal{F} \hookrightarrow f_*\mathcal{F}$, which becomes an isomorphism if f is proper (e.g., a closed inclusion). The assignment $\mathcal{F} \mapsto f_!\mathcal{F}$ extends to a left exact functor

$$f_! : Sh(X) \to Sh(Y),$$

which one can derive, i.e., set $Rf_!\mathcal{F}^\bullet = f_!\mathcal{I}^\bullet$, where $\mathcal{I}^\bullet$ is an injective resolution of $\mathcal{F}^\bullet$.

Definition 5.1.3 A sheaf $\mathcal{F}$ on X is called *c-soft* if the restriction map $\Gamma(X, \mathcal{F}) \to \Gamma(K, \mathcal{F})$ is surjective for all compact subsets K in X.

Injective sheaves are c-soft. Moreover, c-soft sheaves are acyclic for $f_!$, so one may calculate $Rf_!$ by using c-soft resolutions. One also has the following result, see [15, Theorem V.10.6(ii)]:

Proposition 5.1.4 *For continuous maps $f : X \to Y$ and $g : Y \to Z$, one has: $g_! \circ f_! = (g \circ f)_!$ and $Rg_! \circ Rf_! = R(g \circ f)_!$.*

Example 5.1.5 Let $i : X \hookrightarrow Y$ be an open or closed inclusion. If $y \in Y - X$, then (5.2) yields that

$$(i_!\mathcal{F})_y \cong \Gamma_c(i^{-1}(y), \mathcal{F}) = \Gamma_c(\emptyset, \mathcal{F}) = 0,$$

and if $y \in X$ then

$$(i_!\mathcal{F})_y \cong \Gamma_c(y, \mathcal{F}) = \mathcal{F}_y.$$

Thus,

$$i_!\mathcal{F} = \mathcal{F}^Y$$

is just the extension of $\mathcal{F}$ by 0. $\mathcal{F}^Y$ is the unique sheaf on Y that restricts to $\mathcal{F}$ on X and it restricts to the zero sheaf on $Y - X$. If i is a closed inclusion, then i is proper, so $i_! = i_*$ (cf. Example 4.1.25).

Definition 5.1.6 The *higher direct image sheaves with proper support* are

$$R^i f_!(\mathcal{F}^\bullet) := \mathcal{H}^i(Rf_!\mathcal{F}^\bullet).$$

The stalks of the higher direct image sheaves with proper support are given by

$$R^i f_!(\mathcal{F}^\bullet)_y \cong \mathbb{H}^i_c(f^{-1}(y); \mathcal{F}^\bullet). \tag{5.3}$$

We conclude this section with several important results involving the direct image with proper support and its derived functor.

The proof of the following statement can be found, e.g., in [15, V, Proposition 10.7]:

Theorem 5.1.7 (Proper Base Change) *Let*

$$
\begin{array}{ccc}
X' & \xrightarrow{\;g\;} & X \\
{\scriptstyle f'}\downarrow & & \downarrow{\scriptstyle f} \\
Y' & \xrightarrow{\;g'\;} & Y
\end{array}
$$

be a cartesian diagram of spaces. Then

(a) $(g')^* \circ f_! = f'_! \circ g^*$ *in* $Sh(Y')$.
(b) $(g')^* \circ Rf_! = Rf'_! \circ g^*$ *in* $D^b(Y')$.

For the following result, see e.g., [15, V, Proposition 10.8]:

Theorem 5.1.8 (Projection Formula) *Let $f : X \to Y$ be a continuous map.*

(i) Let $\mathcal{F} \in Sh(X)$ be c-soft and let $\mathcal{G} \in Sh(Y)$ be flat. Then

$$f_!(\mathcal{F} \otimes f^*\mathcal{G}) \simeq f_!\mathcal{F} \otimes \mathcal{G} \tag{5.4}$$

in $Sh(Y)$.

(ii) For $\mathcal{F}^\bullet \in D^b(X)$ and $\mathcal{G}^\bullet \in D^b(Y)$, the following identity holds in $D^b(Y)$:

$$Rf_!(\mathcal{F}^\bullet \overset{L}{\otimes} f^*\mathcal{G}^\bullet) \simeq Rf_!\mathcal{F}^\bullet \overset{L}{\otimes} \mathcal{G}^\bullet. \tag{5.5}$$

The following important result is obtained by combining Theorems 5.1.7 and 5.1.8:

Corollary 5.1.9 (Künneth Formula) *Let X_1 and X_2 be topological spaces, with projections $p_i : X_1 \times X_2 \to X_i$, $i = 1, 2$, and let $\mathcal{F}_1^\bullet \in D^b(X_1)$ and $\mathcal{F}_2^\bullet \in D^b(X_2)$ be two bounded complexes. Then*

$$R\Gamma_c(X_1 \times X_2, \mathcal{F}_1^\bullet \overset{L}{\boxtimes} \mathcal{F}_2^\bullet) \cong R\Gamma_c(X_1, \mathcal{F}_1^\bullet) \overset{L}{\otimes} R\Gamma_c(X_2, \mathcal{F}_2^\bullet). \tag{5.6}$$

In particular, if the base ring A is a field, then for all $i \in \mathbb{Z}$ there is a natural isomorphism of A-vector spaces:

$$\mathbb{H}_c^i(X_1 \times X_2; \mathcal{F}_1^\bullet \overset{L}{\boxtimes} \mathcal{F}_2^\bullet) \cong \bigoplus_{p+q=i} \mathbb{H}_c^p(X_1; \mathcal{F}_1^\bullet) \otimes \mathbb{H}_c^q(X_2; \mathcal{F}_2^\bullet). \tag{5.7}$$

Proof Let $a_i : X_i \to pt$, $i = 1, 2$, be the constant map to a point space. By the projection formula (5.5) and the proper base change of Theorem 5.1.7(b), one gets:

$$Rp_{2!}(p_1^*\mathcal{F}_1^\bullet \overset{L}{\otimes} p_2^*\mathcal{F}_2^\bullet) \simeq Rp_{2!}(p_1^*\mathcal{F}_1^\bullet) \overset{L}{\otimes} \mathcal{F}_2^\bullet \simeq a_2^*(Ra_{1!}\mathcal{F}_1^\bullet) \overset{L}{\otimes} \mathcal{F}_2^\bullet. \tag{5.8}$$

By applying $Ra_{2!}$ to the first term of (5.8), one obtains:

$$Ra_{2!}Rp_{2!}(p_1^*\mathcal{F}_1^\bullet \overset{L}{\otimes} p_2^*\mathcal{F}_2^\bullet) \cong R\Gamma_c(X_1 \times X_2, p_1^*\mathcal{F}_1^\bullet \overset{L}{\otimes} p_2^*\mathcal{F}_2^\bullet)$$

$$\cong R\Gamma_c(X_1 \times X_2, \mathcal{F}_1^\bullet \overset{L}{\boxtimes} \mathcal{F}_2^\bullet),$$

which is just the left-hand side of (5.6). On the other hand, by applying $Ra_{2!}$ to the last term of (5.8) and using again the projection formula (5.5) one gets:

$$Ra_{2!}\left(a_2^*(Ra_{1!}\mathcal{F}_1^\bullet) \overset{L}{\otimes} \mathcal{F}_2^\bullet\right) \cong Ra_{1!}\mathcal{F}_1^\bullet \overset{L}{\otimes} Ra_{2!}\mathcal{F}_2^\bullet$$

$$\cong R\Gamma_c(X_1, \mathcal{F}_1^\bullet) \overset{L}{\otimes} R\Gamma_c(X_2, \mathcal{F}_2^\bullet),$$

which is the right-hand side of (5.6). This completes the proof of formula (5.6).

Formula (5.7) follows from (5.6) by an application of the algebraic Künneth formula (e.g., see [79, p.102]).   □

5.2   Inverse Image with Compact Support

One would like to define a functor $f^! : Sh(Y) \to Sh(X)$ so that for $\mathcal{F} \in Sh(X)$ and $\mathcal{G} \in Sh(Y)$ there is a sheaf isomorphism:

$$\mathcal{H}om(f_!\mathcal{F}, \mathcal{G}) \simeq f_*\mathcal{H}om(\mathcal{F}, f^!\mathcal{G}). \tag{5.9}$$

However, as we shall now explain, such a functor does not exist in general. To see why, let us first explore some consequences of (5.9). For $f = j : U \hookrightarrow X$ the open inclusion, $\mathcal{F} = \underline{A}_U$ the constant sheaf on U, and $\mathcal{G} \in Sh(Y)$, one has:

$$\begin{aligned}
\Gamma(U, j^!\mathcal{G}) &= \mathrm{Hom}(\underline{A}_U, j^!\mathcal{G}) \\
&= \Gamma(U, \mathcal{H}om(\underline{A}_U, j^!\mathcal{G})) \\
&= \Gamma(U, j_*\mathcal{H}om(\underline{A}_U, j^!\mathcal{G})) \\
&\overset{(5.9)}{=} \Gamma(U, \mathcal{H}om(j_!\underline{A}_U, \mathcal{G})) \\
&= \mathrm{Hom}(j^*j_!\underline{A}_U, j^*\mathcal{G}) \\
&= \mathrm{Hom}(\underline{A}_U, j^*\mathcal{G}) \\
&= \Gamma(U, j^*\mathcal{G}).
\end{aligned}$$

Performing the same calculation for every open $V \subset U$, one then concludes that

$$j^!\mathcal{G} \simeq j^*\mathcal{G}$$

for open inclusions.

Next, let $f : X \to Y$ be continuous and $j : U \hookrightarrow X$ be an open inclusion. Then, by a similar calculation, one gets from (5.9) that

$$\Gamma(U, f^!\mathcal{G}) = \text{Hom}(f_!(j_!\underline{A_U}), \mathcal{G}),$$

where $j_!\underline{A_U}$ is the extension by zero. This implies that, if $f^!$ as in (5.9) existed, then $f^!\mathcal{G}$ should be given by the assignment

$$U \longmapsto \text{Hom}(f_!(j_!\underline{A_U}), \mathcal{G}).$$

But this is *not* a sheaf in general (as it does not satisfy the gluing condition for sheaves). So one cannot hope for a functor $f^!$ defined on the sheaf level.

To proceed, consider a flat c-soft sheaf $\mathcal{K}$ and replace $\underline{A_U}$ by $\mathcal{K}|_U$, the resulting sheaf $f^!_{\mathcal{K}}\mathcal{G}$ depending on $\mathcal{K}$. To get rid of dependency on $\mathcal{K}$, one passes to the derived category. The final product is a well-defined functor (for complete details see, e.g., [15, V.7.14–V.7.16] or [6, Section 3.2])

$$f^! : D^b(Y) \rightarrow D^b(X).$$

In special cases, $f^!$ can already be defined on the category of sheaves, e.g., (see [15, V.7.19]):

(i) If $j : U \hookrightarrow X$ is an open inclusion, then $j^! \cong j^*$ is exact (and it is also defined on sheaves).
(ii) If $i : Z \hookrightarrow X$ is a closed inclusion, then $i^!$ can be defined on sheaves by the assignment

$$i^!\mathcal{G} : U \cap Z \mapsto \Gamma_{U\cap Z}(U, \mathcal{G})$$

for every open set U of X.

There is also a proper base change property for the inverse image with compact support, see [15, V, Proposition 10.7]:

Theorem 5.2.1 *In the notations of Theorem 5.1.7, the following holds in $D^b(Y')$:*

$$Rf'_* \circ g^! = (g')^! \circ Rf_*.$$

5.3 Dualizing Functor

As usual, the underlying topological spaces are assumed locally compact of finite cohomological dimension, and the base ring for sheaves is a commutative noetherian ring A of finite cohomological dimension.

Definition 5.3.1 Let $f : X \to pt$ be the constant map to a point space. The *dualizing complex of X* is defined by

$$\mathbb{D}_X^\bullet := f^! A_{pt},$$

where A_{pt} is the constant sheaf A on a point space. The *Verdier dual* of a bounded complex $\mathcal{F}^\bullet \in D^b(X)$ is defined as

$$\mathcal{D}_X \mathcal{F}^\bullet := R\mathcal{H}om^\bullet(\mathcal{F}^\bullet, \mathbb{D}_X^\bullet).$$

The dual of a complex was discovered independently by Borel–Moore in [16], at around the same time as the definition of Verdier [235].

In what follows, if there is no danger of confusion, we drop the subscript from the notation of the Verdier dual, and simply write $\mathcal{D}\mathcal{F}^\bullet$ for $\mathcal{D}_X \mathcal{F}^\bullet$.

Lemma 5.3.2 ([15, V.7.6]) *The dualizing complex $\mathbb{D}_X^\bullet$ is injective.*

Exercise 5.3.3 Show that $\mathbb{D}_X^\bullet = \mathcal{D}\underline{A}_X$.

Exercise 5.3.4 Show that $\mathcal{D}(\mathcal{F}^\bullet[n]) \simeq \mathcal{D}(\mathcal{F}^\bullet)[-n]$.

Exercise 5.3.5 Show that if $\mathcal{A}^\bullet \to \mathcal{B}^\bullet \to \mathcal{C}^\bullet \overset{[1]}{\to}$ is a distinguished triangle in $D^b(X)$, then the triangle $\mathcal{D}\mathcal{C}^\bullet \to \mathcal{D}\mathcal{B}^\bullet \to \mathcal{D}\mathcal{A}^\bullet \overset{[1]}{\to}$ obtained by applying the duality functor $\mathcal{D}$ is again a distinguished triangle.

The following result plays an important role in duality statements for manifolds. For a proof, see e.g., [15, V.7.3,V.7.10(4)] or [122, Section 3.3]:

Theorem 5.3.6 *Let X be an n-dimensional topological manifold with orientation sheaf Or_X.[1] There is a canonical isomorphism in $D^b(X)$:*

$$\mathbb{D}_X^\bullet \simeq Or_X[n]. \tag{5.10}$$

More generally, if $\mathcal{L}$ is an A-local system on X so that its stalks $\mathcal{L}_x$ are free A-modules (e.g., if A is a field), then

$$\mathcal{D}_X \mathcal{L} \simeq \mathcal{L}^\vee \otimes Or_X[n], \tag{5.11}$$

where $\mathcal{L}^\vee = \mathcal{H}om(\mathcal{L}, \underline{A}_X)$ is the dual local system, with stalks $\mathcal{L}_x^\vee = \mathrm{Hom}(\mathcal{L}_x, A)$.

The following local Verdier duality theorem was obtained by Verdier in [235], see also [15, V, Theorem 7.17] for a proof:

---

[1] There are various incarnations of the orientation sheaf appearing in these notes; these are all nicely explained in [15, V.7] or [122, Section 3.3].

Theorem 5.3.7 (Verdier Duality) *Let* $f : X \to Y$ *be a continuous map. Then for* $\mathcal{F}^\bullet \in D^b(X)$ *and* $\mathcal{G}^\bullet \in D^b(Y)$ *there is a canonical isomorphism*

$$R\mathcal{H}om^\bullet(Rf_!\mathcal{F}^\bullet, \mathcal{G}^\bullet) \simeq Rf_* R\mathcal{H}om^\bullet(\mathcal{F}^\bullet, f^! \mathcal{G}^\bullet). \tag{5.12}$$

The global version of Verdier duality is obtained from (5.12) by applying $H^0\Gamma(Y, -)$:

Corollary 5.3.8 *Let* $f : X \to Y$ *be a continuous map. Then for* $\mathcal{F}^\bullet \in D^b(X)$ *and* $\mathcal{G}^\bullet \in D^b(Y)$ *there is an isomorphism:*

$$\mathrm{Hom}_{D^b(Y)}(Rf_!\mathcal{F}^\bullet, \mathcal{G}^\bullet) \cong \mathrm{Hom}_{D^b(X)}(\mathcal{F}^\bullet, f^! \mathcal{G}^\bullet). \tag{5.13}$$

In particular, there exist adjunction morphisms

$$\mathcal{F}^\bullet \longrightarrow f^! Rf_!\mathcal{F}^\bullet, \quad Rf_! f^! \mathcal{G}^\bullet \longrightarrow \mathcal{G}^\bullet.$$

An important consequence of Verdier duality is the following:

Proposition 5.3.9 *If* $f : X \to Y$ *is a continuous map, then the following quasi-isomorphisms hold:*

$$\mathcal{D}_Y(Rf_!\mathcal{F}^\bullet) \simeq Rf_*(\mathcal{D}_X\mathcal{F}^\bullet), \quad \mathcal{D}_X(f^*\mathcal{G}^\bullet) \simeq f^!(\mathcal{D}_Y\mathcal{G}^\bullet).$$

Proof First note that $\mathbb{D}_X^\bullet = f^! \mathbb{D}_Y^\bullet$. By using Verdier duality (5.12), one gets the following sequence of isomorphisms:

$$\begin{aligned}
\mathcal{D}_Y(Rf_!\mathcal{F}^\bullet) &= R\mathcal{H}om^\bullet(Rf_!\mathcal{F}^\bullet, \mathbb{D}_Y^\bullet) \\
&\simeq Rf_* R\mathcal{H}om^\bullet(\mathcal{F}^\bullet, f^! \mathbb{D}_Y^\bullet) \\
&= Rf_* R\mathcal{H}om^\bullet(\mathcal{F}^\bullet, \mathbb{D}_X^\bullet) \\
&\simeq Rf_* \mathcal{D}_X\mathcal{F}^\bullet.
\end{aligned}$$

The second identity is proved similarly.   □

5.4 Verdier Dual via the Universal Coefficient Theorem

Assume that the coefficient ring A is a Dedekind domain (e.g., a PID). Then the Verdier dual $\mathcal{D}_X\mathcal{F}^\bullet$ is well defined up to quasi-isomorphism by the following *universal coefficients property*: for all open $U \subseteq X$ there exists a split natural exact sequence

$$0 \longrightarrow \mathrm{Ext}(\mathbb{H}_c^{q+1}(U; \mathcal{F}^\bullet), A) \longrightarrow \mathbb{H}^{-q}(U; \mathcal{D}_X \mathcal{F}^\bullet)$$

$$\longrightarrow \mathrm{Hom}(\mathbb{H}_c^q(U; \mathcal{F}^\bullet), A) \longrightarrow 0. \qquad (5.14)$$

If A is a field, the above sequence simplifies as:

$$\mathbb{H}^{-q}(U; \mathcal{D}_X \mathcal{F}^\bullet) \cong \mathrm{Hom}(\mathbb{H}_c^q(U; \mathcal{F}^\bullet), A). \qquad (5.15)$$

In general, for an arbitrary coefficient ring A, there is a *Universal Coefficient Spectral Sequence* with E_2-term given by:

$$E_2^{p,q} = \mathrm{Ext}^p(\mathbb{H}_c^{-q}(U; \mathcal{F}^\bullet), A),$$

which converges to $\mathbb{H}^{p+q}(U; \mathcal{D}_X \mathcal{F}^\bullet)$, e.g., see [15, V.7.7(3)].

5.5 Poincaré and Alexander Duality on Manifolds

In this section, we explain how to use the dualizing functor and the dualizing complex in order to deduce Poincaré and Alexander duality statements for manifolds.

Proposition 5.5.1 (Poincaré Duality) *Let X be an n-dimensional topological manifold, and assume that the base ring A is a field. Let $\mathcal{L}$ be an A-local system on X. Then there are A-vector space isomorphisms*

$$H^{n-i}(X; \mathcal{L}^\vee \otimes Or_X) \cong H_c^i(X; \mathcal{L})^\vee \qquad (5.16)$$

for all integers i.

Proof Recall that for an A-local system $\mathcal{L}$ on X one has by (5.11) a quasi-isomorphism

$$\mathcal{D}_X \mathcal{L} \simeq \mathcal{L}^\vee \otimes Or_X[n].$$

Applying hypercohomology to it yields:

$$\mathbb{H}^{-i}(X; \mathcal{D}_X \mathcal{L}) \cong \mathbb{H}^{n-i}(X; \mathcal{L}^\vee \otimes Or_X) \cong H^{n-i}(X; \mathcal{L}^\vee \otimes Or_X).$$

On the other hand, by the Universal Coefficient isomorphism (5.15) one gets:

$$\mathbb{H}^{-i}(X; \mathcal{D}_X \mathcal{L}) \cong \mathrm{Hom}(\mathbb{H}_c^i(X; \mathcal{L}), A)$$
$$= H_c^i(X; \mathcal{L})^\vee.$$

Altogether, one obtains the desired Poincaré duality isomorphism:

$$H^{n-i}(X; \mathcal{L}^{\vee} \otimes Or_X) \cong H^i_c(X; \mathcal{L})^{\vee}.$$

$\square$

If the manifold X is moreover *oriented* (i.e., with a chosen isomorphism $Or_X \simeq \underline{A}_X$), then (5.16) yields

$$H^{n-i}(X; \mathcal{L}^{\vee}) \cong H^i_c(X; \mathcal{L})^{\vee}, \tag{5.17}$$

or, equivalently, a non-degenerate Poincaré duality pairing

$$H^{n-i}(X; \mathcal{L}^{\vee}) \otimes H^i_c(X; \mathcal{L}) \to A.$$

The classical Poincaré duality isomorphism and pairing are obtained for the case $\mathcal{L} = \underline{A}_X$ of the constant sheaf on X.

Before discussing the Alexander duality on manifolds, we introduce the notion of relative hypercohomology groups:

Definition 5.5.2 For $f : Y \hookrightarrow X$ an inclusion, define *relative hypercohomology groups* by:

$$\mathbb{H}^*(X, X - Y; \mathcal{F}^{\bullet}) := \mathbb{H}^*(X; Rf_! f^! \mathcal{F}^{\bullet}).$$

Then the following Alexander duality statement holds:

Proposition 5.5.3 (Alexander Duality) *Let X be an n-dimensional oriented topological manifold, $Y \subset X$ a closed subset with inclusion $i : Y \hookrightarrow X$, and assume that the coefficient ring A is a field. Then there are A-vector space isomorphisms:*

$$H^{n-k}(X, X - Y; A) \cong H^k_c(Y; A)^{\vee}, \tag{5.18}$$

for all integers k.

Proof It follows from Proposition 5.3.9 that we have isomorphisms of functors:

$$i_! i^! \mathcal{D}_X \simeq i_! \mathcal{D}_Y i^* \simeq \mathcal{D}_X i_! i^*,$$

where one also uses the fact that $i_! = i_*$, and that $i_!$ is an exact functor on sheaves, so it extends to the derived category without being derived. Apply the above isomorphism to $\underline{A}_X \simeq (\mathcal{D}_X \underline{A}_X)[-n]$ (which follows from the orientation assumption) to get

$$i_! i^! \underline{A}_X[n] \simeq \mathcal{D}_X i_! \underline{A}_Y.$$

Apply $\mathbb{H}^{-k}(X; -)$ on both sides and use (5.15) to obtain:

$$H^{n-k}(X, X - Y; A) \cong \mathbb{H}^{n-k}(X; i_! i^! \underline{A}_X)$$
$$\cong \mathbb{H}^{-k}(X; \mathcal{D}_X i_! \underline{A}_Y)$$
$$\cong \mathbb{H}^k_c(X; i_! \underline{A}_Y)^\vee$$
$$\cong \mathbb{H}^k_c(Y; \underline{A}_Y)^\vee$$
$$\cong H^k_c(Y; A)^\vee.$$

$\square$

5.6 Attaching Triangles. Hypercohomology Long Exact Sequences of Pairs

In this section, we introduce the attaching (adjunction) triangles associated to a closed embedding and its open complement. We also develop the relevant hypercohomology long exact sequences of pairs.

Let $\mathcal{F}$ be an injective sheaf on X. Let Y be a *closed* subset of X with natural inclusions $i : Y \hookrightarrow X$ and $j : X - Y \hookrightarrow X$. For U open in X, applying the functor $\Gamma(U, -)$ to the adjunction map $\mathcal{F} \to j_* j^* \mathcal{F}$ one obtains the restriction of sections $\Gamma(U, \mathcal{F}) \to \Gamma(U - Y, \mathcal{F})$. This map is surjective since $\mathcal{F}$ is an injective sheaf. The kernel of this restriction is

$$\{s \in \Gamma(U, \mathcal{F}) \mid s|_{U-Y} = 0\} = \Gamma_{U \cap Y}(U, \mathcal{F}).$$

So one gets a short exact sequence

$$0 \to \Gamma_{U \cap Y}(U, \mathcal{F}) \to \Gamma(U, \mathcal{F}) \to \Gamma(U - Y, \mathcal{F}) \to 0.$$

Fix $x \in X$ and take the direct limit over neighborhoods U of x (which is an exact functor) to obtain the following exact sequence

$$0 \to (i_* i^! \mathcal{F})_x \to \mathcal{F}_x \to (j_* j^* \mathcal{F})_x \to 0.$$

This shows that the sequence $0 \to i_* i^! \mathcal{F} \to \mathcal{F} \to j_* j^* \mathcal{F} \to 0$ is exact. Passing to the derived category yields a distinguished triangle in $D^b(X)$:

$$i_* i^! \mathcal{F}^\bullet \longrightarrow \mathcal{F}^\bullet \longrightarrow Rj_* j^* \mathcal{F}^\bullet \xrightarrow{[1]} \tag{5.19}$$

called an *attaching (or adjunction) triangle*.

Similarly, there is a distinguished (attaching) triangle

$$j_! j^* \mathcal{F}^\bullet \longrightarrow \mathcal{F}^\bullet \longrightarrow i_* i^* \mathcal{F}^\bullet \overset{[1]}{\longrightarrow}, \tag{5.20}$$

which can be deduced from (5.19) by duality via Proposition 5.3.9.

By considering the associated hypercohomology long exact sequence for the triangle (5.19) and using $i_* = i_!$, one gets the long exact sequence:

$$\cdots \rightarrow \mathbb{H}^k(X; i_! i^! \mathcal{F}^\bullet) \rightarrow \mathbb{H}^k(X; \mathcal{F}^\bullet) \rightarrow \mathbb{H}^k(X; Rj_* j^* \mathcal{F}^\bullet) \rightarrow \cdots \tag{5.21}$$

Since $\mathbb{H}^k(X; i_! i^! \mathcal{F}^\bullet) \cong \mathbb{H}^k(X, X - Y; \mathcal{F}^\bullet)$ and $\mathbb{H}^k(X; Rj_* j^* \mathcal{F}^\bullet) \cong \mathbb{H}^k(X - Y; j^* \mathcal{F}^\bullet)$, in view of Definition 5.5.2 the sequence (5.21) translates into the long exact sequence for the hypercohomology groups of the pair $(X, X - Y)$:

$$\cdots \rightarrow \mathbb{H}^k(X, X - Y; \mathcal{F}^\bullet) \rightarrow \mathbb{H}^k(X; \mathcal{F}^\bullet) \rightarrow \mathbb{H}^k(X - Y; \mathcal{F}^\bullet) \rightarrow \cdots \tag{5.22}$$

Similarly, by applying hypercohomology with compact support to the triangle (5.20), one gets the following long exact sequence for compactly supported hypercohomology:

$$\cdots \rightarrow \mathbb{H}_c^k(X - Y; \mathcal{F}^\bullet) \rightarrow \mathbb{H}_c^k(X; \mathcal{F}^\bullet) \rightarrow \mathbb{H}_c^k(Y; \mathcal{F}^\bullet) \rightarrow \cdots \tag{5.23}$$

In particular, if X and Y are compact, then

$$\mathbb{H}_c^k(X - Y; \mathcal{F}^\bullet) \cong \mathbb{H}^k(X, Y; \mathcal{F}^\bullet) \tag{5.24}$$

for every integer k.

Exercise 5.6.1 (Excision) Show that the relative hypercohomology groups satisfy the *excision property*, i.e., if V is a subset of X so that Y is contained in the interior of V then:

$$\mathbb{H}^*(X, X - Y; \mathcal{F}^\bullet) \cong \mathbb{H}^*(V, V - Y; \mathcal{F}^\bullet). \tag{5.25}$$

Chapter 6
Intersection Homology After Deligne

In this chapter, we explain the sheaf-theoretic approach to intersection homology theory. We introduce here the Deligne intersection cohomology complex [83], whose hypercohomology computes the intersection homology groups. This complex of sheaves can be described axiomatically in a way that is independent of the stratification or any additional geometric structure (such as a piecewise linear structure), leading to a proof of the topological invariance of intersection homology groups.

For a more complete account, see [15], [83] or [6]. For a nicely and compactly written survey, the reader may also consult [201].

6.1 Introduction

Let X^n be an n-dimensional oriented pseudomanifold, and let $\overline{p}, \overline{q}$ be complementary perversities (i.e., $\overline{p}(k) + \overline{q}(k) = k - 2$, for every integer $k \geq 2$). As already indicated in Section 2.6, there exists a bilinear non-degenerate pairing (i.e., *generalized Poincaré duality*):

$$I H_i^{\overline{p}}(X; \mathbb{Q}) \times I^{BM} H_{n-i}^{\overline{q}}(X; \mathbb{Q}) \to \mathbb{Q}. \tag{6.1}$$

The aim in this chapter is to construct a complex of A-sheaves $IC_{\overline{p}}^\bullet$ whose hypercohomology group $\mathbb{H}^{-i}(X; IC_{\overline{p}}^\bullet)$ calculates $I^{BM} H_i^{\overline{p}}(X; A)$, and such that $\mathbb{H}_c^{-i}(X; IC_{\overline{p}}^\bullet)$ calculates $I H_i^{\overline{p}}(X; A)$. This will be done by sheafifying the chain construction of intersection homology. Moreover, for complementary perversities $\overline{p}$ and $\overline{q}$ and field coefficients (e.g., $\mathbb{Q}$), one can show that there exists a quasi-isomorphism

© Springer Nature Switzerland AG 2019
L. G. Maxim, *Intersection Homology & Perverse Sheaves*, Graduate Texts in Mathematics 281, https://doi.org/10.1007/978-3-030-27644-7_6

$$IC_{\overline{p}}^{\bullet} \cong \mathcal{D}_X IC_{\overline{q}}^{\bullet}[n] \qquad (6.2)$$

by checking that the complex $\mathcal{D}_X(IC_{\overline{q}}^{\bullet})[n]$ satisfies a set of axioms that characterize $IC_{\overline{p}}^{\bullet}$ uniquely up to quasi-isomorphism. Finally, the duality statement of (6.1) follows from (6.2) with $A = \mathbb{Q}$, by applying hypercohomology and using the Universal Coefficient Theorem of Section 5.4.

6.2   Intersection Cohomology Complex

Let us now proceed with the construction of the *intersection cohomology complex*, see [83] or [15, Chapter II]. For now we work with coefficients in an arbitrary noetherian ring A (e.g., $\mathbb{Z}$ or a field).

Let X^n be an n-dimensional pseudomanifold with a filtration

$$X = X_n \supseteq X_{n-2} \supseteq \ldots \supseteq X_0 \supseteq \emptyset.$$

As already indicated in Chapter 2, for simplifying the exposition we will assume in this section that X has an underlying PL structure (alternatively, one may use King's singular version of intersection homology, see Remark 2.3.12). Then, if $U \subseteq X$ is an open subset, U has an induced PL structure.

For every integer i, there is a sheaf $IC_{\overline{p}}^{-i} \in Sh(X)$ that is constructed by defining sections on each open subset U of X by

$$X \supseteq U \longmapsto IC_{\overline{p}}^{-i}(U) := IC_i^{\overline{p}}((U)), \qquad (6.3)$$

i.e., the $\overline{p}$-allowable locally finite i-chains on U (with A-coefficients). Moreover, differentials $d^{-i} : IC_{\overline{p}}^{-i} \to IC_{\overline{p}}^{-i+1}$ are induced by the boundary maps $\partial_i : IC_i^{\overline{p}} \to IC_{i-1}^{\overline{p}}$. This defines a bounded complex of sheaves of A-modules

$$IC_{\overline{p}}^{\bullet} \in D^b(X)$$

in the derived category of X, called the *intersection cohomology complex* of X.

Recall that a sheaf $\mathcal{F} \in Sh(X)$ on a paracompact topological space X is called *soft* if $\Gamma(X, \mathcal{F}) \to \Gamma(K, \mathcal{F})$ is onto, for each K closed subset of X. For the following statement, the interested reader may consult [15, Chapter II, Section 5] or [6, Proposition 4.1.19]:

Lemma 6.2.1 $IC_{\overline{p}}^{-i}$ *are soft sheaves.*

Similarly, if $\mathcal{L}$ is an A-local system on $X - X_{n-2}$, the *twisted intersection cohomology complex* $IC_{\overline{p}}^{\bullet}(\mathcal{L})$ is defined by setting

$$U \longmapsto IC_{\overline{p}}^{-i}(\mathcal{L})(U) := IC_i^{\overline{p}}((U; \mathcal{L})),$$

on each open subset $U \subseteq X$. Then it can be shown that $IC_{\overline{p}}^{-i}(\mathcal{L})$ are soft sheaves. Therefore, since by Proposition 4.3.15 soft sheaves are acyclic (hence one is allowed to resolve by soft sheaves in order to compute hypercohomology), one obtains the following:

Proposition 6.2.2 *The intersection homology groups of a pseudomanifold X are computed from the (twisted) intersection cohomology complex via the identifications:*

$$I^{BM}H_i^{\overline{p}}(X; \mathcal{L}) = \mathbb{H}^{-i}(X; IC_{\overline{p}}^{\bullet}(\mathcal{L})) \tag{6.4}$$

and

$$IH_i^{\overline{p}}(X; \mathcal{L}) = \mathbb{H}_c^{-i}(X; IC_{\overline{p}}^{\bullet}(\mathcal{L})). \tag{6.5}$$

Proof Indeed,

$$\mathbb{H}^{-i}(X; IC_{\overline{p}}^{\bullet}(\mathcal{L})) = H^{-i}\Gamma(X, IC_{\overline{p}}^{\bullet}(\mathcal{L}))$$
$$= H^{-i}(IC_{-\bullet}^{\overline{p}}((X; \mathcal{L})))$$
$$= H_i(IC_{\bullet}^{\overline{p}}((X; \mathcal{L})))$$
$$= I^{BM}H_i^{\overline{p}}(X; \mathcal{L}).$$

The second identity is obtained similarly. $\square$

Moreover, using the Künneth formula (Proposition 2.5.1) and the cone formula for intersection homology (Theorem 2.5.2), one gets the following *cohomology stalk calculation* (see, e.g., [15, Lemma V.3.15]):

Proposition 6.2.3 *For $x \in X_{n-k} - X_{n-k-1}$,*

$$\mathcal{H}^i(IC_{\overline{p}}^{\bullet}(\mathcal{L}))_x \cong \begin{cases} 0, & i > \overline{p}(k) - n, \\ IH_{-i-(n-k+1)}^{\overline{p}}(L_x, \mathcal{L}), & i \leq \overline{p}(k) - n, \end{cases} \tag{6.6}$$

with L_x denoting the link of (the stratum containing) x in X.

To fix the notations, set

$$U_k = X - X_{n-k}, \quad k \geq 2.$$

Note that U_2 is the dense open subset of X, and there is an exhausting filtration of X by open subsets

$$U_2 \subseteq U_3 \subseteq \ldots \subseteq U_{n+1} = X.$$

Furthermore, let

$$j_k : U_k \hookrightarrow U_{k+1}$$

be the open inclusion, and denote by

$$i_k : X_{n-k} - X_{n-k-1} = U_{k+1} - U_k \hookrightarrow U_{k+1}$$

the closed (stratum) inclusion.

The following result is an easy consequence of the above stalk calculation and of the cone formula on allowable chains (see, e.g., [83, Section 2.5], [15, Chapter II, Section 6], or [6, Section 4.1.4]):

Proposition 6.2.4 *In the above notations, the following assertions hold:*

(a) *If $x \in X_{n-k} - X_{n-k-1}$ and $U = \mathbb{R}^{n-k} \times \mathring{c} L_x$ is a distinguished neighborhood of x, then*

$$\mathcal{H}^{-i}(IC_{\overline{p}}^{\bullet})_x \cong I^{BM} H_i^{\overline{p}}(U; A).$$

(b) *If $x \in X_{n-k} - X_{n-k-1}$ and U is a distinguished neighborhood of x, then*

$$\mathcal{H}^{-i}(Rj_{k*} IC_{\overline{p}}^{\bullet}|_{U_k})_x \cong I^{BM} H_i^{\overline{p}}(U \cap U_k; A).$$

(c)

$$\mathcal{H}^j(IC_{\overline{p}}^{\bullet}|_{U_{k+1}}) = 0, \quad \text{if } j > \overline{p}(k) - n, \ k \geq 2.$$

(d) *The adjunction*

$$IC_{\overline{p}}^{\bullet}|_{U_{k+1}} \longrightarrow Rj_{k*} j_k^*(IC_{\overline{p}}^{\bullet}|_{U_{k+1}})$$

induces isomorphisms

$$\mathcal{H}^j(i_k^* IC_{\overline{p}}^{\bullet}|_{U_{k+1}}) \xrightarrow{\simeq} \mathcal{H}^j(i_k^* Rj_{k*} j_k^* IC_{\overline{p}}^{\bullet}|_{U_{k+1}})$$

for $j \leq \overline{p}(k) - n$.

Definition 6.2.5 A complex of sheaves $\mathcal{A}^{\bullet} \in C(X)$ is *constructible* with respect to the given pseudomanifold stratification $\mathcal{X}$ on X if the sheaves $\mathcal{H}^j(\mathcal{A}^{\bullet})|_{X_{n-k}-X_{n-k-1}}$ are local systems with finitely generated stalks. Denote by $D_{c,\mathcal{X}}^b(X)$ the full subcategory of $D^b(X)$ consisting of all bounded $\mathcal{X}$-constructible complexes of sheaves.

It then follows from Proposition 6.2.4 that the following property holds:

Proposition 6.2.6 $IC^\bullet_{\overline{p}}$ is $\mathcal{X}$-constructible.

The following set of axioms plays an essential role in understanding the intersection cohomology complex.

Definition 6.2.7 A complex $\mathcal{A}^\bullet \in D^b_{c,\mathcal{X}}(X)$ satisfies the set of axioms $[AX_{\overline{p}}]$ if the following conditions are verified:

(AX0)   *Normalization*: $\mathcal{A}^\bullet|_{U_2} \simeq \underline{A}_{U_2}[n]$.
(AX1)   *Lower bound*: $\mathcal{H}^j(\mathcal{A}^\bullet) = 0$ if $j < -n$.
(AX2)   *Vanishing condition*: $\mathcal{H}^j(\mathcal{A}^\bullet|_{U_{k+1}}) = 0$ if $j > \overline{p}(k) - n$.
(AX3)   *Attaching condition*: $\mathcal{H}^j(i_k^*\mathcal{A}^\bullet|_{U_{k+1}}) \to \mathcal{H}^j(i_k^* R j_{k*} j_k^* \mathcal{A}^\bullet|_{U_{k+1}})$ is an isomorphism if $j \le \overline{p}(k) - n$.

Remark 6.2.8 The constructibility assumption in the above definition is redundant. In the next section, it will become clear that constructibility is a consequence of (AX0)–(AX3), see Remark 6.3.6.

Remark 6.2.9 One may, more generally, replace $(AX0)$ by the requirement that

(AX0') $\mathcal{A}^\bullet|_{U_2} \simeq \mathcal{L}[n]$,

where $\mathcal{L}$ is an A-local coefficient system on $U_2 = X - X_{n-2}$.

Theorem 6.2.10 *If X is an oriented n-dimensional pseudomanifold, then $IC^\bullet_{\overline{p}}$ satisfies $[AX_{\overline{p}}]$.*

Proof (AX2) and (AX3) follow from Proposition 6.2.4 (c) and (d), respectively. For (AX0) and (AX1), let C^i be the sheaf on X defined by the assignment:

$$U \longmapsto C_{-i}((U)),$$

for every open subset U in X. Let $C^\bullet$ be the complex with differentials $d^i : C^i \to C^{i+1}$ induced from $\partial_i : C_{-i+1}((U)) \to C_{-i}((U))$. Then there is a canonical homomorphism $IC^{\overline{p}}_{-i}((U)) \hookrightarrow C_{-i}((U))$, which is an isomorphism if $U \subseteq U_2$ since U_2 has no singularities. So

$$IC^\bullet_{\overline{p}}|_{U_2} \longrightarrow C^\bullet|_{U_2}$$

is a quasi-isomorphism. Hence, for (AX0) it is enough to show that

$$\mathcal{H}^{-i}(C^\bullet|_{U_2}) \simeq \begin{cases} \underline{A}_{U_2}, & i = n, \\ 0, & \text{otherwise.} \end{cases}$$

The sheaf $\mathcal{H}^{-i}(C^\bullet|_{U_2})$ is associated to the presheaf

$$U_2 \supseteq U \longmapsto H^{-i}\Gamma(U, C^\bullet) = H^{BM}_i(U; A).$$

Recall that for $U = \mathbb{R}^n$ a small Euclidean neighborhood, one has that $H_i^{BM}(\mathbb{R}^n; A) = A$ if $i = n$ and 0 otherwise. In other words, $\mathcal{H}^{-n}(C^\bullet|_{U_2})$ is the orientation sheaf Or_{U_2} on U_2. Together with the orientability assumption, this proves (AX0). Finally, (AX1) follows from Proposition 6.2.4 (a) and the cone formula. $\square$

Remark 6.2.11 Similarly, $IC_{\overline{p}}^\bullet(\mathcal{L})$ satisfies $[AX_{\overline{p}}]$ with the normalization condition (AX0').

Remark 6.2.12 The proof of Theorem 6.2.10 shows in fact that, if one drops the orientability assumption for X, then $IC_{\overline{p}}^\bullet$ satisfies $[AX_{\overline{p}}]$ but with (AX0) replaced by:

$$IC_{\overline{p}}^\bullet|_{U_2} \simeq Or_{U_2}[n], \tag{6.7}$$

with $n = \dim X$ and Or_{U_2} the orientation sheaf on U_2; see also [15, II, Theorem 6.1], [15, V.2.9].

6.3   Deligne's Construction of Intersection Homology

In the previous section, we have defined the intersection cohomology complex $IC_{\overline{p}}^\bullet$ of a (PL) pseudomanifold X, and showed that it satisfies a set of axioms $[AX_{\overline{p}}]$. In this section, we define a bounded constructible complex, the *Deligne complex*, on *any* topological pseudomanifold (without assuming a PL structure), which, by its very definition, satisfies $[AX_{\overline{p}}]$. Moreover, we show that the set of axioms $[AX_{\overline{p}}]$ uniquely characterize the Deligne complex up to quasi-isomorphism. It then follows that $IC_{\overline{p}}^\bullet$ is quasi-isomorphic to the Deligne complex, and hence it does not depend on the underlying PL structure of X.

Let X^n be an n-dimensional pseudomanifold with a filtration

$$X = X_n \supseteq X_{n-2} \supseteq \ldots \supseteq X_0 \supseteq \emptyset,$$

and set

$$U_k = X - X_{n-k}, \quad k \geq 2,$$

with inclusion maps $j_k : U_k \hookrightarrow U_{k+1}$ and $i_k : X_{n-k} - X_{n-k-1} \hookrightarrow U_{k+1}$.

Definition 6.3.1 (Deligne Complex)   The *Deligne complex* $\mathfrak{S}_{\overline{p}}^\bullet$ is defined inductively as follows. On U_2, set

$$\mathfrak{S}_{\overline{p}}^\bullet|_{U_2} = \underline{A}_{U_2}[n].$$

By induction, assuming that $\mathfrak{S}_{\overline{p}}^\bullet|_{U_k}$ has been constructed, define

$$\mathfrak{S}_{\overline{p}}^\bullet|_{U_{k+1}} := \tau_{\leq \overline{p}(k)-n} R j_{k*}(\mathfrak{S}_{\overline{p}}^\bullet|_{U_k}).$$

Remark 6.3.2 The construction of Definition 6.3.1 was suggested by Deligne, who also conjectured that it computes the Goresky–MacPherson intersection homology groups. The conjecture was proved by Goresky–MacPherson in [83], who developed sheaf-theoretic axioms for the intersection cohomology complex, which also led to the proof of topological invariance of intersection homology.

Remark 6.3.3 The Deligne complex $\mathfrak{S}_{\overline{p}}^{\bullet}$ is constructible with respect to the given stratification of X (see [15, Proposition V.3.12]), and it satisfies the set of axioms $[AX_{\overline{p}}]$. The constructibility of $\mathfrak{S}_{\overline{p}}^{\bullet}$ is a consequence of the fact that operations of the form Rj_{k*} and $\tau_{\leq \overline{p}(k)-n}$ preserve constructibility (see [15, Corollary V.3.11] for Rj_{k*} where the local normal triviality assumption is used, and obviously for $\tau_{\leq \overline{p}(k)-n}$). See also the comments of [15, Chapter V, Remark 4.20].

Remark 6.3.4 It is very important to note that no PL structure is involved in the construction of the Deligne complex.

The Deligne complex satisfies the following uniqueness property:

Theorem 6.3.5 *If $\mathcal{A}^{\bullet} \in D^b(X)$ satisfies the set of axioms $[AX_{\overline{p}}]$, then $\mathcal{A}^{\bullet}$ is quasi-isomorphic to the Deligne complex $\mathfrak{S}_{\overline{p}}^{\bullet}$.*

Proof The assertion is proved by induction on strata. On U_2, one has

$$\mathcal{A}^{\bullet}|_{U_2} \overset{AX0}{\simeq} \underline{A}_{U_2}[n] = \mathfrak{S}_{\overline{p}}^{\bullet}|_{U_2}.$$

Inductively, assuming that there is a quasi-isomorphism

$$\mathcal{A}^{\bullet}|_{U_k} \simeq \mathfrak{S}_{\overline{p}}^{\bullet}|_{U_k},$$

one can extend it to a morphism over U_{k+1} as follows:

$$\mathcal{A}^{\bullet}|_{U_{k+1}} \overset{AX2}{\simeq} \tau_{\leq \overline{p}(k)-n}(\mathcal{A}^{\bullet}|_{U_{k+1}}) \overset{(1)}{\to} \tau_{\leq \overline{p}(k)-n} Rj_{k*}j_k^*(\mathcal{A}^{\bullet}|_{U_{k+1}})$$

$$\simeq \tau_{\leq \overline{p}(k)-n} Rj_{k*}(\mathcal{A}^{\bullet}|_{U_k})$$

$$\overset{(2)}{\simeq} \tau_{\leq \overline{p}(k)-n} Rj_{k*}(\mathfrak{S}_{\overline{p}}^{\bullet}|_{U_k})$$

$$= \mathfrak{S}_{\overline{p}}^{\bullet}|_{U_{k+1}},$$

where (1) is the adjunction morphism, and (2) follows by the induction hypothesis. It remains to show that the morphism $\mathcal{A}^{\bullet}|_{U_{k+1}} \to \mathfrak{S}_{\overline{p}}^{\bullet}|_{U_{k+1}}$ is an isomorphism over $U_{k+1} - U_k = X_{n-k} - X_{n-k-1}$. Indeed,

$$i_k^*(\mathcal{A}^{\bullet}|_{U_{k+1}}) \overset{AX2}{\simeq} i_k^*(\tau_{\leq \overline{p}(k)-n}\mathcal{A}^{\bullet}|_{U_{k+1}}) \overset{AX3}{\simeq} i_k^* \tau_{\leq \overline{p}(k)-n} Rj_{k*}j_k^*(\mathcal{A}^{\bullet}|_{U_{k+1}}).$$

(Recall that pullback commutes with truncation.) This completes the proof. □

Remark 6.3.6 In view of Remark 6.3.3, the above theorem also shows that the constructibility assumption in the set of axioms $[AX\overline{p}]$ is redundant.

By combining Theorems 6.2.10 and 6.3.5, one gets the following:

Corollary 6.3.7 *The intersection cohomology complex $IC^{\bullet}_{\overline{p}}$ of an oriented pseudo-manifold X is quasi-isomorphic to the Deligne complex $\mathfrak{S}^{\bullet}_{\overline{p}}$.*

Remark 6.3.8 If the oriented n-dimensional pseudomanifold X is also assumed to be *normal*, then

$$IC^{\bullet}_{\overline{0}} \simeq \underline{A}_X[n]. \tag{6.8}$$

Indeed, since $\tau_{\leq 0} \circ Rj_{k*} = j_{k*}$ on $Sh(U_k)$, for every $k \geq 2$, Corollary 6.3.7 implies that $IC^{\bullet}_{\overline{0}} \simeq j_*(\underline{A}_{U_2})[n]$, with $j : U_2 \hookrightarrow X$ the inclusion map. The assertion follows then by the normality assumption, which implies that $j_*(\underline{A}_{U_2}) \simeq \underline{A}_X$ (since intersecting open sets with U_2 preserves the number of connected components). Note that in this case (6.8) yields the isomorphism

$$I^{BM}H^{\overline{0}}_i(X; A) \cong \mathbb{H}^{-i}(X; IC^{\bullet}_{\overline{0}}) \cong \mathbb{H}^{-i}(X; \underline{A}_X[n]) \cong H^{n-i}(X; A),$$

as already mentioned in Theorem 2.4.9.

Remark 6.3.9 One may, more generally, define the *twisted Deligne complex* $\mathfrak{S}^{\bullet}(\mathcal{L})$ by requiring that, on U_2,

$$\mathfrak{S}^{\bullet}_{\overline{p}}|_{U_2} = \mathcal{L}[n],$$

where $\mathcal{L}$ is an A-local coefficient system on U_2. The obtained complex is then quasi-isomorphic to the twisted intersection cohomology complex $IC^{\bullet}_{\overline{p}}(\mathcal{L})$ of an oriented pseudomanifold X.

Without the orientation assumption, Remark 6.2.12 and Theorem 6.3.5 yield that

$$IC^{\bullet}_{\overline{p}} \simeq \mathfrak{S}^{\bullet}_{\overline{p}}(Or_{U_2}), \tag{6.9}$$

and

$$IC^{\bullet}_{\overline{p}}(\mathcal{L}) \simeq \mathfrak{S}^{\bullet}_{\overline{p}}(\mathcal{L} \otimes Or_{U_2}), \tag{6.10}$$

with Or_{U_2} denoting as usual the orientation sheaf on the dense open stratum U_2. Therefore, one shall write $IC^{\bullet}_{\overline{p}}$ (resp., $IC^{\bullet}_{\overline{p}}(\mathcal{L})$) for every incarnation of the (twisted) Deligne complex appearing on the right-hand side of (6.9) (resp., (6.10)).

Furthermore, since no PL structure is involved in the construction of the Deligne complex $\mathfrak{S}^{\bullet}_{\overline{p}}(Or_{U_2})$, by using (6.9) and Proposition 6.2.2 one obtains immediately the following important consequence:

Corollary 6.3.10 *The intersection homology groups $I^{(BM)}H_i^{\overline{p}}(X; A)$ are independent of the underlying PL structure on X.*

For later reference, (AX3) can be reformulated as follows. Recall from Section 5.6 that for a closed subset K of X with natural inclusions $i : K \hookrightarrow X$ and $j : X - K \hookrightarrow X$, there is an attaching triangle in $D^b(X)$:

$$i_*i^! \longrightarrow id \longrightarrow Rj_*j^* \xrightarrow{[1]} .\tag{6.11}$$

Applying i^* to (6.11) and using $i^*i_* = id$ yields the triangle

$$i^! \longrightarrow i^* \longrightarrow i^*Rj_*j^* \xrightarrow{[1]} .$$

Setting $i = i_k$ and $j = j_k$, it can be readily seen that (AX3) is equivalent to the condition:

$$\mathcal{H}^i(i_k^! \mathcal{A}^\bullet|_{U_{k+1}}) = 0, \quad \text{if } i \le \overline{p}(k) - n + 1.\tag{6.12}$$

Condition (6.12) can be further refined by using the following fact (e.g., see [15, V, Proposition 3.7] for a proof):

Lemma 6.3.11 *If M^m is a manifold of dimension m with $f_x : \{x\} \hookrightarrow M$ the point inclusion and $\mathcal{A}^\bullet$ has locally constant cohomology sheaves, then*

$$f_x^! \mathcal{A}^\bullet \simeq f_x^* \mathcal{A}^\bullet[-m].\tag{6.13}$$

($f_x^! \mathcal{A}^\bullet$ is usually called the "co-stalk of $\mathcal{A}$ at x").

Now let $x \in X_{n-k} - X_{n-k-1}, i_x : \{x\} \hookrightarrow U_{k+1}$ and $f_x : \{x\} \hookrightarrow X_{n-k} - X_{n-k-1}$, and recall that $X_{n-k} - X_{n-k-1}$ is a manifold of dimension $n - k$. By (6.13) and using the equality $i_x^! = f_x^! \circ i_k^!$, one gets

$$i_x^! \mathcal{A}^\bullet \simeq f_x^*(i_k^! \mathcal{A}^\bullet)[-n + k].$$

Using (6.12), it follows that (AX3) is equivalent to the following:
(AX3') *Co-stalk vanishing condition*:

$$\mathcal{H}^i(i_x^! \mathcal{A}^\bullet) = 0, \quad i \le \overline{p}(k) - k + 1, \ x \in X_{n-k} - X_{n-k-1}.$$

Remark 6.3.12 Deligne's axiomatic construction of the intersection cohomology complex $IC_{\overline{p}}^\bullet$ applies without change if one uses what is usually referred to as a *super-perversity*, i.e., a function $\overline{p} : \mathbb{Z}_{(\ge 2)} \longrightarrow \mathbb{N}$ satisfying $\overline{p}(2) = 1$ and $\overline{p}(k) \le \overline{p}(k+1) \le \overline{p}(k) + 1$. In particular, the stalk formula of Proposition 6.2.3 remains valid in this context, as it can be derived only from the axiomatic definition of the intersection complex. An important example of super-perversity is the *logarithmic* perversity $\bar{\ell}$, given by $\bar{\ell} = (1, 2, 2, 3, 3, 4, \cdots)$, which plays a fundamental role in works of Cappell–Shaneson [31] and of the author [159, 160]; see also [76] and the references therein.

Exercise 6.3.13 (Witt Spaces) Let X be an n-dimensional oriented pseudomanifold (without boundary), and assume that the coefficient ring A is a field. As in Definition 2.7.1, X is called an A-*Witt space* if for every stratum S of odd codimension $2r + 1$ with link L_S, one has $IH_r^{\bar{m}}(L_S; A) = 0$. Show that X is an A-Witt space if and only if the canonical morphism $IC_{\bar{m}}^\bullet \longrightarrow IC_{\bar{n}}^\bullet$ is a quasi-isomorphism. As a consequence, deduce Theorem 2.7.3. (See [83, Sections 5.5, 5.6].)

Exercise 6.3.14 (Künneth formula) Assume that the base ring A is a field. If X is an A-Witt space, let $IC_X^\bullet$ denote $IC_{\bar{m}}^\bullet \simeq IC_{\bar{n}}^\bullet$.

Let X_1 and X_2 be A-Witt spaces. Show (by verifying axioms) that:

$$IC_{X_1 \times X_2}^\bullet \simeq IC_{X_1}^\bullet \overset{L}{\boxtimes} IC_{X_2}^\bullet. \tag{6.14}$$

Deduce the following Künneth formula:

$$IH_i^{\bar{m}}(X_1 \times X_2; A) \cong \bigoplus_{a+b=i} IH_a^{\bar{m}}(X_1; A) \otimes IH_b^{\bar{m}}(X_2; A). \tag{6.15}$$

More generally, show that if $\mathcal{L}_1$, $\mathcal{L}_2$ are A-local systems defined on dense open subsets of X_1 and X_2, respectively, then (with self-explanatory notations):

$$IC_{X_1 \times X_2}^\bullet(\mathcal{L}_1 \boxtimes \mathcal{L}_2) \simeq IC_{X_1}^\bullet(\mathcal{L}_1) \overset{L}{\boxtimes} IC_{X_2}^\bullet(\mathcal{L}_2). \tag{6.16}$$

Deduce that the following Künneth formula holds:

$$IH_i^{\bar{m}}(X_1 \times X_2; \mathcal{L}_1 \boxtimes \mathcal{L}_2) \cong \bigoplus_{a+b=i} IH_a^{\bar{m}}(X_1; \mathcal{L}_1) \otimes IH_b^{\bar{m}}(X_2; \mathcal{L}_2). \tag{6.17}$$

6.4 Generalized Poincaré Duality

The *generalized Poincaré duality* for intersection homology groups is a consequence of the following sheaf-theoretic result:

Theorem 6.4.1 *Assume the coefficient ring A is a field. Let X be an n-dimensional oriented pseudomanifold, with $\bar{p}$ and $\bar{q}$ complementary perversities. Then there exists a quasi-isomorphism*

$$\mathcal{D}IC_{\bar{q}}^\bullet[n] \simeq IC_{\bar{p}}^\bullet.$$

Proof Start by observing that, for a complex $\mathcal{A}^\bullet \in D^b_c(X)$, $x \in X$ and U_x a small distinguished open neighborhood of x, a hypercohomology spectral sequence argument yields that (e.g., see [15, V, Lemma 8.1]):

$$\mathcal{H}^i(i_x^*\mathcal{A}^\bullet) \cong \mathbb{H}^i(U_x; \mathcal{A}^\bullet), \tag{6.18}$$

and similarly,

$$\mathcal{H}^i(i_x^!\mathcal{A}^\bullet) \cong \mathbb{H}^i_c(U_x; \mathcal{A}^\bullet), \tag{6.19}$$

with $i_x : \{x\} \hookrightarrow X$ denoting the point inclusion (see also Section 7.2).

By Theorem 6.3.5 and Corollary 6.3.7, in order to prove Theorem 6.4.1 it suffices to show that the complex $\mathcal{D}IC_{\overline{q}}^\bullet[n]$ satisfies the set of axioms $[AX_{\overline{p}}]$.

(AX0)   Let $j : U_2 \hookrightarrow X$ denote the inclusion of the regular stratum. Since j is an open inclusion, one has that $j^* = j^!$. Then:

$$j^*\mathcal{D}_X IC_{\overline{q}}^\bullet[n] \simeq \mathcal{D}_{U_2}(j^! IC_{\overline{q}}^\bullet)[n] \simeq \mathcal{D}_{U_2}(\underline{A}_{U_2}[n])[n]$$
$$\simeq \mathcal{D}_{U_2}(\underline{A}_{U_2}) \simeq \mathbb{D}^\bullet_{U_2} \simeq \underline{A}_{U_2}[n], \tag{6.20}$$

where the last isomorphism uses the orientability assumption.

(AX2)   Let $x \in X_{n-k} - X_{n-k-1}$. By (AX3') for $\overline{q}$ one has that

$$\mathcal{H}^i(i_x^! IC_{\overline{q}}^\bullet) = 0, \ \ i \leq \overline{q}(k) - k + 1.$$

Therefore,

$$\mathcal{H}^j(\mathcal{D}IC_{\overline{q}}^\bullet[n])_x \ \cong \ \mathcal{H}^{j+n}(\mathcal{D}IC_{\overline{q}}^\bullet)_x$$
$$\cong \ \mathbb{H}^{j+n}(U_x; \mathcal{D}IC_{\overline{q}}^\bullet)$$
$$\overset{UCT}{\cong} \ \mathbb{H}_c^{-j-n}(U_x; IC_{\overline{q}}^\bullet)^\vee$$
$$\cong \ \mathcal{H}^{-j-n}(i_x^! IC_{\overline{q}}^\bullet)^\vee$$
$$= \ 0$$

for $-j - n \leq \overline{q}(k) - k + 1$, or equivalently, for $j > \overline{p}(k) - n$, since $\overline{p}$ and $\overline{q}$ are complementary perversities. This proves (AX2).

(AX3')   Similarly, the $\overline{q}$-stalk condition for $IC_{\overline{q}}^\bullet$ implies the $\overline{p}$-co-stalk condition for its dual. More precisely,

$$\mathcal{H}^j(i_x^! \mathcal{D}IC_{\overline{q}}^\bullet[n]) \ \cong \ \mathcal{H}^{j+n}(i_x^! \mathcal{D}IC_{\overline{q}}^\bullet)$$
$$\cong \ \mathbb{H}_c^{j+n}(U_x; \mathcal{D}IC_{\overline{q}}^\bullet)$$

$$\overset{UCT}{\cong} \mathbb{H}^{-j-n}(U_x; IC_{\overline{q}}^{\bullet})^{\vee}$$

$$\cong \mathcal{H}^{-j-n}(IC_{\overline{q}}^{\bullet})_x^{\vee}$$

$$= 0$$

for $-j - n > \overline{q}(k) - n$ (from the $\overline{q}$-stalk vanishing for $IC_{\overline{q}}^{\bullet}$), which is equivalent to $j \leq \overline{p}(k) - k + 1$.

(AX1)   This follows from the calculation from (AX2) together with the fact that $\mathcal{H}^j(i_x^! IC_{\overline{q}}^{\bullet}) = 0$ for $j > 0$.

This finishes the proof of Theorem 6.4.1.   □

As a direct consequence of Theorem 6.4.1, one obtains the following duality statement for intersection homology groups:

Corollary 6.4.2 *With the same assumptions as in Theorem 6.4.1, there is a non-degenerate pairing:*

$$I^{BM}H_i^{\overline{p}}(X; A) \otimes IH_{n-i}^{\overline{q}}(X; A) \to A,$$

with A denoting the coefficient field.

Proof Indeed, one has the following sequence of isomorphisms of A-vector spaces:

$$I^{BM}H_i^{\overline{p}}(X; A) \cong \mathbb{H}^{-i}(X; IC_{\overline{p}}^{\bullet})$$

$$\cong \mathbb{H}^{-i}(X; \mathcal{D}IC_{\overline{q}}^{\bullet}[n])$$

$$\cong \mathbb{H}^{n-i}(X; \mathcal{D}IC_{\overline{q}}^{\bullet})$$

$$\overset{(1)}{\cong} \mathrm{Hom}(\mathbb{H}_c^{-n+i}(X; IC_{\overline{q}}^{\bullet}), A)$$

$$\cong \mathrm{Hom}(IH_{n-i}^{\overline{q}}(X; A), A),$$

where (1) follows from the Universal Coefficient Theorem (5.15).   □

Remark 6.4.3 If in Theorem 6.4.1 one assumes, moreover, that X is a $\mathbb{Q}$-Witt space (see Definition 2.7.1 and Exercise 6.3.13), then $IC_{\overline{m}}^{\bullet}$ is self-dual. If X is also compact, then one gets a non-degenerate pairing $IH_i^{\overline{m}}(X; \mathbb{Q}) \otimes IH_{n-i}^{\overline{m}}(X; \mathbb{Q}) \to \mathbb{Q}$ that computes the signature $\sigma(X)$ of X.

Exercise 6.4.4 Show that if X_1 and X_2 are closed $\mathbb{Q}$-Witt spaces in the sense of Exercise 6.3.13, then

$$\sigma(X_1 \times X_2) = \sigma(X_1) \cdot \sigma(X_2). \tag{6.21}$$

More generally, without any orientation assumption on X and with $\mathcal{L}$ an A-local system on the dense open stratum U_2 (where A is a field), one can easily show, by using (5.11) and adapting the argument in (AX0) of Theorem 6.4.1 accordingly, that there is a quasi-isomorphism

$$\mathcal{D}_X \mathfrak{S}_{\overline{q}}^{\bullet}(\mathcal{L})[n] \simeq \mathfrak{S}_{\overline{p}}^{\bullet}(\mathcal{L}^{\vee} \otimes Or_{U_2}), \tag{6.22}$$

(see also [15, V.9.8]). In particular, if the n-dimensional pseudomanifold X is assumed to be *oriented*, then upon applying hypercohomology one gets from Remark 6.3.9 the following Poincaré duality isomorphism, generalizing Corollary 6.4.2:

$$I^{BM} H_i^{\overline{p}}(X; \mathcal{L}^{\vee}) \cong I H_{n-i}^{\overline{q}}(X; \mathcal{L})^{\vee}. \tag{6.23}$$

Poincaré duality pairings over other coefficient rings have been obtained by Goresky–Siegel [86] (for $A = \mathbb{Z}$), and by Cappell–Shaneson [31] (for A a Dedekind domain). For example, it is shown in [86] that integral Poincaré duality for intersection homology theory does hold under the hypothesis that the intersection homology groups of links of strata are torsion free, at least in appropriate dimensions. In [31], Cappell and Shaneson exhibit a duality that holds at the opposite extreme, when the intersection homology of the links is all torsion. The Cappell–Shaneson duality (called *superduality*) is a consequence of the sheaf-theoretic superduality isomorphism (6.24), which is described by Theorem 6.4.5 below.

In view of Remark 6.3.12, two perversities $\overline{p}$ and $\overline{q}$ are called *superdual* if $\overline{p}(k) + \overline{q}(k) = k - 1$ for all $k \geq 2$; in particular, one of the two functions $\overline{p}, \overline{q}$ is a super-perversity. The proof of the following result follows just as in Theorem 6.4.1, by making use of the Universal Coefficient Theorem of Section 5.4; the details are left as an exercise for the interested reader (but see also [31, Theorem 3.2]):

Theorem 6.4.5 *Let A be a Dedekind domain, and let X be an oriented n-dimensional stratified pseudomanifold with singular set $\Sigma = X_{n-2}$. Let $\mathcal{L}$ and $\mathcal{M}$ be local systems of A-modules on $X - \Sigma$ with finitely generated stalks, and let $\overline{p}$ and $\overline{q}$ be superdual perversities. Suppose that for every $x \in \Sigma$ the stalks $\mathcal{H}^j(IC_{\overline{q}}^{\bullet}(\mathcal{M}))_x$ are torsion A-modules for all j. Then a perfect pairing $\mathcal{L} \otimes_A \mathcal{M} \to \underline{A}_{X-\Sigma}$ induces a canonical (superduality) isomorphism:*

$$\mathcal{D}IC_{\overline{q}}^{\bullet}(\mathcal{M})[n] \simeq IC_{\overline{p}}^{\bullet}(\mathcal{L}) \tag{6.24}$$

in $D^b(X)$.

For trivial coefficients, the hypothesis of Theorem 6.4.5 is never satisfied, but for non-trivial local systems this often happens. For example, if $\Sigma \subset S^n = X$ is a knot, $A = \mathbb{Q}[t, t^{-1}]$, $\mathcal{L}$ is defined by the linking number homomorphism on $\pi_1(S^n - \Sigma)$, and $\mathcal{M} = \mathcal{L}^{\vee}$, then the induced pairing on cohomology is the familiar *Blanchfield pairing* of knot theory. For another interesting such situation, with applications to Singularity Theory, see [160], where the Cappell–Shaneson superduality isomorphism (6.24) is used for studying Alexander-type invariants of complex hypersurface complements.

6.5 Topological Invariance of Intersection Homology

Let X^n be a pseudomanifold with a fixed stratification $\mathcal{X}$ and let $\bar{p}$ be a fixed perversity. By making use of the Deligne complex, it was shown in Section 6.3 that the intersection cohomology complex

$$IC^\bullet_{\bar{p},\mathcal{X}} \simeq \mathfrak{S}^\bullet_{\bar{p},\mathcal{X}}(Or_{U_2})$$

is characterized, uniquely up to quasi-isomorphism, by a set of axioms $[AX_{\bar{p},\mathcal{X}}]$. Moreover, intersection homology groups (with A-coefficients) can be computed as:

$$I^{BM}H_i^{\bar{p},\mathcal{X}}(X) \cong \mathbb{H}^{-i}(X; IC^\bullet_{\bar{p},\mathcal{X}}),$$

where the above notations record the fact that the definitions of the objects involved depend on the chosen stratification $\mathcal{X}$ of X. However, in [83, Section 4, Corollary 1], Goresky–MacPherson showed that the following holds:

Theorem 6.5.1 *Intersection homology is a topological invariant, i.e., if $f : X \to Y$ is a homeomorphism, then (with A-coefficients):*

$$I^{BM}H_i^{\bar{p},\mathcal{X}}(X) \cong I^{BM}H_i^{\bar{p},\mathcal{Y}}(Y).$$

(Here $\mathcal{X}$ and $\mathcal{Y}$ are pseudomanifold stratifications of X and Y, respectively.) In particular, for $f = id_X$, this implies that the intersection homology of X does not depend on the chosen stratification.

The above theorem is an immediate consequence of the following result:

Theorem 6.5.2 *Let $\mathcal{X}_1$ and $\mathcal{X}_2$ be two pseudomanifold stratifications on X. Let $IC^\bullet_{\bar{p},\mathcal{X}_1}$ and $IC^\bullet_{\bar{p},\mathcal{X}_2}$ be the intersection cohomology complexes on X defined by using $\mathcal{X}_1$ and $\mathcal{X}_2$, respectively. Then*

$$IC^\bullet_{\bar{p},\mathcal{X}_1} \simeq IC^\bullet_{\bar{p},\mathcal{X}_2}$$

in $D^b(X)$. Equivalently, if $\mathcal{X}$ is a pseudomanifold stratification of X with Or_{U_2} the orientation sheaf on the dense open stratum, then the Deligne complex $\mathfrak{S}^\bullet_{\bar{p},\mathcal{X}}(Or_{U_2})$ is independent of the choice of $\mathcal{X}$.

The proof of Theorem 6.5.2 consists of the following main steps:

a) Introduce a more intrinsic set of axioms $[AX_{\overline{p},intr.}]$ that depends on the stratification X only very weakly.
b) Show that $[AX_{\overline{p},X}] \iff [AX_{\overline{p},intr.}]$.
c) Construct an *intrinsic* sheaf complex $IC^{\bullet}_{\overline{p},intr.}$ defined uniquely up to quasi-isomorphism by $[AX_{\overline{p},intr.}]$.
d) Show that $IC^{\bullet}_{\overline{p},X_1} \simeq IC^{\bullet}_{\overline{p},intr.} \simeq IC^{\bullet}_{\overline{p},X_2}$.

We only indicate a); for complete details, see [15, V.4], [83, Section 4], or [6, Section 4.3]).

For $j = 0, 1, 2, \ldots$, define

$$\overline{p}^{-1}(j) = \begin{cases} \min\{k \mid \overline{p}(k) \geq j\}, & j \leq \overline{p}(n), \\ \infty, & j > \overline{p}(n). \end{cases}$$

So

$$\overline{p}(k) \geq j \iff k \geq \overline{p}^{-1}(j), \text{ for } 2 \leq k \leq n.$$

The *j-th cohomological support* of $A^{\bullet} \in D(X)$ is

$$\mathrm{supp}^j(A^{\bullet}) = \overline{\{x \in X \mid \mathcal{H}^j(i_x^* A^{\bullet}) \neq 0\}},$$

with $i_x : \{x\} \hookrightarrow X$. The *$j$-th cohomological cosupport* is

$$\mathrm{cosupp}^j(A^{\bullet}) = \overline{\{x \in X \mid \mathcal{H}^j(i_x^! A^{\bullet}) \neq 0\}}.$$

Then the set of axioms $[AX_{\overline{p},intr.}]$ consists of the following:

$(AX0)_{intr.}$: (*normalization*) Generically, i.e., on some dense open $U \subset X$, one has:
   $A^{\bullet}|_U \simeq Or_U[n]$.
$(AX1)_{intr.}$: (*lower bound*) $\mathcal{H}^j(A^{\bullet}) = 0$, for all $j < -n$.
$(AX2)_{intr.}$: (*support condition*) $\dim \mathrm{supp}^j(A^{\bullet}) \leq n - \overline{p}^{-1}(j+n)$, for all $j > -n$.
$(AX3)_{intr.}$: (*cosupport condition*) $\dim \mathrm{cosupp}^j(A^{\bullet}) \leq n - \overline{q}^{-1}(-j)$, for all $j < 0$,
   where $\overline{q}$ is the complementary perversity to $\overline{p}$.

Remark 6.5.3 Note that only $(AX0)_{intr.}$ involves the stratification of X. Here the notion of dimension may be taken to be the topological dimension as in [110].

Exercise 6.5.4 (Normally Nonsingular Inclusion)   Generalizing Definition 3.4.5, an inclusion of oriented pseudomanifolds $i : Z \hookrightarrow X$ is said to be *trivial normally nonsingular* of codimension c if Z has a c-dimensional tubular neighborhood in X, that is, an open neighborhood $N \subset X$ and a retraction $\pi : N \to Z$ such that (π, N, Z) is homeomorphic to an $\mathbb{R}^c$-vector bundle over Z, where Z is identified with the zero-section; see [83, Section 5.4.1]. Let $i : Z \hookrightarrow X$ be a normally nonsingular inclusion of codimension c. Fix a perversity $\overline{p}$ and denote by $IC_X^\bullet$ and $IC_Z^\bullet$ the corresponding intersection cohomology complexes. Show that there are canonical quasi-isomorphisms:

$$i^* IC_X^\bullet \simeq IC_Z^\bullet[c] \text{ and } i^! IC_X^\bullet \simeq IC_Z^\bullet. \tag{6.25}$$

6.6   Rational Homology Manifolds

In this section, we give an explicit description of the middle-perversity intersection cohomology complex for a very special class of singular spaces, whose (usual) homology still possesses Poincaré duality over the rationals, see, e.g., [18, Section 1.4].

Definition 6.6.1 A topological space X is called a *rational homology manifold* (or $\mathbb{Q}$-manifold) of real dimension m if, for every $x \in X$, the rational local homology groups at x are computed by

$$H_i(X, X - x; \mathbb{Q}) = \begin{cases} \mathbb{Q}, & i = m, \\ 0, & i \neq m. \end{cases}$$

It is easy to see that locally contractible spaces (e.g., manifolds, pseudomanifolds, or (singular) complex algebraic varieties) for which all points have links that are rational homology spheres (e.g., lens spaces) are rational homology manifolds. For example, if M is a manifold, and G is a *finite* group acting on M, then $X = M/G$ is a $\mathbb{Q}$-manifold. More generally, orbifolds (i.e., varieties that are locally finite quotients) are $\mathbb{Q}$-manifolds. The latter class includes simplicial toric varieties. Conversely, one has the following:

Proposition 6.6.2 *Let X be a complex algebraic variety of complex dimension n, which moreover is assumed to be a rational homology manifold. Then the link of every point in X is a rational homology sphere.*

Proof First note that, by the definition of a rational homology manifold, X is pure-dimensional.

Denote by L_x the link of a point x in a stratum S of X of complex codimension s. Such a point x has a neighborhood U_x in X of type:

$$U_x \cong \mathbb{C}^{n-s} \times \mathring{c} L_x.$$

By excision, there is an isomorphism of rational vector spaces:

$$H_i(X, X - x; \mathbb{Q}) \cong H_i(U_x, U_x - x; \mathbb{Q}).$$

Moreover, the relative Künneth theorem and the long exact sequence of a pair yield the isomorphisms:

$$H_i(U_x, U_x - x; \mathbb{Q}) \cong H_i(\mathbb{C}^{n-s} \times \mathring{c}L_x, \mathbb{C}^{n-s} \times \mathring{c}L_x - \{(0, c)\}; \mathbb{Q})$$

$$\cong \tilde{H}_{i-2n+2s-1}(L_x; \mathbb{Q}).$$

(Here c denotes the cone point in $\mathring{c}L_x$.) Finally, since X is a rational homology manifold,

$$H_i(X, X - x; \mathbb{Q}) \cong \begin{cases} \mathbb{Q}, & i = 2n, \\ 0, & i \neq 2n. \end{cases}$$

Altogether,

$$H_i(L_x; \mathbb{Q}) = \begin{cases} \mathbb{Q}, & i = 2s - 1, \\ 0, & i \neq 2s - 1, \end{cases}$$

showing that L_x is a $\mathbb{Q}$-homology sphere of dimension $2s - 1$. $\qquad\square$

We can now prove the following:

Theorem 6.6.3 *Let X be a complex algebraic variety of complex dimension n, which moreover is assumed to be a rational homology manifold. Then the middle-perversity intersection cohomology complex $IC_{\overline{m}}^\bullet$ with $\mathbb{Q}$-coefficients is quasi-isomorphic to $\underline{\mathbb{Q}}_X[2n]$.*

Proof Since X is assumed to be a rational homology manifold, it is pure-dimensional.

Since X is a complex algebraic variety, it admits a Whitney stratification that makes X into an *oriented* topological pseudomanifold of real dimension $2n$ with all strata of even real dimension.

Let $\mathcal{X}$ be a Whitney stratification of X with strata S_s indexed by complex codimension, i.e., $\operatorname{codim}_{\mathbb{C}} S_s = s$. It suffices to show that the complex $\underline{\mathbb{Q}}_X[2n]$ satisfies the set of axioms $[AX_{\overline{m}}]$, hence it is quasi-isomorphic to $IC_{\overline{m}}^\bullet$.

Since we are in the context of complex algebraic geometry, we take the opportunity to introduce here the *shifted Deligne complex*

$$IC_X := IC_{\overline{m}}^\bullet[-n],$$

and show instead that $\mathcal{A}^\bullet := \underline{\mathbb{Q}}_X[n]$ is quasi-isomorphic to IC_X. For this, we first need to rephrase the set of axioms $[AX_{\overline{m}}]$ in the algebraic geometric convention of Beilinson–Bernstein–Deligne [12], i.e., for IC_X, thus paving the road for the theory of perverse sheaves of the subsequent chapters:

(AX0) $\mathcal{A}^\bullet|_{X_{\text{reg}}} \simeq \underline{\mathbb{Q}}_{X_{\text{reg}}}[n]$, where X_{reg} is the nonsingular locus of X.

(AX1) $\mathcal{H}^j(\mathcal{A}^\bullet) = 0$ for $j < -n$.

(AX2) For all $x \in S_s$: $\mathcal{H}^j(\mathcal{A}^\bullet)_x = 0$ for all $j \geq -n + s$ and $s \geq 1$.

(AX3') For all $x \in S_s$ with $i_x : \{x\} \hookrightarrow X$ the point inclusion: $\mathcal{H}^j(i_x^!\mathcal{A}^\bullet) = 0$ for all $j \leq n - s$ and $s \geq 1$.

The axioms (AX0) and (AX1) are clearly satisfied by $\underline{\mathbb{Q}}_X[n]$.

For (AX2), note that if U_x denotes a small neighborhood of x in X, then:

$$\mathcal{H}^j(\underline{\mathbb{Q}}_X[n])_x \cong \mathbb{H}^j(U_x; \underline{\mathbb{Q}}_X[n]) = H^{j+n}(U_x; \mathbb{Q}) = 0$$

for all $j + n > 0$ since U_x is contractible. In particular, this vanishing holds for $j \geq -n + s$ and $s \geq 1$.

For checking (AX3'), one makes use of the fact that X is a $\mathbb{Q}$-manifold. Let $x \in S_s$. Then the link L_x of x has real dimension $2s - 1$. Moreover, by Proposition 6.6.2, L_x is a $\mathbb{Q}$-homology sphere of dimension $2s - 1$. Then, for $x \in S_s$,

$$\mathcal{H}^j(i_x^!\underline{\mathbb{Q}}_X[n]) \cong \mathbb{H}_c^j(U_x; \underline{\mathbb{Q}}_X[n]) = H_c^{j+n}(U_x; \mathbb{Q}).$$

But $U_x \cong \mathbb{C}^{n-s} \times \mathring{c}L_x$, so the Künneth formula yields that

$$H_c^{j+n}(U_x; \mathbb{Q}) \cong \bigoplus_{k+\ell=j+n} H_c^k(\mathbb{C}^{n-s}; \mathbb{Q}) \otimes H_c^\ell(\mathring{c}L_x; \mathbb{Q}).$$

Recall that

$$H_c^k(\mathbb{C}^{n-s}; \mathbb{Q}) = \begin{cases} \mathbb{Q}, & k = 2n - 2s, \\ 0, & k \neq 2n - 2s. \end{cases}$$

Moreover,

$$H_c^\ell(\mathring{c}L_x; \mathbb{Q}) \cong H^\ell(cL_x, L_x; \mathbb{Q}) \cong \widetilde{H}^{\ell-1}(L_x; \mathbb{Q}) = \begin{cases} \mathbb{Q}, & \ell = 2s, \\ 0, & \ell \neq 2s, \end{cases}$$

where the last equality uses the fact that L_x is a $\mathbb{Q}$-homology sphere of dimension $2s - 1$. Thus $\mathcal{H}^j(i_x^!\underline{\mathbb{Q}}_X[n]) = 0$ if $j \neq n$, and in particular for all $j \leq n - s$ and $s \geq 1$. So $\underline{\mathbb{Q}}_X[n]$ also satisfies (AX3'). $\square$

Remark 6.6.4 The Borel convention [15] for the intersection cohomology complex differs by a shift by the real dimension from the Goresky–MacPherson convention [83]. Specifically, the middle-perversity *Borel intersection cohomology complex* is given by

$$IC_X^B := IC_X[-n] = IC_{\overline{m}}^\bullet[-2n].$$

Borel's indexing convention is convenient for defining (compactly supported) intersection cohomology groups as follows:

Definition 6.6.5 The (compactly supported) rational (middle-perversity) *intersection cohomology groups* of a pure complex n-dimensional complex algebraic variety are defined as:

$$IH^k(X; \mathbb{Q}) := \mathbb{H}^k(X; IC_X^B) \cong I^{BM} H_{2n-k}^{\overline{m}}(X; \mathbb{Q}).$$

$$IH_c^k(X; \mathbb{Q}) := \mathbb{H}_c^k(X; IC_X^B) \cong IH_{2n-k}^{\overline{m}}(X; \mathbb{Q}).$$

In particular, if X is also a $\mathbb{Q}$-manifold as in Theorem 6.6.3, there is a quasi-isomorphism $IC_X^B \simeq \mathbb{Q}_X$. So, in this case, it follows that:

$$
\begin{aligned}
I^{BM} H_{2n-k}^{\overline{m}}(X; \mathbb{Q}) = IH^k(X; \mathbb{Q}) &:= \mathbb{H}^k(X; IC_X^B) = \mathbb{H}^k(X; \mathbb{Q}_X) \\
&= H^k(X; \mathbb{Q}).
\end{aligned}
\tag{6.26}
$$

For completeness, we also introduce here the notion of *relative (middle-perversity) intersection cohomology groups* as follows.

Definition 6.6.6 Let X be a complex algebraic variety of pure complex dimension n, and let $i : Z \hookrightarrow X$ be a closed subvariety. As in Definition 5.5.2, we set:

$$IH^k(X, X - Z; \mathbb{Q}) := \mathbb{H}^k(X, X - Z; IC_X^B) := \mathbb{H}^{k-n}(X; i_! i^! IC_X).$$

Exercise 6.6.7 In the above notations, show that the relative intersection cohomology groups fit into a long exact sequence of intersection cohomology groups for the pair $(X, X - Z)$:

$$\cdots \to IH^k(X, X - Z; \mathbb{Q}) \to IH^k(X; \mathbb{Q}) \to IH^k(X - Z; \mathbb{Q}) \to \cdots$$

Exercise 6.6.8 (Alexander Duality for Intersection Cohomology) Let X be a complex algebraic variety of pure complex dimension n, and let $i : Z \hookrightarrow X$ be a pure-dimensional closed subvariety. Assume that i is a normally nonsingular inclusion (as in Exercise 6.5.4). Show that the following Alexander duality isomorphism holds for every integer k:

$$IH^{2n-k}(X, X - Z; \mathbb{Q}) \cong IH_c^k(Z; \mathbb{Q})^\vee.
\tag{6.27}$$

6.7 Intersection Homology Betti Numbers, I

In this section, we compute the middle-perversity intersection homology groups of complex algebraic varieties in terms of the "size" of their singular locus (e.g., see [67], or [61, Section 5.4]). The results presented here will be refined in Section 11.4, by using Saito's theory of mixed Hodge modules. We work with $\mathbb{Q}$-coefficients, unless stated otherwise.

Recall from the previous section that the (compactly supported) rational (middle-perversity) intersection cohomology groups of a pure complex n-dimensional complex algebraic variety are defined as:

$$I H^k(X; \mathbb{Q}) := \mathbb{H}^{k-n}(X; IC_X), \quad I H_c^k(X; \mathbb{Q}) := \mathbb{H}_c^{k-n}(X; IC_X),$$

with $IC_X := IC_{\overline{m}}^{\bullet}[-n]$.

Exercise 6.7.1 Show that, if X is a pure n-dimensional complex algebraic variety, there exists a natural morphism

$$\alpha_X : \underline{\mathbb{Q}}_X[n] \longrightarrow IC_X,$$

extending the natural quasi-isomorphism on the nonsingular locus of X. (Hint: start by computing $\tau_{\leq -n} IC_X$.)

Remark 6.7.2 Applying the Verdier dualizing functor $\mathcal{D}$ to the morphism α_X of Exercise 6.7.1, and using the isomorphisms $\mathcal{D}(\underline{\mathbb{Q}}_X[n]) \simeq \mathbb{D}_X^{\bullet}[-n]$ (cf. Exercise 5.3.3) and $\mathcal{D}(IC_X) \simeq IC_X$, one gets natural morphisms:

$$\underline{\mathbb{Q}}_X \longrightarrow IC_X[-n] \longrightarrow \mathbb{D}_X^{\bullet}[-2n]. \tag{6.28}$$

Applying the hypercohomology functor to (6.28) yields induced morphisms (with $\mathbb{Q}$-coefficients)

$$H^k(X) \longrightarrow I H^k(X) \longrightarrow H_{2n-k}^{BM}(X),$$

which correspond to the cap product by the fundamental class $[X] \in H_{2n}^{BM}(X)$.

The following statement is a sheaf-theoretic version of Proposition 2.3.18.

Proposition 6.7.3 *Let X be a complex algebraic variety of pure complex dimension n, with only isolated singularities. Let $U = X_{\mathrm{reg}} = X - \mathrm{Sing}(X)$ be the nonsingular locus of X. Then (with $\mathbb{Q}$-coefficients):*

$$I H^k(X) = \begin{cases} H^k(U), & k < n, \\ \mathrm{Image}\,(H^n(X) \to H^n(U)), & k = n, \\ H^k(X), & k > n. \end{cases} \tag{6.29}$$

Proof The partition

$$X = U \sqcup \bigsqcup_{x \in \mathrm{Sing}(X)} \{x\}$$

is a Whitney (hence, a topological pseudomanifold) stratification, so if $j : U \hookrightarrow X$ is the inclusion of the nonsingular locus, it follows by Deligne's construction that

$$IC_X = \tau_{\leq -1}(Rj_*\underline{\mathbb{Q}}_U[n]) = \tau_{\leq n-1}(Rj_*\underline{\mathbb{Q}}_U)[n].$$

In particular, there is a distinguished triangle

$$IC_X[-n] \longrightarrow Rj_*\underline{\mathbb{Q}}_U \longrightarrow \tau_{\geq n}Rj_*\underline{\mathbb{Q}}_U \overset{[1]}{\longrightarrow} . \qquad (6.30)$$

Since $\mathcal{H}^q(\tau_{\geq n}Rj_*\underline{\mathbb{Q}}_U) = 0$ for $q < n$, the hypercohomology spectral sequence together with the long exact sequence of hypercohomology groups associated to the distinguished triangle (6.30) yield the assertion for $k < n$, together with an injective morphism $IH^n(X) \hookrightarrow H^n(U)$.

Let us next embed the canonical morphism $\underline{\mathbb{Q}}_X \longrightarrow IC_X[-n]$ of Exercise 6.7.1 into a distinguished triangle

$$\underline{\mathbb{Q}}_X \longrightarrow IC_X[-n] \longrightarrow \mathcal{F}^\bullet \overset{[1]}{\longrightarrow} \qquad (6.31)$$

and note that $j^!\mathcal{F}^\bullet = j^*\mathcal{F}^\bullet \simeq 0$, i.e., $\mathcal{F}^\bullet$ is supported on the zero-dimensional set $\mathrm{Sing}(X)$. In particular, the obvious adjunction triangle yields that $\mathcal{F}^\bullet \simeq i_*i^*\mathcal{F}^\bullet$, where $i : \mathrm{Sing}(X) \hookrightarrow X$ denotes the inclusion of the singular locus. Therefore,

$$\mathbb{H}^k(X; \mathcal{F}^\bullet) \cong \mathbb{H}^k(\mathrm{Sing}(X); i^*\mathcal{F}^\bullet) \cong \bigoplus_{x \in \mathrm{Sing}(X)} \mathcal{H}^k(\mathcal{F}^\bullet)_x.$$

Note also that, for $x \in \mathrm{Sing}(X)$ and $k \geq 1$, one gets from (6.31) that

$$\mathcal{H}^k(\mathcal{F}^\bullet)_x \cong \mathcal{H}^{k-n}(IC_X)_x.$$

Furthermore, by the axiom (AX2) for IC_X, one has that $\mathcal{H}^\ell(IC_X)_x \cong 0$ for every $x \in \mathrm{Sing}(X)$ and $\ell \geq 0$. Altogether,

$$\mathbb{H}^k(X; \mathcal{F}^\bullet) \cong 0, \quad \text{for } k \geq n. \qquad (6.32)$$

The hypercohomology long exact sequence associated to the distinguished triangle (6.31), together with the vanishing (6.32) yield isomorphisms $H^k(X) \cong IH^k(X)$ for $k > n$, and a surjective morphism $H^n(X) \twoheadrightarrow IH^n(X)$, thus completing the proof.   □

As a generalization of Proposition 6.7.3, one can prove the following result (see [61, Theorem 5.4.12(ii)] and [67, Lemma 1]):

Theorem 6.7.4 *Let X be a complex algebraic variety of pure complex dimension n, and let U be an open subvariety, with $Z = X - U$ a closed subvariety of complex dimension $\leq d$. Then (with $\mathbb{Q}$-coefficients):*

(a) $IH^k(X) \cong IH^k(U)$, for $k < n - d$.

(b) $IH^{n-d}(X) \hookrightarrow IH^{n-d}(U)$.

(c) $IH_c^{n+d}(X) \twoheadleftarrow IH_c^{n+d}(U)$.

(d) $IH_c^k(X) \cong IH_c^k(U)$, for $k > n + d$.

Proof Let $j : U \hookrightarrow X$ and $i : Z \hookrightarrow X$ denote the open and, respectively, closed inclusion.

By applying $\mathbb{H}^{k-n}(X; -)$ to the attaching triangle

$$i_* i^! IC_X \longrightarrow IC_X \longrightarrow Rj_* j^! IC_X \longrightarrow$$

one gets the long exact sequence

$$\cdots \to \mathbb{H}^{k-n}(X; i_* i^! IC_X) \to \mathbb{H}^{k-n}(X; IC_X) \to \mathbb{H}^{k-n}(X; Rj_* j^! IC_X) \to \cdots$$

or, equivalently,

$$\cdots \to \mathbb{H}^{k-n}(Z; i^! IC_X) \to IH^k(X) \to IH^k(U) \to \cdots .$$

Then for (a) and (b) it suffices to show that

$$\mathbb{H}^\ell(Z; i^! IC_X) = 0 \text{ for } \ell \le -d. \tag{6.33}$$

This is proved by induction on d. First note that there is a Zariski-open subset $S \subset Z$ (which is in fact the union of maximal dimensional strata of a Whitney stratification of Z) such that $Z_1 := Z - S$ has $\dim_{\mathbb{C}} Z_1 < \dim_{\mathbb{C}} Z$. If $j_S : S \hookrightarrow Z$ denotes the inclusion and $i_S = i \circ j_S$, then (6.12) yields that $\mathcal{H}^\ell(i_S^! IC_X) = 0$ for $\ell \le -d$. Set $\mathcal{F}^\bullet = i^! IC_X$ and let $i_1 : Z_1 \hookrightarrow Z$ be the inclusion, with $u = i \circ i_1$. Then one has a hypercohomology long exact sequence (derived from the obvious attaching triangle)

$$\cdots \to \mathbb{H}^\ell(Z_1; i_1^! \mathcal{F}^\bullet) \to \mathbb{H}^\ell(Z; \mathcal{F}^\bullet) \to \mathbb{H}^\ell(S; \mathcal{F}^\bullet) \to \cdots$$

But $\mathbb{H}^\ell(Z_1; i_1^! \mathcal{F}^\bullet) = \mathbb{H}^\ell(Z_1; u^! IC_X) = 0$ for $\ell \le -(d-1)$ by the induction hypothesis. Since $\mathcal{H}^\ell(j_S^! \mathcal{F}^\bullet) = 0$ for $\ell \le -d$, and $j_S^! = j_S^*$, the hypercohomology spectral sequence yields that $\mathbb{H}^\ell(Z; \mathcal{F}^\bullet) = \mathbb{H}^\ell(Z; j_S^! \mathcal{F}^\bullet) = 0$ for $\ell \le -d$. This proves (6.33).

By applying $\mathbb{H}_c^{k-n}(X; -)$ to the attaching triangle

$$j_! j^* IC_X \longrightarrow IC_X \longrightarrow i_* i^* IC_X \longrightarrow$$

one gets the long exact sequence

$$\cdots \to \mathbb{H}_c^{k-n}(X; j_! j^* IC_X) \to \mathbb{H}_c^{k-n}(X; IC_X) \to \mathbb{H}_c^{k-n}(X; i_* i^* IC_X) \to \cdots$$

or, equivalently,

$$\cdots \to IH_c^k(U) \to IH_c^k(X) \to \mathbb{H}_c^{k-n}(Z; i^*IC_X) \to \cdots$$

As in the proof of (6.33), or by duality, one has that $\mathcal{H}^\ell(i^*IC_X) = 0$ for $\ell \geq -d$. So, since Z is of complex dimension $\leq d$, the hypercohomology spectral sequence yields that $\mathbb{H}_c^{k-n}(Z; i^*IC_X) = 0$ for $k - n \geq d$. This proves (c) and (d). $\square$

Exercise 6.7.5 Let X be a hypersurface in $\mathbb{C}P^{n+1}$ with only isolated singularities, and denote by $j : X \hookrightarrow \mathbb{C}P^{n+1}$ the inclusion map. Show the following:

(i) $j^k : H^k(\mathbb{C}P^{n+1}; \mathbb{Q}) \to H^k(X; \mathbb{Q})$ is a monomorphism for all k with $0 \leq k \leq 2n$.

(ii) the middle intersection cohomology Betti number of the hypersurface X is given by the formula (with $\mathbb{Q}$-coefficients):

$$\dim IH^n(X) = \dim H^n(X) + \dim H_0^{n+1}(X) - \sum_{x \in \mathrm{Sing}(X)} \dim H^n(L_x),$$

where $H_0^{n+1}(X) := \mathrm{Coker}\,(j^{n+1})$ is the primitive $(n+1)$-st cohomology of X, and L_x denotes the link of the singular point x in X.

(iii) the intersection cohomology Euler characteristic of X can be computed as

$$\chi(IH^*(X)) := \sum_i (-1)^i \dim IH^i(X)$$

$$= \chi(X) - (-1)^n \cdot \sum_{x \in \mathrm{Sing}(X)} \dim H^n(L_x).$$

In particular, $\chi(IH^*(X))$ does not depend on the position of the singularities of the hypersurface X.

Remark 6.7.6 The Betti numbers of complex projective hypersurfaces with only isolated singularities are known to depend on the *position* of singularities. The classical example in this regard goes back to Zariski in the early 1930s and consists of sextic surfaces:

$$X = \{f(x_0, x_1, x_2) + x_3^6 = 0\} \subset \mathbb{C}P^3,$$

where $f(x_0, x_1, x_2) = 0$ is a plane sextic curve $C \subset \mathbb{C}P^2$ having six cusp singularities. There are two possible situations:

(i) The six cusps of the sextic curve C are all situated on a conic, e.g., $f(x_0, x_1, x_2) = (x_0^2 + x_1^2)^3 + (x_1^3 + x_2^3)^2$. In this case it can be shown that $\dim H^2(X) = 2$.

(ii) The six cusps of C are not situated on a conic (this being the generic case). Then it can be shown that dim $H^2(X) = 0$. For more details on these computations see, e.g., [60, Chapter 6].

In light of Example 6.7.5(ii), one can see that for $X \subset \mathbb{C}P^{n+1}$ a complex hypersurface with only isolated singularities, the middle intersection homology Betti number, dim $IH^n(X)$, depends in general on the position of singularities of X in the ambient projective space.

Chapter 7
Constructibility in Algebraic Geometry

Constructible sheaves are the algebraic counterpart of the decomposition of singular spaces into manifold pieces, the strata. These sheaves, which can be seen as generalizations of local systems, have powerful applications to the study of topology of singular spaces, especially in the complex algebraic/analytic context.

Constructible sheaves were introduced by Grothendieck in [228, Exposé IX], where their functorial properties were studied (in the context of étale cohomology). The theory has gained renewed interest after Kashiwara [117] made the connection with holonomic D-modules, and the discovery of intersection cohomology by Goresky–MacPherson [83] and perverse sheaves by Gabber and Beilinson–Bernstein–Deligne [12]. For the derived calculus in the constructible setting, we refer to [83, Section 1] and also [15, V.3].

As usual, we assume that the base ring A is commutative and noetherian, of finite cohomological dimension (e.g., $\mathbb{Z}$ or a field).

7.1 Definition: Properties

It was shown in [242, 233] that a complex algebraic (or analytic) variety X admits a *Whitney stratification*, that is, a (locally) finite partition $\mathcal{X}$ into non-empty, connected, locally closed nonsingular subvarieties X_α of X (called *strata*) that satisfy the following properties:

(a) *frontier condition*: the frontier $\partial X_\alpha := \overline{X_\alpha} - X_\alpha$ is a union of strata of $\mathcal{X}$.
(b) *constructibility*: the closure $\overline{X_\alpha}$ and the frontier ∂X_α are closed complex algebraic (analytic) subspaces in X.

In addition, whenever $X_\alpha \subseteq \overline{X_\beta}$, the pair $(X_\alpha, \overline{X_\beta})$ is required to satisfy certain conditions that guarantee that the variety X is topologically equisingular along each stratum.

© Springer Nature Switzerland AG 2019

L. G. Maxim, *Intersection Homology & Perverse Sheaves*, Graduate Texts in Mathematics 281, https://doi.org/10.1007/978-3-030-27644-7_7

Example 7.1.1 (Whitney Umbrella) Let X be defined by $x^2 - zy^2 = 0$ in $\mathbb{C}^3$. Then $\text{Sing}(X) = \{z - axis\}$, but the origin is "more singular" than any other point on the z-axis. This can be easily seen by considering neighborhoods of (or local homology groups at) points on the z-axis. A Whitney stratification of X is:

$$X \supset \{z - \text{axis}\} \supset \{0\}.$$

This example shows that it is not enough to produce a decomposition of a variety X into nonsingular pieces: one also needs the variety to be uniformly singular (i.e., equisingular) along these pieces.

Definition 7.1.2

(i) A sheaf $\mathcal{F}$ of A-modules on a complex algebraic (or analytic) variety X is *constructible* if there is a partition $X = \bigcup_\alpha X_\alpha$ corresponding to the strata of a Whitney stratification $\mathcal{X}$ of X, so that:

 (a) the restriction $\mathcal{F}|_{X_\alpha}$ is an A-local system for all α,
 (b) the stalks $\mathcal{F}_x$ ($x \in X$) are finite type A-modules.

(ii) A complex $\mathcal{F}^\bullet \in D^b(X)$ is constructible if all its cohomology sheaves $\mathcal{H}^j(\mathcal{F}^\bullet) \in Sh(X)$, $j \in \mathbb{Z}$, are constructible.

Exercise 7.1.3 It is immediate to see that the constant sheaf $\underline{A}_X$ is constructible on X. On the other hand, if $i : \mathfrak{C} \hookrightarrow \mathbb{C}^1$ denotes the closed inclusion of the *Cantor set* into the complex affine line, show that the direct image sheaf $i_*\underline{A}_{\mathfrak{C}}$ is not constructible.

Exercise 7.1.4 Show that the category $Constr(X)$ of constructible sheaves of A-modules on X is an abelian category.

Let $D^b_c(X)$ be the full triangulated subcategory of the derived category $D^b(X)$ consisting of all constructible bounded complexes of sheaves of A-modules on X. By slight abuse of terminology, $D^b_c(X)$ is usually referred to as the *bounded derived category of constructible sheaves*. However, as indicated by the result below, terminology is not completely abusive. In fact, in the notations of Exercise 7.1.4, there is a natural morphism $D^b(Constr(X)) \to D^b_c(X)$, and the following result holds (see [11, 188]):

Theorem 7.1.5 *Let X be a complex algebraic variety, and assume that A is a field. Then the morphism*

$$D^b(Constr(X)) \longrightarrow D^b_c(X)$$

is an equivalence of categories.

Constructibility is preserved under many natural sheaf-theoretic operations. For example, the following result holds (for a proof, see [15, V.8.7] or [214, Corollary 4.2.2]):

Theorem 7.1.6 *Let X be a complex algebraic (or analytic) variety, let $\mathcal{F}^\bullet \in D^b(X)$, and assume that the coefficient ring A is a Dedekind domain (e.g., a field or $\mathbb{Z}$). Then $\mathcal{F}^\bullet$ is constructible if and only if its dual $\mathcal{DF}^\bullet$ is constructible. In particular, the dualizing complex $\mathbb{D}_X^\bullet = \mathcal{D}\underline{A}_X$ is constructible.*

Moreover, one has the following (e.g., see [15] or the unified treatment of [214, Theorem 4.0.2]):

Theorem 7.1.7 *Let $f : X \to Y$ be a morphism of complex algebraic (or analytic) varieties.*

(a) *If $\mathcal{G}^\bullet \in D_c^b(Y)$, then $f^*\mathcal{G}^\bullet, f^!\mathcal{G}^\bullet \in D_c^b(X)$.*

(b) *If $\mathcal{F}^\bullet \in D_c^b(X)$ and f is an algebraic map, then $Rf_*\mathcal{F}^\bullet, Rf_!\mathcal{F}^\bullet \in D_c^b(Y)$. If $\mathcal{F}^\bullet \in D_c^b(X)$ and f is an analytic map so that the restriction of f to $\mathrm{supp}(\mathcal{F}^\bullet)$ is proper (e.g., f is proper), then $Rf_*\mathcal{F}^\bullet, Rf_!\mathcal{F}^\bullet \in D_c^b(Y)$.*

(c) *If $\mathcal{F}^\bullet, \mathcal{G}^\bullet \in D_c^b(X)$, then $\mathcal{F}^\bullet \overset{L}{\otimes} \mathcal{G}^\bullet, R\mathcal{H}om^\bullet(\mathcal{F}^\bullet, \mathcal{G}^\bullet) \in D_c^b(X)$.*

In other words, the derived category $D_c^b(X)$ of bounded constructible complexes is closed under Grothendieck's six operations: $Rf_*, Rf_!, f^*, f^!, R\mathcal{H}om^\bullet$, and $\overset{L}{\otimes}$.

We conclude this section with the following important result, whose proof can be found in [214, Corollary 2.0.4] (see also [61, Theorem 4.3.14]):

Theorem 7.1.8 (Künneth Formula for Constructible Complexes) *Let A be a field, and let X_1 and X_2 be complex algebraic varieties. For $\mathcal{F}_1^\bullet \in D_c^b(X_1)$ and $\mathcal{F}_2^\bullet \in D_c^b(X_2)$, there are natural isomorphisms of A-vector spaces for every integer $i \in \mathbb{Z}$:*

$$\mathbb{H}^i(X_1 \times X_2; \mathcal{F}_1^\bullet \overset{L}{\boxtimes} \mathcal{F}_2^\bullet) \cong \bigoplus_{p+q=i} \mathbb{H}^p(X_1; \mathcal{F}_1^\bullet) \otimes \mathbb{H}^q(X_2; \mathcal{F}_2^\bullet). \tag{7.1}$$

7.2 Local Calculus

Our local (stalk) calculations are based on the following *Morse Lemma for Constructible complexes* (see, e.g., [61, Corollary 4.3.11] and the references therein):

Lemma 7.2.1 (Homotopy Invariance) *Let X be a complex analytic space with a Whitney stratification $\mathcal{X}$. Let $r : X \to [0, a)$ be a proper, $\mathbb{R}$-analytic map such that for every stratum $S \in \mathcal{X}$, $r|_S$ has no critical values except at 0. Then the inclusion $r^{-1}(0) \hookrightarrow X$ induces an isomorphism*

$$\mathbb{H}^i(X; \mathcal{F}^\bullet) \overset{\cong}{\longrightarrow} \mathbb{H}^i(r^{-1}(0); \mathcal{F}^\bullet)$$

for all $i \in \mathbb{Z}$ and every complex $\mathcal{F}^\bullet \in D_\mathcal{X}^b(X)$ (i.e., constructible with respect to $\mathcal{X}$).

We apply this homotopy invariance to show the following:

Proposition 7.2.2 *Let $\mathcal{F}^\bullet \in D^b_c(X)$, $x \in X$ and $i_x : \{x\} \hookrightarrow X$ the inclusion. Then*

$$\mathcal{H}^j(\mathcal{F}^\bullet)_x \cong H^j(i_x^*\mathcal{F}^\bullet) \cong \mathbb{H}^j(\mathring{B}_\epsilon(x); \mathcal{F}^\bullet), \tag{7.2}$$

where $\mathring{B}_\epsilon(x)$ is the intersection of X with an open small ϵ-ball neighborhood of x in some local embedding of X in $\mathbb{C}^N$.

Proof Define $r : \mathring{B}_\epsilon(x) \to [0, \epsilon^2)$ by

$$r(y) = d^2(y, x).$$

Then 0 is the only critical value of r for $\epsilon > 0$ small enough, and $r^{-1}(0) = \{x\}$. The result follows by applying Lemma 7.2.1. $\square$

Corollary 7.2.3 *If A is a field, one has:*

$$\mathcal{H}^j(\mathcal{D}\mathcal{F}^\bullet)_x \cong \mathbb{H}_c^{-j}(\mathring{B}_\epsilon(x); \mathcal{F}^\bullet)^\vee.$$

Proof The assertion follows by combining Proposition 7.2.2, which yields

$$\mathcal{H}^j(\mathcal{D}\mathcal{F}^\bullet)_x \cong \mathbb{H}^j(\mathring{B}_\epsilon(x); \mathcal{D}\mathcal{F}^\bullet),$$

with the Universal Coefficient Theorem (5.15), from which one obtains:

$$\mathbb{H}^j(\mathring{B}_\epsilon(x); \mathcal{D}\mathcal{F}^\bullet) \cong \mathbb{H}_c^{-j}(\mathring{B}_\epsilon(x); \mathcal{F}^\bullet)^\vee.$$

$\square$

Before discussing the result dual to (7.2), we need the following:

Lemma 7.2.4 *In the notations of Proposition 7.2.2, one has:*

$$\mathbb{H}_c^j(\mathring{B}_\epsilon(x); \mathcal{F}^\bullet) \cong \mathbb{H}^j(\mathring{B}_\epsilon(x), \mathring{B}_\epsilon(x) - x; \mathcal{F}^\bullet).$$

Proof Consider the following commutative diagram

$$
\begin{array}{ccccc}
\mathbb{H}^j(\mathring{B}_\epsilon(x), \mathring{B}_\epsilon(x) - x; \mathcal{F}^\bullet) & \longrightarrow & \mathbb{H}^j(\mathring{B}_\epsilon(x); \mathcal{F}^\bullet) & \longrightarrow & \mathbb{H}^j(\mathring{B}_\epsilon(x) - x; \mathcal{F}^\bullet) \\
\uparrow & & \uparrow{\scriptstyle\cong} & & \uparrow{\scriptstyle\cong} \\
\mathbb{H}_c^j(\mathring{B}_\epsilon(x); \mathcal{F}^\bullet) & \longrightarrow & \mathbb{H}^j(B_\epsilon(x); \mathcal{F}^\bullet) & \longrightarrow & \mathbb{H}^j(\partial B_\epsilon(x); \mathcal{F}^\bullet)
\end{array}
$$

where the middle and right vertical arrows are isomorphisms, and the existence of the left arrow and the fact that it is an isomorphism is a consequence of the five-lemma. For the middle vertical isomorphism, see e.g., [61, Corollary 4.3.11(i)]. The right vertical map is given by the composition of isomorphisms

$$\mathbb{H}^j(\partial B_\epsilon(x); \mathcal{F}^\bullet) \cong \mathbb{H}^j(B_\epsilon(x) - x; \mathcal{F}^\bullet) \tag{7.3}$$

and

$$\mathbb{H}^j(B_\epsilon(x) - x; \mathcal{F}^\bullet) \cong \mathbb{H}^j(\mathring{B}_\epsilon(x) - x; \mathcal{F}^\bullet) \tag{7.4}$$

induced by inclusions. For instance, to show the isomorphism (7.3), define $r : B_\epsilon(x) - x \to [0, \epsilon^2)$ by

$$r(y) = \epsilon^2 - d^2(y, x).$$

Then $r^{-1}(0) = \partial B_\epsilon(x)$ and we apply Lemma 7.2.1. Isomorphism (7.4) is left as an exercise.   □

One can now easily prove the result dual to (7.2), namely in the notations of Proposition 7.2.2 one has the following:

Proposition 7.2.5

$$H^j(i_x^! \mathcal{F}^\bullet) \cong \mathbb{H}_c^j(\mathring{B}_\epsilon(x); \mathcal{F}^\bullet) \cong \mathbb{H}^j(\mathring{B}_\epsilon(x), \mathring{B}_\epsilon(x) - x; \mathcal{F}^\bullet). \tag{7.5}$$

Proof First, note that by the definition of relative hypercohomology and excision, one has:

$$H^j(i_x^! \mathcal{F}^\bullet) \cong \mathbb{H}^j(X, X - \{x\}; \mathcal{F}^\bullet) \cong \mathbb{H}^j(B_\epsilon(x), B_\epsilon(x) - \{x\}; \mathcal{F}^\bullet).$$

Then, by (7.3) and the compactness of $B_\epsilon(x)$ and of its boundary, one gets:

$$\mathbb{H}^j(B_\epsilon(x), B_\epsilon(x) - \{x\}; \mathcal{F}^\bullet) \cong \mathbb{H}^j(B_\epsilon(x), \partial B_\epsilon(x); \mathcal{F}^\bullet) \cong \mathbb{H}_c^j(\mathring{B}_\epsilon(x); \mathcal{F}^\bullet).$$

The second part of (7.5) is just Lemma 7.2.4.   □

All of these local computations will play an important role in Theorem 8.3.12 of the next chapter.

We conclude this section with an example that shows that even if the variety X is a *contractible* manifold, it is still possible to have $H^k(X; \mathcal{G}) \neq 0$ if $k > 0$ for some constructible sheaf $\mathcal{G}$. (This cannot happen if $\mathcal{G}$ is a local system, since then $\mathcal{G}$ is a constant sheaf $\underline{A}_X^{\otimes l}$, for some l, so its higher cohomology vanishes.) This also shows that cohomology with sheaf coefficients is not a homotopy invariant.

Example 7.2.6 Let $X = \{(x, y) \in \mathbb{C}^2 \mid x^2 + y^2 = 1\}$. Then X has the homotopy type of S^1. Consider the projection

$$f : X \to \mathbb{C}, \ (x, y) \mapsto x.$$

It follows that

$$(R^q f_* \underline{\mathbb{C}}_X)_p \cong H^q(f^{-1}(D_p); \mathbb{C}) = 0 \quad \text{if } q > 0,$$

where D_p is a small enough neighborhood of p in $\mathbb{C}$, and note that $f^{-1}(D_p)$ is a union of contractible sets. Therefore, $R^q f_* \underline{\mathbb{C}}_X = 0$ if $q > 0$ and $R^0 f_* \underline{\mathbb{C}}_X \neq 0$.

Consider now the Leray spectral sequence of the map f, that is,

$$E_2^{p,q} = H^p(\mathbb{C}; R^q f_* \underline{\mathbb{C}}_X) \Longrightarrow \mathbb{H}^{p+q}(X; \underline{\mathbb{C}}_X) = H^{p+q}(X; \mathbb{C}).$$

Since $R^q f_* \underline{\mathbb{C}}_X = 0$ if $q > 0$, this spectral sequence degenerates at E_2. Therefore,

$$\mathbb{C} = H^1(X; \mathbb{C}) = H^1(\mathbb{C}; R^0 f_* \underline{\mathbb{C}}_X) \neq 0.$$

The constructible sheaf $\mathcal{G} = R^0 f_* \underline{\mathbb{C}}_X$ on $\mathbb{C}$ provides the desired example.

7.3 Euler Characteristics of Constructible Complexes. Applications

We begin this section with the following easy consequence of Theorem 7.1.7:

Corollary 7.3.1 *Assume that $\mathcal{F}^\bullet \in D_c^b(X)$ and that either*

(a) X is a complex algebraic variety, or
(b) X is an analytic space and $\mathrm{supp}(\mathcal{F}^\bullet)$ is compact.

Then $\mathbb{H}^i(X, \mathcal{F}^\bullet)$ and $\mathbb{H}_c^i(X, \mathcal{F}^\bullet)$ are finite type A-modules for every $i \in \mathbb{Z}$.

With $\mathcal{F}^\bullet \in D_c^b(X)$ satisfying the assumptions of the above corollary, we make the following definition:

Definition 7.3.2 Assume A is a field. The *(compactly supported) Euler characteristic of $\mathcal{F}^\bullet \in D_c^b(X)$* is defined as:

$$\chi_{(c)}(X, \mathcal{F}^\bullet) := \sum_{i \in \mathbb{Z}} (-1)^i \dim_A \mathbb{H}_{(c)}^i(X; \mathcal{F}^\bullet).$$

(Here, we use the notation $\chi_{(c)}$ and $\mathbb{H}_{(c)}^i$ to indicate that the definition applies to the compactly supported Euler characteristic χ_c by using $\mathbb{H}_c^i$, as well as to the usual Euler characteristic χ by using $\mathbb{H}^i$.)

From here on, we will be assuming throughout this chapter that A is a field. We also need the following construction, e.g., see the discussion in [61, Example 2.3.18] and the references therein. If X is a complex quasi-projective variety and $Z \subset X$ is a closed algebraic subvariety, then Z has a closed tubular neighborhood T in X

such that the inclusion $Z \hookrightarrow T$ is a homotopy equivalence. In fact, the subvariety Z has a fundamental system of neighborhoods $(T_k)_{k \in K}$ of this type, such that the inclusions $T_k \hookrightarrow T_\ell$ are homotopy equivalences. The homotopy type of

$$T_k^0 := T_k - Z$$

is independent of k, and is called the *link of Z in X*, denoted by $L_X(Z)$. Now let $U = X - Z$ and $j : U \hookrightarrow X, i : Z \hookrightarrow X$ be the inclusion maps. Then

$$H^s(L_X(Z); A) \cong \mathbb{H}^s(Z; i^* Rj_* \underline{A}_U), \tag{7.6}$$

for all $s \in \mathbb{Z}$. With these notations, we have the following well-known result of Sullivan [226]:

Proposition 7.3.3 *If $L_X(Z)$ is the link of a closed subvariety Z in a complex quasi-projective variety X, then:*

$$\chi(L_X(Z)) = 0.$$

Proof Let $U = X - Z$. By the additivity of the Euler characteristic, we have that $\chi(X) = \chi(Z) + \chi(U)$. On the other hand, by the Mayer–Vietoris sequence for the open cover $X = U \cup T$, we get:

$$\chi(X) + \chi(L_X(Z)) = \chi(U) + \chi(T) = \chi(U) + \chi(Z),$$

where we also used the fact that the inclusion $Z \hookrightarrow T$ is a homotopy equivalence. The claim follows. □

Exercise 7.3.4 Show that in the above notations, for every $k \in \mathbb{Z}$ one has:

$$IH^k(L_X(Z); A) \cong \mathbb{H}^{k-n}(Z; i^* Rj_* IC_U),$$

where IC_U denotes as before the middle-perversity intersection cohomology sheaf complex in algebraic geometric (BBD) conventions of [12], i.e., $IC_U := IC_{\overline{m}}^\bullet[-\dim_{\mathbb{C}}(U)]$. More generally, if $\mathcal{F}^\bullet \in D_c^b(X)$, and $L_X(Z)$ is the link of a closed subvariety Z of X, then one has:

$$\mathbb{H}^k(L_X(Z); \mathcal{F}^\bullet) \cong \mathbb{H}^k(Z; i^* Rj_* j^* \mathcal{F}^\bullet), \tag{7.7}$$

for all $k \in \mathbb{Z}$.

The result of Proposition 7.3.3 can be generalized as follows (see [61, Theorem 4.1.21]):

Theorem 7.3.5

(i) *Let $\mathcal{F}^\bullet$ be a bounded $\mathcal{X}$-constructible complex on the complex algebraic variety X, and let V be a stratum in the Whitney stratification $\mathcal{X}$ with inclusion i_V : $V \hookrightarrow X$. Then:*

$$\chi(V, i_V^* \mathcal{F}^\bullet) = \chi(V, i_V^! \mathcal{F}^\bullet). \qquad (7.8)$$

(ii) *If, moreover, V is closed in X, and $K := L_X(V)$ denotes the link of V in X, one has:*

$$\chi(K, \mathcal{F}^\bullet) = 0. \qquad (7.9)$$

Proof Since the stratum V is locally closed in X, one can assume, without any loss of generality, that V is closed; otherwise replace X by an open subset containing V as a closed subset. Let $U = X - V$ with inclusion map $j : U \hookrightarrow X$. For simplicity, write i for the inclusion i_V.

By applying the functor i^* to the attaching triangle

$$i_! i^! \mathcal{F}^\bullet \longrightarrow \mathcal{F}^\bullet \longrightarrow Rj_* j^* \mathcal{F}^\bullet \xrightarrow{[1]},$$

one gets a distinguished triangle

$$i^! \mathcal{F}^\bullet \longrightarrow i^* \mathcal{F}^\bullet \longrightarrow i^* Rj_* j^* \mathcal{F}^\bullet \xrightarrow{[1]},$$

from which, by using (7.7), one obtains the following Euler characteristic identity:

$$\chi(V, i_V^! \mathcal{F}^\bullet) + \chi(K, \mathcal{F}^\bullet) = \chi(V, i_V^* \mathcal{F}^\bullet). \qquad (7.10)$$

So (7.8) is equivalent to the vanishing $\chi(K, \mathcal{F}^\bullet) = 0$.

Since taking the Verdier dual preserves constructibility, the complex $\mathcal{G}^\bullet = \mathcal{D}\mathcal{F}^\bullet$ is also $\mathcal{X}$-constructible. Since A is a field, by using the Universal Coefficient Theorem (5.15) one gets that

$$\mathbb{H}^m(K; \mathcal{D}\mathcal{G}^\bullet) \cong \mathbb{H}_c^{-m}(K; \mathcal{G}^\bullet)^\vee.$$

Therefore,

$$\chi(K, \mathcal{F}^\bullet) = \chi_c(K, \mathcal{D}\mathcal{F}^\bullet) = \chi_c(K, \mathcal{G}^\bullet),$$

where links are regarded as complex analytic spaces. So it suffices to show that $\chi_c(K, \mathcal{G}^\bullet) = 0$ for every $\mathcal{X}$-constructible bounded complex $\mathcal{G}^\bullet$.

Let X^m denote the union of all strata in $\mathcal{X}$ of complex dimension at most m. This is a closed subvariety of X. Note that K is filtered by its intersections with these subvarieties X^m. More precisely, if $\dim_{\mathbb{C}} V = s$, then K has a filtration

$$\emptyset = K_s \subset K_{s+1} \subset \cdots \subset K,$$

with

$$K_m := K \cap X^m$$

the link of V in X^m. Let $K' \subset K''$ be two consecutive terms in the above stratification of the link K, with $X' \subset X''$ the corresponding pair of subvarieties in X. By using the exact sequence

$$\cdots \longrightarrow \mathbb{H}_c^k(K'' - K'; \mathcal{G}^\bullet) \longrightarrow \mathbb{H}_c^k(K''; \mathcal{G}^\bullet) \longrightarrow \mathbb{H}_c^k(K'; \mathcal{G}^\bullet)$$
$$\longrightarrow \mathbb{H}_c^{k+1}(K'' - K'; \mathcal{G}^\bullet) \longrightarrow \cdots,$$

it follows by induction on dimension that it is enough to show that

$$\chi_c(K'' - K', \mathcal{G}^\bullet) = 0. \tag{7.11}$$

Furthermore, by using the hypercohomology spectral sequence

$$E_2^{p,q} = H_c^p(K'' - K'; \mathcal{H}^q(\mathcal{G}^\bullet)) \Longrightarrow \mathbb{H}_c^{p+q}(K'' - K'; \mathcal{G}^\bullet),$$

it is enough to prove the vanishing (7.11) for $\mathcal{G}^\bullet$ a local system $\mathcal{L}$ on $K'' - K'$. If $K'' - K'$ is connected, the desired vanishing follows from the identity

$$\chi_c(K'' - K', \mathcal{L}) = \chi(K'' - K', \mathcal{L}) = \chi(K'' - K') \cdot \mathrm{rank}(\mathcal{L}),$$

(where the first equality uses Poincaré duality for the stratum $K'' - K'$, while for the second see Exercise 4.2.15) together with $\chi(K'' - K') = \chi_c(K'' - K') = \chi(K'') - \chi(K') = 0$, cf. Proposition 7.3.3. (Here links are regarded as complex analytic spaces, which explains the first equality in the previous equation.) When $K'' - K'$ is not connected, the above argument applies to each connected component of $K'' - K'$. Altogether, one gets:

$$\chi_c(K, \mathcal{G}^\bullet) = 0,$$

as desired. $\qquad\qquad\qquad\square$

As a consequence, one obtains the following additivity property for the Euler characteristic of a constructible complex (see [61, Theorem 4.1.22]):

Theorem 7.3.6 *Let X be a complex algebraic variety with a fixed Whitney stratification $\mathfrak{X}$, and let $\mathcal{F}^{\bullet}$ be a $\mathfrak{X}$-constructible bounded complex on X. Then:*

$$\chi(X, \mathcal{F}^{\bullet}) = \sum_{V \in \mathfrak{X}} \chi(V, i_V^! \mathcal{F}^{\bullet}) = \sum_{V \in \mathfrak{X}} \chi(V, i_V^* \mathcal{F}^{\bullet}) = \sum_{V \in \mathfrak{X}} \chi(V) \cdot \chi(\mathcal{F}_{xV}^{\bullet}),$$

where $i_V : V \hookrightarrow X$ denotes the stratum inclusion, $x_V \in V$ is a point in V, and $\chi(\mathcal{F}_{xV}^{\bullet}) := \sum_{i \in \mathbb{Z}} (-1)^i \dim_A \mathcal{H}^i(\mathcal{F}^{\bullet})_{xV}$.

Proof The second equality follows from Theorem 7.3.5(i). For proving the first equality, we proceed by induction on $\dim_{\mathbb{C}} X$. For $\dim_{\mathbb{C}} X = 0$, the result is immediate. Let U be the union of open strata in X, with $j : U \hookrightarrow X$ the open inclusion, and let $i : Z = X - U \hookrightarrow X$ be the inclusion of the closed complement. The hypercohomology long exact sequence associated to the attaching triangle

$$i_! i^! \mathcal{F}^{\bullet} \longrightarrow \mathcal{F}^{\bullet} \longrightarrow Rj_* j^* \mathcal{F}^{\bullet} \xrightarrow{[1]}$$

yields the Euler characteristic identity:

$$\chi(X, \mathcal{F}^{\bullet}) = \chi(U, \mathcal{F}^{\bullet}) + \chi(Z, i^! \mathcal{F}^{\bullet}).$$

Since $\dim_{\mathbb{C}} Z < \dim_{\mathbb{C}} X$, one has by the induction hypothesis applied to the pair $(Z, i^! \mathcal{F}^{\bullet})$ that:

$$\chi(Z, i^! \mathcal{F}^{\bullet}) = \sum_{V \in \mathfrak{X}, \, \dim_{\mathbb{C}} V < \dim_{\mathbb{C}} X} \chi(V, i_V^! \mathcal{F}^{\bullet}).$$

On the other hand,

$$\chi(U, \mathcal{F}^{\bullet}) = \sum_{V \in \mathfrak{X}, \, \dim_{\mathbb{C}} V = \dim_{\mathbb{C}} X} \chi(V, i_V^* \mathcal{F}^{\bullet})$$

$$= \sum_{V \in \mathfrak{X}, \, \dim_{\mathbb{C}} V = \dim_{\mathbb{C}} X} \chi(V, i_V^! \mathcal{F}^{\bullet}),$$

since for an open embedding j one has that $j^* = j^!$. This proves the first equality.

To get the third equality, we use the hypercohomology spectral sequence to compute $\mathbb{H}^*(V; i_V^* \mathcal{F}^\bullet)$, together with the fact stated in Exercise 4.2.15, that if $\mathcal{L}$ is an A-local system on a topological space V, the twisted Euler characteristic $\chi(V, \mathcal{L}) := \sum_i (-1)^i \dim_A H_i(V, \mathcal{L})$ is computed by:

$$\chi(V, \mathcal{L}) = \mathrm{rank}(\mathcal{L}) \cdot \chi(V).$$

$\square$

Corollary 7.3.7 *With the above notations, one has the identity:*

$$\chi(X, \mathcal{F}^\bullet) = \chi_c(X, \mathcal{F}^\bullet). \tag{7.12}$$

Proof By the additivity of the previous theorem, it suffices to prove the statement for a complex $\mathcal{F}^\bullet$ whose cohomology sheaves are local systems (i.e., $X = V$ is a stratum). Then the assertion follows by using the (compactly supported) hypercohomology spectral sequence. We leave the details as an exercise. $\square$

Remark 7.3.8 The equality (7.12) is a generalization of the classical formula stating that for a complex algebraic (or analytic) variety, one has:

$$\chi(X) = \chi_c(X). \tag{7.13}$$

Note that this identification is equivalent to the additivity of the Euler characteristic in the complex algebraic (analytic) setting, already used in the proof of Proposition 7.3.3. It should also be noted that formula (7.13) is not true outside of the complex context. For example, if M is an oriented m-dimensional topological manifold, then Poincaré duality yields that $\chi_c(M) = (-1)^m \chi(M)$.

Theorems 7.3.5 and 7.3.6 hold in other contexts as well, e.g., for compact analytic varieties or for certain spaces obtained from complex algebraic varieties by real algebraic constructions, see [61, Remark 4.1.24]. For example, if $f : X \to Y$ is a morphism of complex algebraic varieties, $y \in Y$ is a point and B_y is a small open ball neighborhood of y in Y constructed by using a local embedding of (Y, y) in a smooth germ, then Theorems 7.3.5 and 7.3.6 also hold for $T_y := f^{-1}(B_y)$, the *tube* of f at the point y. As a consequence, if $X_y = f^{-1}(y)$ is the fiber of f at y, the following holds:

Corollary 7.3.9 *For $\mathcal{F}^\bullet \in D_c^b(X)$ one has the equality:*

$$\chi(T_y, \mathcal{F}^\bullet) = \chi(X_y, \mathcal{F}^\bullet).$$

Proof First note that by using (7.12), one has that

$$\chi(T_y, \mathcal{F}^\bullet) = \chi(T_y^*, \mathcal{F}^\bullet) + \chi(X_y, \mathcal{F}^\bullet),$$

where $T_y^* = T_y - X_y$ denotes the punctured tube of f at y. Furthermore,

$$\chi(T_y^*, \mathcal{F}^\bullet) = \chi(B_y^*, Rf_*\mathcal{F}^\bullet),$$

with $B_y^* = B_y - \{y\}$ the link of y in Y. Then Theorem 7.3.5(ii) implies that $\chi(T_y^*, \mathcal{F}^\bullet) = 0$, thus proving the desired result.   □

Chapter 8
Perverse Sheaves

Perverse sheaves are fundamental objects of study in topology, algebraic geometry, analysis, and differential equations, with a plethora of applications, including in adjacent fields such as number theory, representation theory, combinatorics, and algebra. In this chapter, we overview the relevant definitions and results of the theory of perverse sheaves, with an emphasis on examples and applications (see also Chapters 9 and 10 for more applications of perverse sheaves).

Perverse sheaves are an important class of constructible complexes, introduced in [12] as a formalization of the celebrated Riemann–Hilbert correspondence of Kashiwara [118, 119] (see also [172, 173]), which relates the topology of algebraic varieties (intersection homology) and the algebraic theory of differential equations (microlocal calculus and holonomic D-modules).

Perverse sheaves on a variety form an abelian category, and they are close in spirit to local systems (i.e., locally constant sheaves) on nonsingular varieties. However, perverse sheaves are not sheaves. The use of "sheaves" in "perverse sheaves" is partly motivated by the fact that the objects and morphisms in the category of perverse sheaves can be glued from local data, just like in the case of sheaves. For more explanation on terminology, see also [12], [129, p. 10] or [52].

For comprehensive references on perverse sheaves and their many applications, the reader is referred to [6], [12], [61], [107], [122], or [51].

From now on, we work in the complex algebraic (or analytic) category and we only consider the middle-perversity $\overline{m}$. Moreover, in the complex analytic context all spaces are assumed to be finite dimensional. Intersection homology (both PL and topological versions) can be adapted to this setup because of the existence of Whitney stratifications and of triangulations compatible with such stratifications.

Note that for a complex algebraic (or analytic) variety X, the middle-perversity function $\overline{m}$ is self-dual, since all strata have even real codimensions. We call

$$IC_{\overline{m}}^{\bullet} =: IC_X^{top}$$

© Springer Nature Switzerland AG 2019

L. G. Maxim, *Intersection Homology & Perverse Sheaves*, Graduate Texts in Mathematics 281, https://doi.org/10.1007/978-3-030-27644-7_8

the *topological Deligne complex* (this notation follows the normalization condition of Goresky–MacPherson [83]). As already mentioned, in complex algebraic geometry it is more common to use the complex

$$IC_X := IC_X^{top}[-n]$$

that satisfies the normalization axiom as in the BBD convention [12]. Generalized Poincaré duality (with X as above and a field of coefficients) takes in this case the form:

$$\mathcal{D}_X IC_X \simeq IC_X,$$

i.e., IC_X is *Verdier self-dual*. As we will see later, IC_X is the main example of a *perverse sheaf* on X.

8.1 Definition, Examples

The original construction of perverse sheaves in [12] proceeds through the machinery of triangulated categories and homological algebra. We overview here the main constructions and properties.

Definition 8.1.1 (*t*-Structure) A *t-structure* on a triangulated category D consists of two strictly full[1] subcategories $D^{\leq 0}$ and $D^{\geq 0}$ of D, so that by setting $D^{\leq n} = D^{\leq 0}[-n]$ and $D^{\geq n} = D^{\geq 0}[-n]$ one has:

(1) $\mathrm{Hom}_D(P, R) = 0$ if $P \in D^{\leq 0}$ and $R \in D^{\geq 1}$.
(2) $D^{\leq 0} \subset D^{\leq 1}$ and $D^{\geq 1} \subset D^{\geq 0}$.
(3) for every $P \in D$, there exists a triangle

$$P' \longrightarrow P \longrightarrow P'' \overset{[1]}{\longrightarrow} P'[1],$$

with $P' \in D^{\leq 0}, P'' \in D^{\geq 1}$.

If the above conditions are satisfied, we say that $(D^{\leq 0}, D^{\geq 0})$ is a *t-structure on* D.

Definition 8.1.2 The full subcategory

$$\mathcal{C} := D^{\leq 0} \cap D^{\geq 0}$$

of D is called the *heart* (or *core*) of the given *t*-structure.

---

[1] *Strictly full* means that if $F \in D$ and $G \in D^{\leq 0}$, and $F \cong G$ in D, then $F \in D^{\leq 0}$.

Definition 8.1.3 A t-structure is *non-degenerate* if

$$\bigcap_n D^{\leq n} = \bigcap_n D^{\geq n} = \{0\},$$

where 0 is the isomorphism class of the zero object in D.

Proposition 8.1.4 *Let D be a triangulated category with a t-structure. The inclusion $D^{\leq n} \hookrightarrow D$ has a right adjoint functor $\tau_{\leq n} : D \to D^{\leq n}$, and the inclusion $D^{\geq n} \hookrightarrow D$ has a left adjoint functor $\tau_{\geq n} : D \to D^{\geq n}$.*

Proposition 8.1.5 *For every object P in a triangulated category D with a t-structure, there is a distinguished triangle*

$$\tau_{\leq 0} P \longrightarrow P \longrightarrow \tau_{\geq 1} P \xrightarrow{[1]}$$

Exercise 8.1.6 Let D be a triangulated category with a t-structure, and let $P \in D$. Show that the following are equivalent:

(i) $P \in D^{\leq n}$ (resp., $P \in D^{\geq n}$).
(ii) the canonical morphism $\tau_{\leq n} P \to P$ (resp., $P \to \tau_{\geq n} P$) is an isomorphism.
(iii) $\tau_{\geq n+1} P = 0$ (resp., $\tau_{\leq n-1} P = 0$).

Exercise 8.1.7 Let $P' \to P \to P'' \xrightarrow{[1]}$ be a distinguished triangle in a triangulated category D with t-structure. Show that if $P', P'' \in D^{\leq 0}$ (resp., $D^{\geq 0}$), then $P \in D^{\leq 0}$ (resp., $D^{\geq 0}$). In particular, if $P', P'' \in C$, then $P \in C$.

Proposition 8.1.8 *For every object P in a triangulated category D with a t-structure, the following assertions hold:*

(i) for integers n, m:

$$\tau_{\leq n}(P[m]) \simeq \tau_{\leq n+m}(P)[m] \quad , \quad \tau_{\geq n}(P[m]) \simeq \tau_{\geq n+m}(P)[m].$$

(ii) for integers $m \leq n$, there is a unique isomorphism

$$\tau_{\geq m} \tau_{\leq n} P \simeq \tau_{\leq n} \tau_{\geq m} P.$$

Proposition 8.1.9 *The heart C of a t-structure is an abelian category, and it is stable by extensions.*

One can extend the notion of cohomology groups to any t-category by using the above *truncation functors* $\tau_{\leq 0}$ and $\tau_{\geq 0}$ as follows:

Definition 8.1.10 The functor

$$^t H^0 := \tau_{\geq 0} \tau_{\leq 0} = \tau_{\leq 0} \tau_{\geq 0} : D \to C$$

is called the *cohomology functor* of the given *t*-structure. We also set

$$^t H^i := {}^t H^0 \circ [i],$$

i.e., $^t H^i(P) = {}^t H^0(P[i]) = (\tau_{\geq i}\tau_{\leq i} P)[i]$, for $P \in D$.

Proposition 8.1.11 *The functor* $^t H^0$ *is cohomological, that is, by applying* $^t H^0$ *to a distinguished triangle* $P' \to P \to P'' \overset{[1]}{\to}$, *one gets an exact sequence*

$$^t H^0(P') \longrightarrow {}^t H^0(P) \longrightarrow {}^t H^0(P'').$$

Furthermore, by turning the triangle[2] *repeatedly, the functor* $^t H^0$ *induces in fact a long exact sequence*

$$\cdots \to {}^t H^{i-1}(P'') \to {}^t H^i(P') \to {}^t H^i(P) \to {}^t H^i(P'') \to {}^t H^{i+1}(P') \to \cdots$$

Example 8.1.12 (Standard t-Structure) Let $Sh(X)$ be the abelian category of sheaves of A-modules on X, and let $D(X)$ be the corresponding derived category. Then

$$D^{\leq 0}(X) := \{\mathcal{F}^\bullet \in D(X) \mid \mathcal{H}^i(\mathcal{F}^\bullet) = 0, \ i > 0\},$$

$$D^{\geq 0}(X) := \{\mathcal{F}^\bullet \in D(X) \mid \mathcal{H}^i(\mathcal{F}^\bullet) = 0, \ i < 0\}$$

yields a *t*-structure on $D(X)$ (e.g., see [6, Example 7.1.3]), called the *standard t-structure*. The corresponding truncations are defined by

$$\tau_{\leq 0}(\mathcal{F}^\bullet) = \{\cdots \to \mathcal{F}^{-2} \to \mathcal{F}^{-1} \to ker(d^0) \to 0 \to \dots\},$$

$$\tau_{\geq 0}(\mathcal{F}^\bullet) = \{\cdots \to 0 \to coker(d^{-1}) \to \mathcal{F}^1 \to \mathcal{F}^2 \to \dots\}.$$

The heart

$$D^{\leq 0}(X) \cap D^{\geq 0}(X) = \{\cdots \to 0 \to \mathcal{H}^0(\mathcal{F}^\bullet) \to 0 \to \dots\}$$

of this *t*-structure is equivalent to $Sh(X)$.

Definition 8.1.13 (Perverse t-Structure) Let X be a complex algebraic (or analytic) variety, and fix as before a noetherian and commutative coefficient ring A of finite cohomological dimension. Denote by $D_c^b(X)$ the full triangulated subcategory of the derived category $D^b(X)$ of bounded complexes of sheaves of A-modules,

---

[2] *Turning a triangle* refers to the fact that $P' \to P \to P'' \overset{[1]}{\to}$ is a distinguished triangle if and only if $P \to P'' \to P'[1] \overset{[1]}{\to}$ is a distinguished triangle.

consisting of constructible complexes (i.e., constructible with respect to some Whitney stratification).

(i) The *perverse t-structure* on $D_c^b(X)$ is defined by:

$$^p D^{\leq 0}(X) = \{\mathcal{F}^\bullet \in D_c^b(X) \mid \dim_{\mathbb{C}} \operatorname{supp}^{-j}(\mathcal{F}^\bullet) \leq j, \forall j \in \mathbb{Z}\},$$

$$^p D^{\geq 0}(X) = \{\mathcal{F}^\bullet \in D_c^b(X) \mid \dim_{\mathbb{C}} \operatorname{cosupp}^{j}(\mathcal{F}^\bullet) \leq j, \forall j \in \mathbb{Z}\},$$

where we recall that, for $i_x : \{x\} \hookrightarrow X$ denoting the point inclusion, we define the *support* and, respectively, the *cosupport* of $\mathcal{F}^\bullet$ by:

$$\operatorname{supp}^{j}(\mathcal{F}^\bullet) = \overline{\{x \in X \mid \mathcal{H}^j(i_x^* \mathcal{F}^\bullet) \neq 0\}},$$

$$\operatorname{cosupp}^{j}(\mathcal{F}^\bullet) = \overline{\{x \in X \mid \mathcal{H}^j(i_x^! \mathcal{F}^\bullet) \neq 0\}}.$$

(For a constructible complex $\mathcal{F}^\bullet$, the sets $\operatorname{supp}^{j}(\mathcal{F}^\bullet)$ and $\operatorname{cosupp}^{j}(\mathcal{F}^\bullet)$ are closed algebraic (or analytic) subvarieties of X, hence their dimensions are well defined.)

We say that a complex $\mathcal{F}^\bullet \in {}^p D^{\leq 0}(X)$ satisfies the *condition of support*, whereas $\mathcal{F}^\bullet \in {}^p D^{\geq 0}(X)$ is said to satisfy the *condition of cosupport*.

(ii) A complex $\mathcal{F}^\bullet \in D_c^b(X)$ is called a *perverse sheaf* if

$$\mathcal{F}^\bullet \in Perv(X) := {}^p D^{\leq 0}(X) \cap {}^p D^{\geq 0}(X).$$

Remark 8.1.14 The perverse t-structure is non-degenerate.

The above generalities about t-structures can be translated into the following facts about perverse sheaves:

(1) There exist (perverse) truncations $^p \tau_{\leq 0}, {}^p \tau_{\geq 0}$ that are adjoint to the inclusions $^p D^{\leq 0}(X) \hookrightarrow D_c^b(X) \hookleftarrow {}^p D^{\geq 0}(X)$, i.e., for every $m \in \mathbb{Z}$,

$$\operatorname{Hom}_{D_c^b(X)}(\mathcal{F}^\bullet, \mathcal{G}^\bullet) = \operatorname{Hom}_{{}^p D^{\leq m}(X)}(\mathcal{F}^\bullet, {}^p \tau_{\leq m} \mathcal{G}^\bullet)$$

if $\mathcal{F}^\bullet \in {}^p D^{\leq m}(X)$ and, similarly,

$$\operatorname{Hom}_{D_c^b(X)}(\mathcal{F}^\bullet, \mathcal{G}^\bullet) = \operatorname{Hom}_{{}^p D^{\geq m}(X)}({}^p \tau_{\geq m} \mathcal{F}^\bullet, \mathcal{G}^\bullet)$$

if $\mathcal{G}^\bullet \in {}^p D^{\geq m}(X)$. In particular, there are adjunction maps $\mathcal{F}^\bullet \longrightarrow {}^p \tau_{\geq m} \mathcal{F}^\bullet$ and $^p \tau_{\leq m} \mathcal{F}^\bullet \longrightarrow \mathcal{F}^\bullet$.

(2) The i-th *perverse cohomology* of $\mathcal{F}^\bullet$ is defined as

$$^p \mathcal{H}^i(\mathcal{F}^\bullet) := {}^p \tau_{\leq 0} {}^p \tau_{\geq 0}(\mathcal{F}^\bullet[i]).$$

(3) $\mathcal{F}^\bullet \in D_c^b(X)$ satisfies the condition of support if, and only if, $^p\mathcal{H}^i(\mathcal{F}^\bullet) = 0$ for all $i > 0$. Similarly, $\mathcal{F}^\bullet \in D_c^b(X)$ satisfies the condition of cosupport if, and only if, $^p\mathcal{H}^i(\mathcal{F}^\bullet) = 0$ for all $i < 0$. In particular, $\mathcal{F}^\bullet \in Perv(X)$ if, and only if, $^p\mathcal{H}^0(\mathcal{F}^\bullet) = \mathcal{F}^\bullet$ and $^p\mathcal{H}^i(\mathcal{F}^\bullet) = 0$ for all $i \neq 0$.

(4) For every distinguished triangle $\mathcal{A}^\bullet \to \mathcal{B}^\bullet \to \mathcal{C}^\bullet \overset{[1]}{\to}$ in $D_c^b(X)$, there is an associated long exact sequence in $Perv(X)$:

$$\cdots \to {}^p\mathcal{H}^i(\mathcal{A}^\bullet) \to {}^p\mathcal{H}^i(\mathcal{B}^\bullet) \to {}^p\mathcal{H}^i(\mathcal{C}^\bullet) \to {}^p\mathcal{H}^{i+1}(\mathcal{A}^\bullet) \to \cdots$$

(5) If $\mathcal{A}^\bullet, \mathcal{C}^\bullet \in Perv(X)$ and $\mathcal{A}^\bullet \to \mathcal{B}^\bullet \to \mathcal{C}^\bullet \overset{[1]}{\to}$ is a distinguished triangle in $D_c^b(X)$, then $\mathcal{B}^\bullet \in Perv(X)$.

(6) For $\mathcal{F}^\bullet \in D_c^b(X)$, there is a *perverse cohomology spectral sequence*

$$E_2^{i,j} = \mathbb{H}^i(X; {}^p\mathcal{H}^j(\mathcal{F}^\bullet)) \Longrightarrow \mathbb{H}^{i+j}(X; \mathcal{F}^\bullet). \tag{8.1}$$

Exercise 8.1.15 Let $\mathcal{F}^\bullet \in D_c^b(X)$. Show that $\mathcal{F}^\bullet \simeq 0$ if, and only if, $^p\mathcal{H}^j(\mathcal{F}^\bullet) = 0$ for every $j \in \mathbb{Z}$.

Exercise 8.1.16 Let $u : \mathcal{F}^\bullet \to \mathcal{G}^\bullet$ be a morphism in $D_c^b(X)$. Show that u is a quasi-isomorphism if, and only if, $^p\mathcal{H}^j(u) : {}^p\mathcal{H}^j(\mathcal{F}^\bullet) \to {}^p\mathcal{H}^j(\mathcal{G}^\bullet)$ is an isomorphism of perverse sheaves for every $j \in \mathbb{Z}$.

Exercise 8.1.17 Let $\Phi : \mathcal{A}^\bullet \to \mathcal{B}^\bullet$ be a morphism of perverse sheaves, with mapping cone $\mathcal{C}_\Phi^\bullet$. Show that $\mathrm{Ker}\,\Phi = {}^p\mathcal{H}^{-1}(\mathcal{C}_\Phi^\bullet)$ and $\mathrm{Coker}\,\Phi = {}^p\mathcal{H}^0(\mathcal{C}_\Phi^\bullet)$.

Remark 8.1.18 The perverse cohomology sheaf construction provides a way to get perverse sheaves out of any constructible complex. Another important method for constructing perverse sheaves, the *intermediate extension*, will be discussed later on, in Section 8.4.

In the case when the base ring is a field, the Universal Coefficient Theorem (5.15) yields the following:

Lemma 8.1.19 *If the base ring A is a field, then*

$$\mathrm{cosupp}^j(\mathcal{F}^\bullet) = \mathrm{supp}^{-j}(\mathcal{D}\mathcal{F}^\bullet).$$

Proof Let $\mathcal{G}^\bullet = \mathcal{D}\mathcal{F}^\bullet$, hence $\mathcal{F}^\bullet = \mathcal{D}\mathcal{G}^\bullet$. Since we work over a field, we have:

$$\mathcal{H}^j(i_x^!\mathcal{F}^\bullet) = \mathcal{H}^j(i_x^!\mathcal{D}\mathcal{G}^\bullet) = \mathcal{H}^j(\mathcal{D}i_x^*\mathcal{G}^\bullet) \overset{UCT}{=} \mathcal{H}^{-j}(i_x^*\mathcal{G}^\bullet)^\vee = \mathcal{H}^{-j}(i_x^*\mathcal{D}\mathcal{F}^\bullet)^\vee.$$

$\square$

Corollary 8.1.20 *If the coefficient ring A is a field, then $\mathcal{F}^\bullet \in {}^pD^{\leq 0}(X)$ if and only if $\mathcal{D}\mathcal{F}^\bullet \in {}^pD^{\geq 0}(X)$. In particular, the duality functor $\mathcal{D} : D_c^b(X) \to D_c^b(X)$ preserves perverse sheaves, i.e., it restricts to a functor*

$$\mathcal{D} : Perv(X) \to Perv(X).$$

Remark 8.1.21 Corollary 8.1.20 can be rephrased by saying that, if the base ring A is a field, then

$$^{p}\tau_{\leq 0}(\mathcal{D}\mathcal{F}^{\bullet}) \simeq \mathcal{D}(^{p}\tau_{\geq 0}\mathcal{F}^{\bullet}) \,, \quad ^{p}\tau_{\geq 0}(\mathcal{D}\mathcal{F}^{\bullet}) \simeq \mathcal{D}(^{p}\tau_{\leq 0}\mathcal{F}^{\bullet}) \,,$$

whence:

$$\mathcal{D}(^{p}\mathcal{H}^{j}(\mathcal{F}^{\bullet})) \simeq {}^{p}\mathcal{H}^{-j}(\mathcal{D}(\mathcal{F}^{\bullet})). \tag{8.2}$$

8.2 Gluing of t-Structures

Definition 8.2.1 Let D_i $(i = 1, 2)$ be two triangulated categories with t-structures $(D_i^{\leq 0}, D_i^{\geq 0})$. Let $F : D_1 \to D_2$ be a functor of triangulated categories. Then F is *left t-exact* if $F(D_1^{\geq 0}) \subseteq D_2^{\geq 0}$, *right t-exact* if $F(D_1^{\leq 0}) \subseteq D_2^{\leq 0}$, and *$t$-exact* if F is both left and right t-exact.

Definition 8.2.2 If $\mathcal{C}_1, \mathcal{C}_2$ are the hearts of the above t-structures, with $k_i : \mathcal{C}_i \hookrightarrow D_i$, then

$$^{p}F := {}^{t}H^{0} \circ F \circ k_1 : \mathcal{C}_1 \to \mathcal{C}_2$$

is called the *perverse functor associated to F*.

Remark 8.2.3 If $F : D_1 \to D_2$ is a t-exact functor, then F restricts to a functor $F : \mathcal{C}_1 \to \mathcal{C}_2$ on the corresponding hearts that is naturally isomorphic to the perverse functor ^{p}F.

Exercise 8.2.4 Let $F : D_1 \to D_2$ be a functor of triangulated categories with t-structures. Show that if F is left (resp., right) t-exact and P is an object in $D_1^{\geq 0}$ (resp., $D_1^{\leq 0}$), then:

$$^{t}H^{0}(F(P)) \simeq {}^{p}F(^{t}H^{0}(P)). \tag{8.3}$$

Exercise 8.2.5 Show that if $F : D_1 \to D_2$ is left/right t-exact, then $^{p}F : \mathcal{C}_1 \to \mathcal{C}_2$ is left/right exact.

Exercise 8.2.6 Let D_1 and D_2 be two triangulated categories with t-structures. Assume that $F : D_1 \to D_2$ and $G : D_2 \to D_1$ are functors of triangulated categories, and F is the left adjoint functor of G. Show that F is right t-exact if and only if G is left t-exact.

Let X be a complex algebraic (or analytic) variety with a stratification $\mathcal{X}$. Let $Z \subseteq X$ be a closed subset that is a union of strata. Let $U = X - Z$ and $i : Z \hookrightarrow$

X, $j : U \hookrightarrow X$ be the inclusions maps. Then one has the following diagrams of functors of triangulated categories

$$D^b(Z) \xrightarrow{i_* = i_!} D^b(X) \xrightarrow{j^* = j^!} D^b(U)$$

$$D^b(U) \xrightarrow{Rj_*, j_!} D^b(X) \xrightarrow{i^*, i^!} D^b(Z),$$

so that $j^* \circ i_* = 0$, $i^* \circ j_! = 0$, and $i^! \circ Rj_* = 0$. Moreover, for $\mathcal{F}^\bullet \in D^b_c(X)$ these functors fit into the attaching triangles:

$$i_! i^! \mathcal{F}^\bullet \longrightarrow \mathcal{F}^\bullet \longrightarrow Rj_* j^* \mathcal{F}^\bullet \xrightarrow{[1]}$$

$$j_! j^! \mathcal{F}^\bullet \longrightarrow \mathcal{F}^\bullet \longrightarrow i_* i^* \mathcal{F}^\bullet \xrightarrow{[1]}.$$

The adjunction morphisms

$$i^* i_* \longrightarrow id \longrightarrow i^! i_!$$

$$j^* Rj_* \longrightarrow id \longrightarrow j^! j_!$$

are isomorphisms (equivalently, the functors i_*, j_*, and $j_!$ are fully faithful). The functors i_*, $j^* = j^!$, $i^!$, i^* preserve constructibility in both the algebraic and the analytic setting. The functors $j_!$ and Rj_* preserve constructibility in the algebraic setting; the same holds in the analytic setting provided that one fixes a Whitney stratification $\mathcal{X}$ of the pair (X, Z) and constructibility is taken with respect to $\mathcal{X}$ (indeed, in this case, X, U, Z inherit Whitney stratifications as well). See [214, Proposition 4.2.1] for more details.

A fundamental result in the theory of perverse sheaves (see [12, Theorem 1.4.10], or [6, Section 7.2]) is that the perverse t-structure on $D^b_c(X)$ can be obtained by "gluing" the perverse t-structures on $D^b_c(U)$ and $D^b_c(Z)$ as follows:

Theorem 8.2.7 (Gluing of t-Structures) *Let X be a complex algebraic variety. Then the perverse t-structure on $D^b_c(X)$ can be obtained by gluing the perverse t-structures on $D^b_c(U)$ and $D^b_c(Z)$, as follows:*

$$^p D^{\leq 0}(X) = \{\mathcal{F}^\bullet \in D^b_c(X) \mid j^* \mathcal{F}^\bullet \in {}^p D^{\leq 0}(U), \; i^* \mathcal{F}^\bullet \in {}^p D^{\leq 0}(Z)\},$$

$$^p D^{\geq 0}(X) = \{\mathcal{F}^\bullet \in D^b_c(X) \mid j^! \mathcal{F}^\bullet \in {}^p D^{\geq 0}(U), \; i^! \mathcal{F}^\bullet \in {}^p D^{\geq 0}(Z)\}.$$

Corollary 8.2.8 *The functors $j_!$, i^* are right t-exact, the functors $j^! = j^*$, $i_* = i_!$ are t-exact, and Rj_*, $i^!$ are left t-exact.*

Remark 8.2.9 In particular, Corollary 8.2.8 implies that restriction to open subsets preserves perverse sheaves. Similar results also hold in the complex analytic context, provided one works with fixed Whitney stratifications.

Corollary 8.2.10 *Let* $\mathcal{F}^\bullet \in D^b_c(X)$ *with* $\mathrm{supp}(\mathcal{F}^\bullet) := \bigcup_j \mathrm{supp}^j(\mathcal{F}^\bullet) \subseteq Z$, *for* $i : Z \hookrightarrow X$ *a closed subset. Then* $\mathcal{F}^\bullet \in Perv(X)$ *if and only if* $i^*\mathcal{F}^\bullet \in Perv(Z)$.

Proof Since $\mathcal{F}^\bullet \in Perv(X) = {}^p D^{\leq 0}(X) \cap {}^p D^{\geq 0}(X)$, one has by Theorem 8.2.7 that $i^*\mathcal{F}^\bullet \in {}^p D^{\leq 0}(Z)$ and $i^!\mathcal{F}^\bullet \in {}^p D^{\geq 0}(Z)$. We claim that

$$i^*\mathcal{F}^\bullet \simeq i^!\mathcal{F}^\bullet, \tag{8.4}$$

so then $i^*\mathcal{F}^\bullet \in {}^p D^{\geq 0}(Z) \cap {}^p D^{\leq 0}(Z) = Perv(Z)$. In order to prove the claim, note that the assumption $\mathrm{supp}(\mathcal{F}^\bullet) \subseteq Z$ implies that $j^*\mathcal{F}^\bullet \simeq 0$, where $j : U = X - Z \hookrightarrow X$ is the open inclusion. From the attaching triangles, we then get that $\mathcal{F}^\bullet \simeq i_! i^! \mathcal{F}^\bullet \simeq i_* i^* \mathcal{F}^\bullet$, hence

$$i_! i^! \mathcal{F}^\bullet \simeq i_* i^* \mathcal{F}^\bullet.$$

Applying i^* and using the fact that $i^* i_! \simeq id \simeq i^* i_*$, yields that $i^!\mathcal{F}^\bullet \simeq i^*\mathcal{F}^\bullet$, as claimed. For the converse statement, since $j^*\mathcal{F}^\bullet = j^!\mathcal{F}^\bullet \simeq 0$, Theorem 8.2.7 yields that: $\mathcal{F}^\bullet \in {}^p D^{\leq 0}(X)$ if and only if $i^*\mathcal{F}^\bullet \in {}^p D^{\leq 0}(Z)$, and $\mathcal{F}^\bullet \in {}^p D^{\geq 0}(X)$ if and only if $i^!\mathcal{F}^\bullet \in {}^p D^{\geq 0}(Z)$. Then the equivalence in the statement follows via (8.4), which only uses the assumption on support. □

Remark 8.2.11 As already seen above, perverse sheaves are in general not sheaves, but rather complexes of sheaves. The reason for using the terminology perverse "sheaf" is that the functor

$$U \longmapsto Perv(U),$$

for U open subset of X, behaves like a sheaf with respect to gluing local data into a global object. In other words, the category $Perv(X)$ is a *stack*, see [12, Corollary 2.1.23]. This means that given an open covering $X = \cup_i U_i$, perverse sheaves $\mathcal{F}^\bullet_i \in Perv(U_i)$, and isomorphisms $\mathcal{F}^\bullet_i|_{U_i \cap U_j} \simeq \mathcal{F}^\bullet_j|_{U_i \cap U_j}$ satisfying the usual compatibility conditions, one can glue the $\mathcal{F}^\bullet_i$'s uniquely to get a perverse sheaf $\mathcal{F}^\bullet \in Perv(X)$. Moreover, morphisms of perverse sheaves can also be glued (see [12, Corollary 2.1.22]).

8.3 Examples of Perverse Sheaves

Before discussing examples of perverse sheaves, it is important to note that the categories ${}^p D^{\leq 0}(X), {}^p D^{\geq 0}(X)$ can also be described in terms of a fixed Whitney stratification $\mathcal{X}$ of X. Indeed, by applying the gluing theorem (Theorem 8.2.7) inductively on strata, the perverse t-structure can be characterized as follows:

Theorem 8.3.1 *Say $X = \bigcup_{\alpha} X_\alpha$, with X_α the connected components of the strata in a Whitney stratification $\mathfrak{X}$, with inclusions $i_{X_\alpha} : X_\alpha \hookrightarrow X$. If $\mathcal{F}^\bullet$ is constructible with respect to $\mathfrak{X}$, then:*

$$\mathcal{F}^\bullet \in {}^p D^{\leq 0}(X) \iff \mathcal{H}^j(i_{X_\alpha}^* \mathcal{F}^\bullet) = 0, \ \forall \alpha, \ \forall j > -\dim_{\mathbb{C}} X_\alpha,$$

$$\mathcal{F}^\bullet \in {}^p D^{\geq 0}(X) \iff \mathcal{H}^j(i_{X_\alpha}^! \mathcal{F}^\bullet) = 0, \ \forall \alpha, \ \forall j < -\dim_{\mathbb{C}} X_\alpha.$$

Exercise 8.3.2 Show that $\mathcal{F}^\bullet \in Perv(\{x\})$ if and only if $H^i(\mathcal{F}^\bullet) = 0$ for all $i \neq 0$.

Exercise 8.3.3 Show that if X is a nonsingular variety of complex dimension n and $\mathcal{F}^\bullet$ is constructible with respect to the trivial stratification, then $\mathcal{F}^\bullet$ is perverse if and only if

$$\mathcal{F}^\bullet \simeq \mathcal{H}^{-n}(\mathcal{F}^\bullet)[n].$$

Example 8.3.4 If X is a nonsingular variety of complex dimension n with the trivial stratification, and $\mathcal{L}$ is a local system on X, then $\mathcal{L}[n] \in Perv(X)$, e.g., $\underline{A}_X[n] \in Perv(X)$ (see Theorem 8.3.12 below for an instance where this fact remains true even in the singular case). On the other hand, if X is singular and $\mathcal{F}$ is a constructible sheaf on X, it is not true in general that $\mathcal{F}[\dim_{\mathbb{C}} X]$ is perverse on X.

Exercise 8.3.5 Show that if $\mathcal{F}^\bullet \in Perv(X)$ then $\mathcal{H}^i(\mathcal{F}^\bullet) = 0$ for $i \notin [-\dim_{\mathbb{C}} X, 0]$.

Exercise 8.3.6 Show that if $\mathcal{F}^\bullet \in Perv(X)$ is supported on a closed d-dimensional stratum of X, then

$$\mathcal{F}^\bullet \simeq \mathcal{H}^{-d}(\mathcal{F}^\bullet)[d].$$

Example 8.3.7 If X is a pure-dimensional variety then $IC_X \in Perv(X)$.

Example 8.3.8 Because the support conditions for perverse sheaves are relaxed slightly from those of IC_X, the shifted logarithmic intersection cohomology complex $IC_{\overline{\ell}}^\bullet[-\dim_{\mathbb{C}} X]$ from Remark 6.3.12 is also perverse.

Exercise 8.3.9 Let $f : X \to Y$ be a finite map (i.e., proper, with finite fibers). Then $Rf_* = f_!$ preserves perverse sheaves. If, moreover, f is generically bijective, then $Rf_* IC_X \simeq f_* IC_X \simeq IC_Y$. The latter fact applies, in particular, to the case when X is the (algebraic) normalization of Y.

Before giving another important example of perverse sheaves, let us recall the following:

Definition 8.3.10

(i) A germ of a complex analytic set $(X, 0)$ at the origin $0 \in \mathbb{C}^n$ is a *complete intersection singularity* if all irreducible components of $(X, 0)$ have the same dimension, say m, and $(X, 0)$ can be defined set-theoretically as the zero set of

$n - m$ analytic functions. (This means that in a small enough open ball B_ε at the origin in $\mathbb{C}^n$,

$$X \cap B_\varepsilon = \{x \in B_\varepsilon \mid f_1(x) = \cdots = f_{n-m}(x) = 0\}$$

for some analytic functions $f_1, \cdots, f_{n-m}$.)

(ii) Let Y be a nonsingular variety of complex dimension n, and let $X \subset Y$ be a closed subvariety. We say that X is a *local complete intersection* if for every $x \in X$, the germ (X, x) is a complete intersection singularity in $(Y, x) \cong (\mathbb{C}^n, 0)$.

Example 8.3.11

(i) A hypersurface singularity is a complete intersection singularity.

(ii) If $(X, 0)$ is the germ of a smooth complex subvariety in $\mathbb{C}^n$, then $(X, 0)$ is a complete intersection singularity.

The following result of Lê [138] provides an extension of Example 8.3.4 in the singular context:

Theorem 8.3.12 *Let X be a complex algebraic (or analytic) space of pure complex dimension n, which is a local complete intersection. Then $\underline{A}_X[n] \in Perv(X)$. More generally, $\mathcal{L}[n]$ is a perverse sheaf on X for every local system $\mathcal{L}$ on X.*

Proof For simplicity of the exposition, we prove the assertion only in the case when A is a field and X has only isolated singularities, that is, $X = X_n \sqcup X_0$, where $X_n = X_{reg}$ is the (top) n-dimensional stratum and $X_0 = Sing(X)$ is the 0-dimensional stratum. (The proof of the general case is not much harder, and can be found, e.g., in Dimca's book [61, Theorem 5.1.20].) Let i_n, i_0 denote the inclusion maps of the strata X_n, X_0, respectively, into X. Then

$$\mathcal{H}^j(\underline{A}_X[n])_{x \in X} = \begin{cases} A, & j = -n, \\ 0, & j \neq -n. \end{cases}$$

So, $\mathcal{H}^j(i_0^* \underline{A}_X[n]) = 0$ if $j > 0$, and $\mathcal{H}^j(i_n^* \underline{A}_X[n]) = 0$ if $j > -n$, which implies $\underline{A}_X[n] \in {}^p D^{\leq 0}(X)$.

For proving that $\underline{A}_X[n] \in {}^p D^{\geq 0}(X)$, we make use of the local calculations of Section 7.2 as follows. Recall that we work under the assumption that X has only isolated singularities. Note that, since A is a field, we have

$$\underline{A}_X[n] \in {}^p D^{\geq 0}(X) \iff \mathcal{D}(\underline{A}_X[n]) \in {}^p D^{\leq 0}(X).$$

Moreover, for $x \in X$, by the local calculations of Section 7.2 we have the following sequence of isomorphisms:

$$
\begin{aligned}
\mathcal{H}^j(\mathcal{D}(\underline{A}_X[n]))_x &\cong \mathcal{H}^{j-n}(\mathcal{D}\underline{A}_X)_x \\
&\cong H_c^{n-j}(\mathring{B}_\epsilon(x); A)^\vee \\
&\cong H^{n-j}(\mathring{B}_\epsilon(x), \mathring{B}_\epsilon(x) - x; A)^\vee \\
&\cong \tilde{H}^{n-j-1}(K_x; A)^\vee
\end{aligned}
\tag{8.5}
$$

where K_x denotes the link of x in X, and the last isomorphism uses the local contractibility of X and the long exact sequence for the cohomology of a pair.

If x is a smooth point, i.e., $x \in X_n$, then K_x is homeomorphic to the sphere S^{2n-1}, so $\mathcal{H}^j(\mathcal{D}(\underline{A}_X[n]))_x \cong 0$ if $n - j - 1 < 2n - 1$, i.e., $j > -n$.

Next, we need the following classical fact from Singularity Theory, which asserts that the link of a singular point x in an n-dimensional local complete intersection is $(n - 2)$-connected (see [94]). So, at an isolated singular point $x \in X$ we have that

$$
\mathcal{H}^j(\mathcal{D}(\underline{A}_X[n]))_x \cong \tilde{H}^{n-j-1}(K_x; A)^\vee \cong 0
$$

if $n - j - 1 < n - 1$, or $j > 0$.

Altogether $\mathcal{D}(\underline{A}_X[n]) \in {}^p D^{\leq 0}(X)$.   □

Exercise 8.3.13 For the special case of local complete intersections with isolated singularities, prove the condition of cosupport directly, without reference to the Verdier dual (the proof should apply to any coefficient ring A).

8.4 Intermediate Extension

Let $j : U \hookrightarrow X$ be the inclusion of an open constructible subset of the complex algebraic (or analytic) variety X, with $i : Z = X - U \hookrightarrow X$ the closed inclusion.

Definition 8.4.1 A sheaf complex $\mathcal{F}^\bullet \in D_c^b(X)$ is an *extension* of $\mathcal{G}^\bullet \in D_c^b(U)$ if $j^*\mathcal{F}^\bullet \simeq \mathcal{G}^\bullet$.

Remark 8.4.2 If $\mathcal{F}^\bullet \in D_c^b(X)$ is an extension of $\mathcal{G}^\bullet \in D_c^b(U)$, then by adjunction one gets morphisms:

$$
j_!\mathcal{G}^\bullet \longrightarrow \mathcal{F}^\bullet \longrightarrow Rj_*\mathcal{G}^\bullet.
\tag{8.6}
$$

Since $j^* j_! \simeq id \simeq j^* Rj_*$, $j_!\mathcal{G}^\bullet$ can be regarded as the "smallest" extension of $\mathcal{G}^\bullet$ to X, and $Rj_*\mathcal{G}^\bullet$ can be viewed as the "largest" such extension.

Assume now that $\mathcal{G}^{\bullet} \in Perv(U)$, and we look for extensions $\mathcal{F}^{\bullet} \in Perv(X)$. Applying the functor ${}^{p}\mathcal{H}^{0}$ to the above morphisms of (8.6), and using ${}^{p}\mathcal{H}^{0}(\mathcal{F}^{\bullet}) \simeq \mathcal{F}^{\bullet}$, we get the diagram:

$$ {}^{p}j_{!}\mathcal{G}^{\bullet} \longrightarrow \mathcal{F}^{\bullet} \longrightarrow {}^{p}j_{*}\mathcal{G}^{\bullet}. $$

Definition 8.4.3 (Intermediate Extension) The *intermediate extension* $j_{!*}\mathcal{G}^{\bullet}$ of the perverse sheaf $\mathcal{G}^{\bullet} \in Perv(U)$ is the unique extension $\mathcal{F}^{\bullet} \in Perv(X)$ of $\mathcal{G}^{\bullet}$ satisfying the following equivalent properties:

(i) $\mathcal{F}^{\bullet}$ is the image in $Perv(X)$ of the morphism ${}^{p}j_{!}\mathcal{G}^{\bullet} \to {}^{p}j_{*}\mathcal{G}^{\bullet}$.
(ii) $\mathcal{F}^{\bullet}$ is the unique extension of the perverse sheaf $\mathcal{G}^{\bullet}$ in $D_{c}^{b}(X)$ so that $i^{*}\mathcal{F}^{\bullet} \in {}^{p}D^{\leq -1}(Z)$ and $i^{!}\mathcal{F}^{\bullet} \in {}^{p}D^{\geq 1}(Z)$.

(For the equivalence of the above two properties in Definition 8.4.3, see [12, 1.4.22, 1.4.24].)

Using the description of the perverse t-structure from Theorem 8.3.1, the definition of the intermediate extension can be reformulated as follows:

Proposition 8.4.4 *If* $\mathcal{G}^{\bullet} \in Perv(U)$, *the intermediate extension* $j_{!*}\mathcal{G}^{\bullet}$ *is, up to isomorphism, the unique extension* $\mathcal{F}^{\bullet}$ *of* $\mathcal{G}^{\bullet}$ *in* $D_{c}^{b}(X)$ *such that for every stratum* $V \subset X - U$ *(in a stratification with respect to which* $\mathcal{F}^{\bullet}$ *is constructible) and inclusion* $i_{V} : V \hookrightarrow X$, *one has:*

(i) $\mathcal{H}^{k}(i_{V}^{*}\mathcal{F}^{\bullet}) = 0$, *for all* $k \geq - \dim_{\mathbb{C}} V$.
(ii) $\mathcal{H}^{k}(i_{V}^{!}\mathcal{F}^{\bullet}) = 0$, *for all* $k \leq - \dim_{\mathbb{C}} V$.

The intermediate extension can also be described inductively on strata, using iterated truncations. Let us denote by U_{ℓ} the union of strata of pure complex dimension at least ℓ, with open inclusion

$$ v = v_{\ell} : U_{\ell+1} \hookrightarrow U_{\ell}. $$

With these notations, the following result holds, see [12, Proposition 2.1.11]:

Proposition 8.4.5 *Given* $\mathcal{G}^{\bullet} \in Perv(U_{\ell+1})$, *one has*

$$ v_{!*}\mathcal{G}^{\bullet} = \tau_{\leq -\ell-1} v_{*}\mathcal{G}^{\bullet} \in Perv(U_{\ell}). $$

Exercise 8.4.6 Recall that if X is of pure dimension n, and $j : U = X_{\mathrm{reg}} \hookrightarrow X$ is the inclusion of nonsingular locus, then for a local system $\mathcal{L}$ on U one has that $\mathcal{L}[n] \in Perv(U)$. Show that:

$$ IC_{X}(\mathcal{L}) \simeq j_{!*}(\mathcal{L}[n]). $$

Exercise 8.4.7 Show that for every open subvariety U of X with inclusion map $j : U \hookrightarrow X$, one has:

$$IC_X \simeq j_{!*}IC_U.$$

Exercise 8.4.8 Let X be an irreducible nonsingular algebraic curve and let $j : U \hookrightarrow X$ be the inclusion of a Zariski-open and dense subset. Show that if $\mathcal{L}$ is a local system on U, then:

$$IC_X(\mathcal{L}) \simeq j_{!*}(\mathcal{L}[1]) \simeq (j_*\mathcal{L})[1].$$

Exercise 8.4.9 Show that the functor $j_{!*}$ is not exact (i.e., it does not necessarily send a short exact sequence in $Perv(U)$ into a short exact sequence in $Perv(X)$).

Proposition 8.4.10 *Let* $\mathcal{G}^\bullet \in Perv(U)$ *and* $Z = X - U$. *Then:*

(i) $^P j_* \mathcal{G}^\bullet$ *has no non-trivial sub-object whose support is contained in* Z.
(ii) $^P j_! \mathcal{G}^\bullet$ *has no non-trivial quotient object whose support is contained in* Z.

Proof Let $\mathcal{P}^\bullet \hookrightarrow {}^P j_* \mathcal{G}^\bullet$ be a sub-object of $^P j_* \mathcal{G}^\bullet$ with support contained in Z. It follows from Corollary 8.2.10 that $i^! \mathcal{P}^\bullet \simeq i^* \mathcal{P}^\bullet \in Perv(Z)$, thus $^P i^! \mathcal{P}^\bullet \simeq i^! \mathcal{P}^\bullet$. Since $\mathcal{P}^\bullet \simeq i_* i^! \mathcal{P}^\bullet$, it suffices to show that $^P i^! \mathcal{P}^\bullet = 0$. Apply the left exact functor $^P i^!$ to the exact sequence $0 \to \mathcal{P}^\bullet \to {}^P j_* \mathcal{G}^\bullet$ to obtain an exact sequence

$$0 \longrightarrow {}^P i^! \mathcal{P}^\bullet \longrightarrow {}^P i^! {}^P j_* \mathcal{G}^\bullet$$

Moreover, since $i^! R j_* \simeq 0$, it follows from (8.3) that

$$^P i^! {}^P j_* \mathcal{G}^\bullet \simeq {}^P \mathcal{H}^0(i^! R j_* \mathcal{G}^\bullet) \simeq 0.$$

Therefore, $^P i^! \mathcal{P}^\bullet = 0$. The proof of (ii) is similar, and is left as an exercise. □

Corollary 8.4.11 *The intermediate extension* $j_{!*}\mathcal{G}^\bullet$ *of* $\mathcal{G}^\bullet \in Perv(U)$ *has no non-trivial sub-object and no non-trivial quotient object whose supports are contained in* Z.

Exercise 8.4.12 Let X be a local complete intersection of pure complex dimension n. Show that the natural map $\alpha_X : \underline{A}_X[n] \to IC_X$ is a surjection in the category $Perv(X)$. The kernel of α_X is a perverse sheaf on X supported on the singular locus, which is often referred to as the *comparison complex*.

Corollary 8.4.13 *Assume* $\mathcal{G}^\bullet \in Perv(U)$ *is a simple object. Then* $j_{!*}\mathcal{G}^\bullet \in Perv(X)$ *is also a simple object.*

Proof Let $\mathcal{P}^\bullet \hookrightarrow j_{!*}\mathcal{G}^\bullet$ be a sub-object of $j_{!*}\mathcal{G}^\bullet$, and consider the associated short exact sequence in $Perv(X)$:

$$0 \longrightarrow \mathcal{P}^\bullet \longrightarrow j_{!*}\mathcal{G}^\bullet \longrightarrow \mathcal{Q}^\bullet \to 0.$$

Apply the t-exact functor $j^! = j^*$ to it to get the following short exact sequence in $Perv(U)$:

$$0 \longrightarrow j^*\mathcal{P}^\bullet \longrightarrow \mathcal{G}^\bullet \longrightarrow j^*\mathcal{Q}^\bullet \to 0.$$

Since $\mathcal{G}^\bullet$ is simple, $j^*\mathcal{P}^\bullet$ or $j^*\mathcal{Q}^\bullet$ is zero, i.e., $\mathcal{P}^\bullet$ or $\mathcal{Q}^\bullet$ has support contained in Z. By Corollary 8.4.11, it follows that $\mathcal{P}^\bullet = 0$ or $\mathcal{Q}^\bullet = 0$. □

As suggested by the previous corollary, the intermediate extension functor plays an important role in describing the simple objects in the abelian category $Perv(X)$. More precisely, the following result holds (see [12, Theorem 4.3.1]):

Theorem 8.4.14 *Let X be a complex algebraic variety, and assume that the coefficient ring A is a field.*

(a) *The category of perverse sheaves $Perv(X)$ is artinian and noetherian, i.e., every perverse sheaf on X admits an increasing finite filtration with quotients simple perverse sheaves.*

(b) *The simple A-perverse sheaves on X are the twisted intersection complexes $IC_{\overline{V}}(\mathcal{L})$ (regarded as complexes on X via extension by zero), where V runs through the family of smooth algebraic subvarieties of X, $\mathcal{L}$ is a simple (i.e., irreducible) A-local system on V, and $\overline{V}$ is the closure of V in X.*

We conclude this section with a result describing the behavior of the intermediate extension with respect to the dualizing functor.

Proposition 8.4.15 *Assume the coefficient ring A is a field, and let $\mathcal{G}^\bullet \in Perv(U)$. Then*

$$\mathcal{D}_X(j_{!*}\mathcal{G}^\bullet) \simeq j_{!*}\mathcal{D}_U(\mathcal{G}^\bullet). \tag{8.7}$$

Proof By applying the dualizing functor $\mathcal{D}_X$ to the sequence

$$^p j_! \mathcal{G}^\bullet \twoheadrightarrow j_{!*}\mathcal{G}^\bullet \hookrightarrow {}^p j_* \mathcal{G}^\bullet$$

of morphisms in $Perv(X)$, one gets by Corollary 8.1.20 a sequence

$$\mathcal{D}_X(^p j_* \mathcal{G}^\bullet) \twoheadrightarrow \mathcal{D}_X(j_{!*}\mathcal{G}^\bullet) \hookrightarrow \mathcal{D}_X(^p j_! \mathcal{G}^\bullet)$$

of morphisms in $Perv(X)$. Moreover, Corollary 8.1.20 also yields that

$$\mathcal{D}_X(^p j_* \mathcal{G}^\bullet) \simeq {}^p\mathcal{H}^0 \mathcal{D}_X(Rj_* \mathcal{G}^\bullet) \simeq {}^p\mathcal{H}^0(Rj_!(\mathcal{D}_U \mathcal{G}^\bullet)) \simeq {}^p j_!(\mathcal{D}_U \mathcal{G}^\bullet)$$

$$\mathcal{D}_X(^p j_! \mathcal{G}^\bullet) \simeq {}^p\mathcal{H}^0 \mathcal{D}_X(Rj_! \mathcal{G}^\bullet) \simeq {}^p\mathcal{H}^0(Rj_*(\mathcal{D}_U \mathcal{G}^\bullet)) \simeq {}^p j_*(\mathcal{D}_U \mathcal{G}^\bullet).$$

Therefore, one obtains a sequence

$$^p j_!(\mathcal{D}_U \mathcal{G}^\bullet) \twoheadrightarrow \mathcal{D}_X(j_{!*}\mathcal{G}^\bullet) \hookrightarrow {}^p j_*(\mathcal{D}_U \mathcal{G}^\bullet)$$

of morphisms in $Perv(X)$, which proves the assertion (by uniqueness).   □

8.5   A Splitting Criterion for Perverse Sheaves

In this section, we discuss a splitting criterion for perverse sheaves (see [49, Lemma 4.1.3]) that plays an important role in de Cataldo–Migliorini's proof of the BBDG decomposition theorem (see Section 9.3). We assume that the base ring A is a field.

Let X be a complex algebraic variety of dimension n, $\mathcal{F}^\bullet \in Perv(X)$, and $\mathcal{X}$ a Whitney stratification of X with respect to which $\mathcal{F}^\bullet$ is constructible. To set the notations, recall that $\mathcal{X}$ yields a filtration

$$X = X_n \supseteq X_{n-1} \supseteq \cdots \supseteq X_1 \supseteq X_0 \supseteq X_{-1} = \emptyset$$

by closed algebraic subsets such that

$$Z_i := X_i - X_{i-1}$$

is either empty or a locally closed algebraic subset of pure dimension i, and whose connected components are the i-dimensional strata of X. Set

$$U_i := X - X_{i-1} = \bigsqcup_{j \geq i} Z_j,$$

so that $U_i = U_{i+1} \sqcup Z_i$. In particular, U_n is a Zariski-dense open subset of X, and $U_0 = X$.

Let $\ell \in \mathbb{Z}$ be fixed. Assume, moreover, that

$$X = U \sqcup Z,$$

where $U = U_{\ell+1} = \bigsqcup_{i > \ell} Z_i$ and $Z = Z_\ell$ is a *closed* ℓ-dimensional stratum. Let $u : Z \hookrightarrow X$ and $v : U \hookrightarrow X$ be the corresponding closed and open inclusions.

By the condition of support for $\mathcal{F}^\bullet \in Perv(X)$ (that is, the first condition in Theorem 8.3.1), the map $\tau_{\leq -\ell}\mathcal{F}^\bullet \to \mathcal{F}^\bullet$ is an isomorphism. Similarly, the condition of support for $v^*\mathcal{F}^\bullet \in Perv(U)$ implies that the map $\tau_{\leq -\ell-1}v^*\mathcal{F}^\bullet \to v^*\mathcal{F}^\bullet$ is an isomorphism. Since $\mathcal{F}^\bullet \in D^{\leq -\ell}(X)$, we have that

$$\mathrm{Hom}_{D^b_c(X)}(\mathcal{F}^\bullet, \mathcal{G}^\bullet) \cong \mathrm{Hom}_{D^{\leq -\ell}(X)}(\mathcal{F}^\bullet, \tau_{\leq -\ell}\mathcal{G}^\bullet)$$

for every $\mathcal{G}^\bullet \in D_c^b(X)$. For $\mathcal{G}^\bullet = Rv_*v^*\mathcal{F}^\bullet$, it then follows that the adjunction map $\mathcal{F}^\bullet \to Rv_*v^*\mathcal{F}^\bullet$ admits a natural lifting

$$t : \mathcal{F}^\bullet \longrightarrow \tau_{\leq -\ell}Rv_*v^*\mathcal{F}^\bullet.$$

We also have a map

$$c : \mathcal{F}^\bullet \to \tau_{\geq -\ell}\mathcal{F}^\bullet \simeq \mathcal{H}^{-\ell}(\mathcal{F}^\bullet)[\ell],$$

where the last isomorphism uses again the fact that $\mathcal{F}^\bullet \in D^{\leq -\ell}(X)$. Moreover, the support condition for $\mathcal{F}^\bullet$ implies that $\mathrm{supp}\mathcal{H}^{-\ell}(\mathcal{F}^\bullet) \subseteq Z$, hence, by constructibility, $\mathcal{H}^{-\ell}(\mathcal{F}^\bullet)$ is a local system on the stratum Z. In particular, the complexes

$$\mathcal{H}^{-\ell}(\mathcal{F}^\bullet)[\ell] \simeq u_*u^*\mathcal{H}^{-\ell}(\mathcal{F}^\bullet)[\ell] \simeq \mathcal{H}^{-\ell}(u_*u^*\mathcal{F}^\bullet)[\ell]$$

are perverse. Note also that, since $v^*\mathcal{F}^\bullet \in Perv(U)$, it follows from the inductive construction of the intermediate extension (see Proposition 8.4.5) that

$$\tau_{\leq -\ell-1}Rv_*v^*\mathcal{F}^\bullet \simeq v_{!*}v^*\mathcal{F}^\bullet \in Perv(X).$$

In this setup, the following splitting result holds, see [49, Lemma 4.1.3] (compare also with Proposition 4.6.3):

Theorem 8.5.1 *Assume*

$$\dim_A \mathcal{H}^{-\ell}(u_!u^!\mathcal{F}^\bullet)_z = \dim_A \mathcal{H}^{-\ell}(u_*u^*\mathcal{F}^\bullet)_z \qquad (8.8)$$

for some (or, equivalently, every) point $z \in Z$. Then the following statements are equivalent:

(i) The natural map $\mathcal{H}^{-\ell}(u_!u^!\mathcal{F}^\bullet) \to \mathcal{H}^{-\ell}(\mathcal{F}^\bullet)$ is an isomorphism.
(ii) The map $t : \mathcal{F}^\bullet \to \tau_{\leq -\ell}Rv_*v^*\mathcal{F}^\bullet$ has a unique lifting $\tilde{t} : \mathcal{F}^\bullet \to \tau_{\leq -\ell-1}Rv_*v^*\mathcal{F}^\bullet$ and

$$(\tilde{t}, c) : \mathcal{F}^\bullet \longrightarrow v_{!*}v^*\mathcal{F}^\bullet \oplus \mathcal{H}^{-\ell}(\mathcal{F}^\bullet)[\ell]$$

is an isomorphism of perverse sheaves.

Remark 8.5.2 Theorem 8.5.1 can be applied inductively, by attaching one stratum at a time. Recall now that the artinian property of $Perv(X)$ implies that every perverse sheaf can be written as a finite extension of twisted intersection cohomology complexes. Theorem 8.5.1 asserts that, if certain conditions on strata (as in part (i)) are satisfied, all these extensions are trivial.

Exercise 8.5.3 Show that if $\mathcal{F}^\bullet \in Perv(X)$ is Verdier self-dual (e.g., $\mathcal{F}^\bullet = IC_X$ or $\mathcal{F}^\bullet = Rf_*IC_Y$ for $f : Y \to X$ a proper algebraic map), then the equal-rank condition (8.8) is satisfied.

8.6   Artin's Vanishing Theorem

In this section we assume that the base ring A is a field.

Let us recall that a closed algebraic subvariety in an affine space $\mathbb{C}^n$ (i.e., the common zero-locus of a finite family of complex polynomials in n variables) is called a *complex affine variety*. An analytic variety X is called a *Stein variety* if every coherent $\mathcal{O}_X$-sheaf $\mathcal{F}$ is acyclic on X. For example, a closed analytic subvariety in $\mathbb{C}^n$ is a Stein variety. Clearly, an affine variety is Stein.

We begin with a formulation of Artin's vanishing theorem for constructible sheaves, which was first proved in [228, Exposé XIV, Corollary 3.2] (see also [133, Theorem 3.1.13] or [214, Corollary 6.1.2]).

Theorem 8.6.1 *Let X be an affine n-dimensional complex algebraic variety, and let $\mathcal{F}$ be a constructible sheaf on X. Then $H^i(X; \mathcal{F}) = 0$ for all $i > n$.*

Note that if $\mathcal{F}$ is a constructible sheaf, then $\mathcal{F}[\dim_{\mathbb{C}} X]$ satisfies the condition of support from the definition of perverse sheaves. In fact, Artin's proof also applies to a constructible complex satisfying the condition of support. Since perverse sheaves satisfy the conditions of support and cosupport, they satisfy a stronger version of the Artin vanishing theorem, as we shall next indicate.

Definition 8.6.2 A morphism $f : X \to Y$ of complex algebraic (resp. analytic) varieties is an *affine* (resp., *Stein*) *morphism* if for all $y \in Y$ there exists an open neighborhood U_y of y in the Zariski (resp., analytic) topology such that $f^{-1}(U_y)$ is an affine variety (resp., a Stein space).

For the following important result, see [12, Section 4.1] (and also [214, Corollary 6.0.8]):

Theorem 8.6.3 *Let $f : X \to Y$ be an affine morphism. Then Rf_* is right t-exact and $Rf_!$ is left t-exact. If f is quasi-finite (i.e., $\dim_{\mathbb{C}} f^{-1}(y) \leq 0$, for all $y \in Y$), then the functors Rf_* and $Rf_!$ are t-exact. (Here, t-exactness is with respect to the perverse t-structure.)*

Example 8.6.4 If $j : U \hookrightarrow X$ is an affine open immersion, then j is quasi-finite, so $Rj_!$ and Rj_* preserve perverse sheaves. For example, this applies if U is obtained from $\mathbb{C}^n$ by removing a finite union of hypersurfaces, with $j : U \hookrightarrow \mathbb{C}^n$ the inclusion map.

Remark 8.6.5 A similar property holds for $f : X \to Y$ a Stein morphism, provided that one restricts only to constructible complexes $\mathcal{F}^\bullet$ such that $Rf_*\mathcal{F}^\bullet$ and/or $Rf_!\mathcal{F}^\bullet$ are constructible, see [122, Proposition 10.3.17] and also [214, Corollary 6.0.8].

Corollary 8.6.6 (Artin's Vanishing for Perverse Sheaves) *Let X be a complex affine variety and $\mathcal{F}^\bullet \in Perv(X)$. Then*

$$\mathbb{H}^i(X; \mathcal{F}^\bullet) = 0, \quad \text{for } i > 0,$$

and

$$\mathbb{H}^i_c(X; \mathcal{F}^\bullet) = 0, \text{ for } i < 0.$$

Proof Since X is affine, the morphism $f : X \to pt$ is affine. Then Rf_* is right t-exact, so $Rf_*\mathcal{F}^\bullet \in {}^p D^{\leq 0}(pt) = D^{\leq 0}(pt)$. This implies that $\mathbb{H}^i(X; \mathcal{F}^\bullet) \cong H^i(Rf_*\mathcal{F}^\bullet) = 0$ for $i > 0$. Similarly, $Rf_!$ is left t-exact, so $Rf_!\mathcal{F}^\bullet \in {}^p D^{\geq 0}(pt) = D^{\geq 0}(pt)$, which implies that $\mathbb{H}^i_c(X; \mathcal{F}^\bullet) \cong H^i(Rf_!\mathcal{F}^\bullet) = 0$ for $i < 0$. □

Remark 8.6.7 There is no *sheaf* analogue of Artin vanishing (Theorem 8.6.1) for compactly supported cohomology, as the Verdier dual of a constructible sheaf is not a sheaf.

Remark 8.6.8 Assuming Theorem 8.6.1, a different proof of Artin's vanishing theorem for perverse sheaves (Corollary 8.6.6) can be given as follows. Let X be a complex affine variety and $\mathcal{F}^\bullet \in Perv(X)$. Since the supports of cohomology sheaves of $\mathcal{F}^\bullet$ are closed affine subvarieties, the hypercohomology spectral sequence

$$H^p(X; \mathcal{H}^q(\mathcal{F}^\bullet)) \Longrightarrow \mathbb{H}^{p+q}(X; \mathcal{F}^\bullet)$$

together with Artin's vanishing theorem for constructible sheaves (Theorem 8.6.1) yield readily the vanishing of $\mathbb{H}^i(X; \mathcal{F}^\bullet)$ in the desired range. The vanishing assertion for the hypercohomology with compact support of a perverse sheaf follows by duality, using the fact that $\mathcal{F}^\bullet \in Perv(X)$ if and only if $\mathcal{D}\mathcal{F}^\bullet \in Perv(X)$.

Corollary 8.6.9 *Let X be an affine pure n-dimensional complex algebraic variety. Assume that X is a local complete intersection. Then*

$$H^i(X; \mathbb{Q}) = 0, \text{ for } i > n,$$

and

$$H^i_c(X; \mathbb{Q}) = 0, \text{ for } i < n.$$

Proof Theorem 8.3.12 yields that $\underline{\mathbb{Q}}_X[n] \in Perv(X)$. The assertion follows now from Corollary 8.6.6. □

Remark 8.6.10 A complex affine variety of complex dimension n has the homotopy type of a finite CW complex of real dimension at most n, see [115] and also [60, (1.6.10)]. (The Stein version of this fact is due to Hamm [96], though in this case the CW complex may be infinite; e.g., take $X = \{x \in \mathbb{C} \mid \sin(x) \neq 0\}$.) The above corollary is a cohomological version of this result.

If we drop the assumption that X is affine, we have the following weaker vanishing result (see, e.g., [61, Proposition 5.2.20]):

Proposition 8.6.11 *Let X be a pure n-dimensional complex algebraic (or analytic) variety, and let $\mathcal{F}^\bullet \in Perv(X)$. Then*

$$\mathbb{H}^i(X; \mathcal{F}^\bullet) = \mathbb{H}^i_c(X; \mathcal{F}^\bullet) = 0$$

for all $i \notin [-n, n]$.

Proof The assertion follows from the definition of $Perv(X)$ (via Exercise 8.3.5), together with the hypercohomology spectral sequence

$$E_2^{p,q} = H^p_{(c)}(X; \mathcal{H}^q(\mathcal{F}^\bullet)) \Longrightarrow \mathbb{H}^{p+q}_{(c)}(X; \mathcal{F}^\bullet).$$

$\square$

Remark 8.6.12 For Artin-type vanishing results for perverse sheaves over a more general base ring, see [214, Corollary 6.0.4] and the references therein.

Chapter 9
The Decomposition Package and Applications

In this chapter, we discuss how to recast Lefschetz-type results for complex projective varieties, by using intersection cohomology groups (Sections 9.1 and 9.2). In the nonsingular context, one also recovers the classical statements mentioned in Section 1.2. Section 9.3 is devoted to the BBDG decomposition theorem for proper algebraic maps, one of the most important results of the theory of perverse sheaves. A sample of the numerous applications of the decomposition package is presented in Section 9.4.

Since we are transitioning towards geometric results with Hodge-theoretic implications, in this chapter we assume that the base ring is $A = \mathbb{Q}$.

9.1 Lefschetz Hyperplane Section Theorem

We begin with a general Lefschetz hyperplane section theorem (also referred to as the "Weak Lefschetz theorem") for perverse sheaves:

Theorem 9.1.1 (Weak Lefschetz Theorem for Perverse Sheaves) *If X is a complex projective variety and $i : D \hookrightarrow X$ denotes the inclusion of a hyperplane section, then for every $\mathcal{F}^\bullet \in Perv(X)$ the following hold:*

(i) *the restriction map $\mathbb{H}^k(X; \mathcal{F}^\bullet) \to \mathbb{H}^k(D; i^*\mathcal{F}^\bullet)$ is an isomorphism for $k < -1$ and is injective for $k = -1$.*
(ii) *the pushforward map $\mathbb{H}^k(D; i^!\mathcal{F}^\bullet) \to \mathbb{H}^k(X; \mathcal{F}^\bullet)$ is an isomorphism for $k > 1$ and is surjective for $k = 1$.*

Proof Let $i : D \hookrightarrow X$, $j : U = X - D \hookrightarrow X$ be the inclusion maps. Note that U is an affine complex n-dimensional variety. Consider the compactly supported hypercohomology long exact sequence associated to the attaching triangle:

$$j_! j^! \mathcal{F}^\bullet \longrightarrow \mathcal{F}^\bullet \longrightarrow i_* i^* \mathcal{F}^\bullet \xrightarrow{[1]}$$

© Springer Nature Switzerland AG 2019
L. G. Maxim, *Intersection Homology & Perverse Sheaves*, Graduate Texts in Mathematics 281, https://doi.org/10.1007/978-3-030-27644-7_9

namely,

$$\cdots \longrightarrow \mathbb{H}^k_c(U; j^*\mathcal{F}^\bullet) \longrightarrow \mathbb{H}^k(X; \mathcal{F}^\bullet) \longrightarrow \mathbb{H}^k(D; i^*\mathcal{F}^\bullet) \longrightarrow \cdots$$

Since restriction to open subsets preserves perverse sheaves, we have that $j^*\mathcal{F}^\bullet \in Perv(U)$. So by Artin's vanishing theorem (Theorem 8.6.6), we get that

$$\mathbb{H}^k_c(U; j^*\mathcal{F}^\bullet) = 0 \text{ for } k < 0.$$

Together with the above long exact sequence, this yields (i).

The assertion (ii) is proved similarly, by using the hypercohomology long exact sequence associated to the dual attaching triangle

$$i_! i^! \mathcal{F}^\bullet \longrightarrow \mathcal{F}^\bullet \longrightarrow Rj_* j^*\mathcal{F}^\bullet \xrightarrow{[1]} ,$$

together with Artin's vanishing theorem (Theorem 8.6.6). $\square$

Exercise 9.1.2 Let X be a complex quasi-projective variety and let $i : D \hookrightarrow X$ denote the inclusion of a generic hyperplane section of X relative to a fixed embedding $X \hookrightarrow \mathbb{C}P^N$. Show that if $\mathcal{F}^\bullet \in Perv(X)$, then $i^*\mathcal{F}^\bullet[-1] \in Perv(D)$. (Hint: check directly that $i^*\mathcal{F}^\bullet[-1]$ and its Verdier dual satisfy the conditions of support.)

The following consequence of Theorem 9.1.1 was originally obtained in [84, Section 5.4] by using stratified Morse theory, and then in [83, Section 7] via Artin's vanishing:

Theorem 9.1.3 (Lefschetz Hyperplane Section Theorem for Intersection Cohomology) *Let $X^n \subset \mathbb{C}P^N$ be a pure n-dimensional closed algebraic subvariety with a Whitney stratification $\mathcal{X}$. Let $H \subset \mathbb{C}P^N$ be a generic hyperplane (i.e., transversal to all strata of $\mathcal{X}$). Then the natural homomorphism*

$$IH^i(X; \mathbb{Q}) \longrightarrow IH^i(X \cap H; \mathbb{Q})$$

is an isomorphism for $0 \leq i \leq n - 2$ and a monomorphism for $i = n - 1$.

Proof Let $D = X \cap H$ and denote by $i : D \hookrightarrow X$ the inclusion map. By Theorem 9.1.1,

$$IH^{k+n}(X; \mathbb{Q}) := \mathbb{H}^k(X; IC_X) \longrightarrow \mathbb{H}^k(D; i^*IC_X)$$

is an isomorphism for $k < -1$ and a monomorphism for $k = -1$. Furthermore, by the transversality assumption, the inclusion i is locally *normally nonsingular*, hence

$$i^*IC_X \simeq IC_D[1].$$

For the last claim, one can easily check that $i^* IC_X[-1]$ satisfies $[AX_{\overline{m}}]$ on D (see Exercise 6.5.4). Then

$$\mathbb{H}^k(D; i^* IC_X) \cong \mathbb{H}^k(D; IC_D[1]) =: IH^{k+n}(D; \mathbb{Q}),$$

thus finishing the proof.   □

9.2 Hard Lefschetz Theorem for Intersection Cohomology

There are by now several approaches for obtaining the Hard Lefschetz theorem and purity of intersection cohomology groups of complex projective varieties, e.g., by positive characteristic methods [12, Theorem 6.2.10], by using Saito's theory of mixed Hodge modules [205, 207] (see Chapter 11 for an overview), or the more geometric approach of de Cataldo–Migliorini [49].

The Hard Lefschetz theorem for intersection cohomology can be deduced from a more general sheaf-theoretic statement, which we now describe. Let $f : X \to Y$ be a projective[1] morphism and let $L \in H^2(X; \mathbb{Q})$ be the first Chern class of an f-ample line bundle on X.[2,3] Recall from (4.11) that L corresponds to a map of complexes

$$L : \underline{\mathbb{Q}}_X \longrightarrow \underline{\mathbb{Q}}_X[2].$$

Tensoring with IC_X yields a map

$$L : IC_X \longrightarrow IC_X[2].$$

This induces $L : Rf_* IC_X \longrightarrow Rf_* IC_X[2]$, and after taking perverse cohomology one gets a map of perverse sheaves on Y:

$$L : {}^p\mathcal{H}^i(Rf_* IC_X) \longrightarrow {}^p\mathcal{H}^{i+2}(Rf_* IC_X).$$

Iterating, one obtains maps of perverse sheaves

$$L^i : {}^p\mathcal{H}^{-i}(Rf_* IC_X) \longrightarrow {}^p\mathcal{H}^i(Rf_* IC_X)$$

---

[1] $f : X \to Y$ is projective if it can be factored as $X \overset{i}{\hookrightarrow} Y \times \mathbb{C}P^N \overset{p}{\to} Y$ for some N, with i a closed embedding, and p a projection.

[2] A *very ample* line bundle is one with enough global sections to set up an embedding of its base variety into projective space. An *ample* line bundle is one such that some positive power is very ample. For a morphism $f : X \to Y$, an f-ample line bundle on X is a line bundle that is ample on every fiber of f.

[3] An example of such an f-ample line bundle on X can be obtained as follows: pull back the hyperplane bundle from $\mathbb{C}P^N$ to $Y \times \mathbb{C}P^N$, and then restrict to X.

for every $i \geq 0$. In the above notations, the following result holds:

Theorem 9.2.1 (Relative Hard Lefschetz) *Let $f : X \to Y$ be a projective morphism of complex algebraic varieties with X pure-dimensional, and let $L \in H^2(X; \mathbb{Q})$ be the first Chern class of an f-ample line bundle on X. For every $i > 0$, one has isomorphisms of perverse sheaves*

$$L^i : {}^p\mathcal{H}^{-i}(Rf_*IC_X) \xrightarrow{\simeq} {}^p\mathcal{H}^i(Rf_*IC_X).$$

Remark 9.2.2 The special case of Theorem 9.2.1 for a smooth (i.e., submersive) projective morphism $f : X \to Y$ of complex algebraic manifolds was proved by Deligne, and it gives sheaf isomorphisms

$$R^{n-m-i}f_*\underline{\mathbb{Q}}_X \simeq R^{n-m+i}f_*\underline{\mathbb{Q}}_X \qquad (9.1)$$

for every $i > 0$, where $n = \dim_{\mathbb{C}} X$ and $m = \dim_{\mathbb{C}} Y$. Stalkwise, (9.1) is just the classical Hard Lefschetz theorem on the (nonsingular projective) fibers of the map f. In particular, if X is a complex projective manifold and one considers f to be the constant map $f : X \to pt$ to a point space, then (9.1) yields the classical Hard Lefschetz theorem on $H^*(X; \mathbb{Q})$.

By taking f in Theorem 9.2.1 to be the constant map $f : X \to pt$ to a point space, one obtains as a consequence the Hard Lefschetz theorem for intersection cohomology groups:

Corollary 9.2.3 (Hard Lefschetz Theorem for Intersection Cohomology) *Let X be a complex projective variety of pure complex dimension n, with $L \in H^2(X; \mathbb{Q})$ the first Chern class of an ample line bundle on X. Then there are isomorphisms*

$$L^i : IH^{n-i}(X; \mathbb{Q}) \xrightarrow{\cong} IH^{n+i}(X; \mathbb{Q}) \qquad (9.2)$$

for every integer $i > 0$, induced by the cup product by L^i.

Remark 9.2.4 Recall that intersection cohomology is not a ring. However, as seen above, the cup product with a cohomology class is well defined, and intersection cohomology is a module over cohomology.

One of the nice consequences of the Hard Lefschetz theorem for intersection cohomology is the *unimodality* of the corresponding intersection homology Betti numbers, generalizing Corollary 1.2.9 to the singular context.

Remark 9.2.5 The perverse cohomology sheaves appearing in Theorem 9.2.1 underlie a more complicated structure, that of pure mixed Hodge modules, e.g., see Chapter 11. In particular, the isomorphisms (9.2) hold (up to a Tate twist) in the category of pure Hodge structures.

9.3 The BBDG Decomposition Theorem and Applications

One of the most important results in the theory of perverse sheaves is the *BBDG decomposition theorem* [12, Theorem 6.2.5]. It was originally conjectured by S. Gelfand and R. MacPherson, and it was proved soon after by Beilinson, Bernstein, Deligne, and Gabber by reduction to positive characteristic. The proof given in [12] ultimately rests on Deligne's proof of the Weil conjectures. In late 1980s, Morihiko Saito gave another proof of the decomposition theorem, as a consequence of his theory of mixed Hodge modules [205, 206]. More recently, de Cataldo and Migliorini [49] gave a more geometric proof, involving only classical Hodge theory.[4]

In this section, we explain the statement and motivation of the decomposition theorem, along with a few important applications. We follow here the approach from [49]. (See also [245] for a detailed account of de Cataldo–Migliorini's proof, as well as Chapter 11, where Saito's method of proof is briefly reviewed.) While the decomposition theorem and the relative Hard Lefschetz theorem (Theorem 9.2.1) are proved as a package in [49], here we focus only on the decomposition statement, and steer away from any of its Hodge-theoretic aspects.

Stratifications of Algebraic Maps

Let us first recall some basic facts about complex algebraic maps.

First, every algebraic map $f : X \to Y$ of complex algebraic varieties can be *stratified*, e.g., see [233]. This means that there exist algebraic Whitney stratifications $\mathcal{X}$ of X and $\mathcal{Y}$ of Y such that, given any connected component S of a $\mathcal{Y}$-stratum on Y one has the following properties:

(a) $f^{-1}(S)$ is a union of connected components of strata of $\mathcal{X}$, each of which is mapped submersively to S by f; in particular, every fiber $f^{-1}(y)$ of f is stratified by its intersection with the strata of $\mathcal{X}$.

(b) For every point $y \in S$, there is an Euclidean neighborhood U of y in S and a stratum-preserving homeomorphism $h : U \times f^{-1}(y) \to f^{-1}(U)$ such that $f|_{f^{-1}(U)} \circ h$ is the projection to U.

Property (b) is the celebrated *Thom's isotopy lemma*: for every stratum S in Y, the restriction $f|_{f^{-1}(S)} : f^{-1}(S) \to S$ is a topologically locally trivial fibration.

Example 9.3.1 If $f : X \to Y$ is an open immersion, then a Whitney stratification $\mathcal{Y}$ on Y induces a Whitney stratification $\mathcal{X}$ on X with respect to which f is stratified.

---

[4]A more general decomposition theorem (for semi-simple coefficients) has been obtained by Mochizuki [183, 182] (with substantial contributions of Sabbah [204]), in relation to a conjecture of Kashiwara [121].

If $f : X \to Y$ is a closed immersion, one can choose a Whitney stratification $\mathcal{X}$ on X so that every stratum of $\mathcal{X}$ is the intersection of X with a stratum of $\mathcal{Y}$ of the same dimension.

Secondly, complex algebraic varieties and maps can be *compactified*, i.e., given an algebraic map $f : X \to Y$ of complex algebraic varieties, there are compact varieties X', Y' containing X and Y as Zariski-dense open subvarieties, and a proper map $f' : X' \to Y'$ such that $f'(X) \subseteq Y$ and $f'|_X = f$. Moreover, if f is proper, then $f'(X' - X) \subseteq Y' - Y$.

Exercise 9.3.2 Use the fact that complex algebraic varieties are compactifiable to give a proof of Corollary 7.3.1(a). Namely if $(X, \mathcal{X})$ is a complex algebraic variety with a Whitney stratification, then $\mathbb{H}^i(X; \mathcal{F}^\bullet)$ and $\mathbb{H}^i_c(X; \mathcal{F}^\bullet)$ are finite dimensional for every bounded complex $\mathcal{F}^\bullet$ that is constructible with respect to $\mathcal{X}$.

Deligne's Decomposition Theorem

The BBDG decomposition theorem is a generalization of the following theorem of Deligne, from a smooth projective morphism to an arbitrary proper morphism.

Theorem 9.3.3 (Deligne) *Let $f : X \to Y$ be a smooth projective map of complex algebraic manifolds. Then*

$$Rf_*\underline{\mathbb{Q}}_X \simeq \bigoplus_{i \geq 0} \mathcal{H}^i(Rf_*\underline{\mathbb{Q}}_X)[-i] \in D^b_c(Y), \tag{9.3}$$

and the local systems $R^i f_*\mathbb{Q}_X = \mathcal{H}^i(Rf_*\underline{\mathbb{Q}}_X)$ are semi-simple[5] on Y.

Since in the decomposition (9.3) the map $f : X \to Y$ is assumed to be smooth and projective, every fiber of f is a nonsingular complex projective variety. Deligne deduced his decomposition (9.3) by applying the Hard Lefschetz theorem to the cohomology of the fibers of f. (This idea is also used repeatedly in the proof of the BBDG decomposition theorem by de Cataldo–Migliorini [49].) For a detailed proof of Theorem 9.3.3, see [54] and [55, Theorem 4.2.6].

It should be noted that Ehresmann's fibration theorem implies that if $f : X \to Y$ is a smooth proper map of complex algebraic manifolds, then f is a locally trivial topological fibration. In particular, under the hypotheses of Deligne's theorem, each sheaf $R^i f_*\underline{\mathbb{Q}}_X$ is a local system on Y, with fiber at $y \in Y$ given by $\mathcal{H}^i(Rf_*\underline{\mathbb{Q}}_X)_y \cong H^i(f^{-1}(y); \mathbb{Q})$.

---

[5]A local system $\mathcal{L}$ on Y is *semi-simple* if every local subsystem $\mathcal{L}'$ of $\mathcal{L}$ admits a complement, i.e., a local subsystem $\mathcal{L}''$ of $\mathcal{L}$ such that $\mathcal{L} \simeq \mathcal{L}' \oplus \mathcal{L}''$.

Remark 9.3.4 The semi-simplicity part of Theorem 9.3.3 fails in general if the map f is not smooth or non-proper. Indeed, in such cases, the sheaves $R^i f_* \underline{\mathbb{Q}}_X$ are not necessarily local systems on the target, and the category of $\mathbb{Q}$-sheaves has too few simple objects (it is not "artinian"). If the projective map f has singular fibers, it is also not difficult to find examples (e.g., resolutions of singularities of a complex projective variety), where the decomposition statement (9.3) fails to hold.

Deligne's decomposition theorem has the following immediate cohomological consequence:

Corollary 9.3.5 *Let* $f : X \to Y$ *be a smooth projective map of complex algebraic manifolds. Then*

$$H^*(X; \mathbb{Q}) \cong \bigoplus_{i \geq 0} H^{*-i}(Y; R^i f_* \underline{\mathbb{Q}}_X). \tag{9.4}$$

Deligne's theorem (Theorem 9.3.3) also yields that the Leray spectral sequence

$$E_2^{p,q} = H^p(Y; R^q f_* \underline{\mathbb{Q}}_X) \Longrightarrow H^{p+q}(X; \mathbb{Q}) \tag{9.5}$$

of the smooth projective map f degenerates on the E_2-page. This fact is specific to the realm of complex algebraic geometry, and it fails in the non-algebraic or non-proper situation. It also fails in real algebraic geometry or complex geometry. The simplest example is provided by the Hopf fibration $S^3 \to S^2$ with fiber S^1 (this is a real algebraic proper submersion), or its complex algebraic version given by $\mathbb{C}^2 - \{0\} \to \mathbb{C}P^1$. Indeed, if the Leray spectral sequence degenerated, then (with rational coefficients) one would have that $H^*(S^3) = H^*(S^1) \otimes H^*(S^2)$, which is obviously not true.

If $f : X \to Y$ is a smooth projective map of complex algebraic manifolds (i.e., a family of complex projective manifolds, the fibers of f), the degeneration of the Leray spectral sequence for f yields a surjective map

$$H^q(X; \mathbb{Q}) \twoheadrightarrow E_\infty^{0,q} = E_2^{0,q} = H^0(Y; R^q f_* \underline{\mathbb{Q}}_X) \cong H^q(F; \mathbb{Q})^{\pi_1(Y)},$$

where F denotes the fiber of the fibration f, and $H^q(F; \mathbb{Q})^{\pi_1(Y)} \subseteq H^q(F; \mathbb{Q})$ is the subspace of monodromy invariants. Hence *every monodromy invariant class is global*, i.e., it comes from the total space X of the family. With help from the theory of mixed Hodge structures (see Section 11.1 for a short introduction), one can moreover show the following:

Theorem 9.3.6 (Global Invariant Cycle Theorem) *Under the above notations, if* $\overline{X}$ *is a smooth compactification of the complex algebraic manifold* X*, then*

$$\text{Image} (H^*(\overline{X}; \mathbb{Q}) \to H^*(F; \mathbb{Q})) = \text{Image} (H^*(X; \mathbb{Q}) \to H^*(F; \mathbb{Q}))$$

$$= H^*(F; \mathbb{Q})^{\pi_1(Y)}.$$

In particular, since $H^*(\overline{X}; \mathbb{Q})$ carries a pure Hodge structure, it follows that the subspace $H^*(F; \mathbb{Q})^{\pi_1(Y)}$ of monodromy invariants is a Hodge substructure (see Chapter 11 for a brief overview of Hodge theory). This fact is quite striking, because the space of monodromy invariants is obtained via a topological construction.

Semi-Small Maps

Semi-small maps are a class of maps that behave especially nicely with respect to pushforward of perverse sheaves. They were defined and first analyzed by Borho–MacPherson in [17] and [18].

For simplicity, we work with proper surjective maps $f : X \to Y$, with X nonsingular and pure-dimensional. Let us fix a stratification $Y = \bigsqcup_\lambda S_\lambda$ of Y with respect to which f is stratified. For any given stratum S_λ, let $s_\lambda \in S_\lambda$ denote any of its points. Recall that $f^{-1}(S_\lambda) \to S_\lambda$ is a topologically locally trivial fibration, so fibers of f over such a stratum S_λ have constant dimension.

Definition 9.3.7 (Defect of Semi-Smallness, Semi-Small Maps) The *defect of semi-smallness* $r(f)$ of f is defined by

$$r(f) = \max_\lambda \{2 \dim_{\mathbb{C}} f^{-1}(s_\lambda) + \dim_{\mathbb{C}} S_\lambda - \dim_{\mathbb{C}} X\} \geq 0.$$

The map f is called *semi-small* if $r(f) = 0$, i.e., if for every λ,

$$\dim_{\mathbb{C}} S_\lambda + 2 \dim_{\mathbb{C}} f^{-1}(s_\lambda) \leq \dim_{\mathbb{C}} X.$$

A stratum S_λ of Y is called *relevant* if $\dim_{\mathbb{C}} S_\lambda + 2 \dim_{\mathbb{C}} f^{-1}(s_\lambda) = \dim_{\mathbb{C}} X$. A semi-small map with no relevant strata of positive codimension is called a *small map*.

Remark 9.3.8 A semi-small map is finite on every open stratum of Y, hence $\dim_{\mathbb{C}} Y = \dim_{\mathbb{C}} X$. In particular, open strata are relevant, and all relevant strata have even complex codimension in Y. Note also that the complex dimension of every fiber of a semi-small map $f : X \to Y$ is at most half of the complex dimension of X, and that equality can only occur at finitely many points in Y.

Example 9.3.9 If Y has dimension at most 2, then a small map $f : X \to Y$ must be finite. The blowup of a point in $\mathbb{C}^2$ is semi-small. In fact, surjective maps between complex surfaces are always semi-small. One the other hand, the blowup of a point in $\mathbb{C}^3$ is not semi-small. But the blowup of a line in $\mathbb{C}^3$ is semi-small. A surjective map of complex threefolds is semi-small if and only if no divisor (i.e., codimension one subvariety) is contracted to a point.

Exercise 9.3.10 Show that the blowup of $\mathbb{C}^k \subset \mathbb{C}^n$, $k \leq n - 2$, is semi-small if and only if $k = n - 2$. Moreover, none of these maps is small.

Example 9.3.11 An important class of semi-small maps (due to Kaledin [114]) consists of projective birational maps from a *holomorphic symplectic* nonsingular variety. Here, a nonsingular quasi-projective complex variety is holomorphic symplectic if it is even- dimensional and admits a closed holomorphic 2-form ω that is non-degenerate, i.e., $\omega^{\frac{\dim_{\mathbb{C}} X}{2}}$ is nowhere vanishing.

Recall that an algebraic map $f : X \to Y$ is a *resolution of singularities* of a complex algebraic variety Y if X is a nonsingular complex variety and f is proper and an isomorphism away from a proper closed subset of Y (i.e., f is a proper *birational* morphism). Such resolutions always exist by a fundamental result of Hironaka [99].

Example 9.3.12 (Springer Resolution) Let $Fl(n)$ denote the *complete flag variety* of $\mathbb{C}^n$, whose points parameterize sequences of nested subspaces (flags)

$$\{0\} \subset V_1 \subset \cdots \subset V_{n-1} \subset V_n \subset \mathbb{C}^n.$$

The general linear group $GL(n, \mathbb{C})$ acts transitively on $Fl(n)$, and the standard flag

$$\{0\} \subset \langle e_1 \rangle \subset \langle e_1, e_2 \rangle \subset \cdots \subset \langle e_1, \cdots, e_{n-1} \rangle \subset \mathbb{C}^n$$

(with $\{e_1, \cdots, e_n\}$ the standard basis of $\mathbb{C}^n$) is fixed by the Borel subgroup B_n of all upper triangular invertible matrices with $\mathbb{C}$-coefficients. Therefore, $Fl(n)$ is a homogeneous manifold (in fact, a smooth projective variety), which can be written as $Fl(n) = GL(n, \mathbb{C})/B_n$. Let $X = T^* Fl(n)$ be its cotangent bundle. Let $\mathbb{C}[X]$ be the ring of regular functions on X and $\mathcal{N} := Spec\, \mathbb{C}[X]$. The affine variety $\mathcal{N}$ can be identified with the *nilpotent cone*, i.e., the variety of nilpotent $n \times n$ matrices. There is a map $\pi : X \to \mathcal{N}$, so that the fiber $\pi^{-1}(A)$ over a nilpotent matrix A is the set of A-invariant flags. For example, the fiber over the zero-matrix is the whole flag variety, viewed as the zero-section of its cotangent bundle. In fact, X can be identified with

$$\{(A, F) \in \mathcal{N} \times Fl(n) \mid A \text{ stabilizes } F\},$$

and π is the first projection map. It's not hard to see that the map π is birational. Indeed, consider the subset $\mathcal{N}_{\text{reg}}$ of nilpotent matrices that are conjugate to the Jordan matrix of order n, i.e., matrices A for which there is a basis $\{v_1, \cdots, v_n\}$ with $Av_i = v_{i+1}$ for $i \leq n-1$ and $Av_n = 0$. Then $\mathcal{N}_{\text{reg}}$ is a Zariski-open dense subset of $\mathcal{N}$, and the restriction $\pi_| : \pi^{-1}(\mathcal{N}_{\text{reg}}) \to \mathcal{N}_{\text{reg}}$ is an isomorphism since the only flag preserved by such a matrix is

$$\{0\} \subset \text{Span}(v_n) \subset \text{Span}(v_n, v_{n-1}) \subset \cdots \subset \text{Span}(v_n, \cdots, v_2) \subset \mathbb{C}^n.$$

Therefore, π is a resolution of singularities of $\mathcal{N}$, called the *Springer resolution*. Moreover, using properties of the Jordan form of a nilpotent matrix, it can be shown that π is a semi-small map. In fact, since the cotangent bundle X has a natural symplectic structure, the map π is a symplectic resolution, so it can also be seen from Example 9.3.11 that π is semi-small. We can partition $\mathcal{N}$ according to the Jordan canonical form. This gives rise to a stratification of $\mathcal{N}$ and π. Every stratum of this stratification turns out to be relevant for the semi-small map π.

It is easy to see (by checking the axioms) that the following holds (compare also with Proposition 9.3.18 below):

Proposition 9.3.13 *Let $f : X \to Y$ be a small map, with X nonsingular of complex pure dimension n. Let $U \subseteq Y$ be the nonsingular dense open subset over which f is a covering map. Then*

$$Rf_*\underline{\mathbb{Q}}_X[n] \simeq IC_Y(\mathcal{L}),$$

where $\mathcal{L}$ is the local system $(f_*\underline{\mathbb{Q}}_X)|_U$.

In particular, Proposition 9.3.13 yields the following:

Corollary 9.3.14 *If $f : X \to Y$ is a small resolution (i.e., a resolution of singularities that is also a small map), then*

$$Rf_*\underline{\mathbb{Q}}_X[\dim_{\mathbb{C}} X] \simeq IC_Y.$$

In particular, in this case there is an isomorphism:

$$IH^*(Y; \mathbb{Q}) \cong H^*(X; \mathbb{Q}).$$

Remark 9.3.15 Small resolutions do not always exist, and they are not necessarily unique when they do exist. If Y has several small resolutions, their cohomologies are isomorphic as groups, but not necessarily isomorphic as rings. (This shows that there is no natural product on intersection cohomology.) However, two small resolutions must have the same Euler characteristic and the same signature.

For the following example, see [39, Section 5.2]:

Example 9.3.16 For positive integers $a \leq b$, the Grassmann variety $\mathbf{G}_a(\mathbb{C}^b)$ of a-dimensional subspaces of $\mathbb{C}^b$ is a nonsingular complex projective variety of complex dimension $a(b - a)$. If M is a fixed subspace of $\mathbb{C}^b$ and $c \leq a$ is a positive integer, then

$$S = \{V \in \mathbf{G}_a(\mathbb{C}^b) \mid \dim_{\mathbb{C}} V \cap M \geq c\}$$

is a projective subvariety of $\mathbf{G}_a(\mathbb{C}^b)$, called a *single condition Schubert variety*. There is a resolution of singularities $f : \widetilde{S} \to S$, where

$$\widetilde{S} = \{(V, W) \in \mathbf{G}_a(\mathbb{C}^b) \times \mathbf{G}_c(\mathbb{C}^b) \mid W \subseteq V \cap M\}$$

and $f(V, W) = V$. It can be shown that the resolution f is in fact a small map. Therefore, Corollary 9.3.14 yields that

$$IH^*(S; \mathbb{Q}) \cong H^*(\widetilde{S}; \mathbb{Q}). \tag{9.6}$$

If we choose an isomorphism $M \cong \mathbb{C}^d$, where $d = \dim_{\mathbb{C}} M$, then we can define

$$\pi : \widetilde{S} \longrightarrow \mathbf{G}_c(\mathbb{C}^b), \ (V, W) \mapsto W.$$

It is a routine check to show that π is a locally trivial fibration with fiber $\mathbf{G}_{a-c}(\mathbb{C}^{b-c})$. It then follows from Corollary 9.3.5 that we have a decomposition:

$$H^j(\widetilde{S}; \mathbb{Q}) \cong \bigoplus_{p+q=j} H^p(\mathbf{G}_c(\mathbb{C}^b); H^q(\mathbf{G}_{a-c}(\mathbb{C}^{b-c}); \mathbb{Q})). \tag{9.7}$$

Since Grassmann varieties are simply connected, it follows that every local system on $\mathbf{G}_c(\mathbb{C}^b)$ is trivial. Hence, one gets from (9.6) and (9.7) the following computation of the intersection cohomology groups of S:

$$IH^j(S; \mathbb{Q}) \cong \bigoplus_{p+q=j} H^p(\mathbf{G}_c(\mathbb{C}^b); \mathbb{Q}) \otimes H^q(\mathbf{G}_{a-c}(\mathbb{C}^{b-c}); \mathbb{Q}). \tag{9.8}$$

The (co)homology of Grassmann manifolds is well-known, see, e.g., [91, Chapter 1, Section 5]. For example, $\mathbf{G}_a(\mathbb{C}^b)$ has an algebraic cell decomposition (i.e., all cells are complex affine spaces), so all of its cells are in even real dimensions. This implies, in particular, that all of its odd Betti numbers vanish.

Exercise 9.3.17 Let X be the blowup of $\mathbb{C}P^2$ at a point, with blowdown map $f : X \to \mathbb{C}P^2$. Show that $Rf_*\mathbb{Q}_X[2]$ is perverse on $\mathbb{C}P^2$.

More generally, one has the following:

Proposition 9.3.18 *Let $f : X \to Y$ be a proper surjective map, with X nonsingular of pure complex dimension n. Then*

$$Rf_*(\underline{\mathbb{Q}}_X[n]) \in {}^pD^{\leq r(f)}(Y) \cap {}^pD^{\geq -r(f)}(Y),$$

i.e., ${}^p\mathcal{H}^i(Rf_*(\underline{\mathbb{Q}}_X[n])) = 0$ for $i \notin [-r(f), r(f)]$.
In particular, if f is semi-small then $Rf_*(\underline{\mathbb{Q}}_X[n]) \in Perv(Y)$.

Proof Since f is proper, $Rf_! = Rf_*$. Hence

$$\mathcal{D}(Rf_*(\underline{\mathbb{Q}}_X[n])) \simeq Rf_*(\mathcal{D}(\underline{\mathbb{Q}}_X[n])) \simeq Rf_*(\underline{\mathbb{Q}}_X[n]),$$

i.e., $Rf_*(\underline{\mathbb{Q}}_X[n])$ is Verdier self-dual. By (8.2) it then suffices to check that $Rf_*(\underline{\mathbb{Q}}_X[n]) \in {}^pD^{\leq r(f)}(Y)$. Equivalently, by Theorem 8.3.1, one has to show that, for $s_\lambda \in S_\lambda$, the following support (stalk vanishing) condition is satisfied: $\mathcal{H}^i(Rf_*(\underline{\mathbb{Q}}_X[n]))_{s_\lambda} = 0$ for all $i > r(f) - \dim_{\mathbb{C}} S_\lambda$. Indeed, for such a point $s_\lambda \in S_\lambda$, we have that $\mathcal{H}^i(Rf_*(\underline{\mathbb{Q}}_X[n]))_{s_\lambda} = H^{i+n}(f^{-1}(s_\lambda); \mathbb{Q})$, since f is proper. Moreover,

$$H^{i+n}(f^{-1}(s_\lambda); \mathbb{Q}) = 0, \text{ if } i+n > 2\dim_{\mathbb{C}} f^{-1}(s_\lambda).$$

The desired vanishing follows readily since, by definition, $r(f) - \dim_{\mathbb{C}} S_\lambda \geq 2\dim_{\mathbb{C}} f^{-1}(s_\lambda) - n$.   □

It was noted in [48, Proposition 2.2.7] that semi-smallness is essentially equivalent to the Hard Lefschetz phenomenon, namely the following result holds:

Theorem 9.3.19 *Let* $f : X \to Y$ *be a projective surjective map of complex projective varieties with X nonsingular and n-dimensional. Let $L := f^*L' \in H^2(X; \mathbb{Q})$ be the first Chern class of the pullback to X of an ample line bundle L' on Y. The iterated cup product maps*

$$L^i : H^{n-i}(X; \mathbb{Q}) \longrightarrow H^{n+i}(X; \mathbb{Q})$$

are isomorphisms for every $i \geq 0$ if and only if f is semi-small.

The Decomposition Theorem for Semi-Small Maps

In general, if f is not a smooth proper map, or if X and/or Y are singular, Deligne's decomposition theorem (Theorem 9.3.3) fails, see, e.g., [50, Section 3.1]. In this section, we discuss a decomposition result for semi-small maps (see [48] for complete details).

The decomposition theorem for a semi-small map $f : X \to Y$ has a particularly simple form: the perverse sheaf $Rf_*\underline{\mathbb{Q}}_X[n]$ splits as a direct sum of twisted intersection cohomology complexes $IC_{\overline{S}}(\mathcal{L}_S)$, one for each relevant stratum S, and with corresponding local systems $\mathcal{L}_S$ of finite monodromy.

Let $f : X \to Y$ be a proper surjective semi-small map with X nonsingular of complex dimension n. Let S be a relevant stratum of f, and let $s \in S$. Define

$$\mathcal{L}_S := (R^{n-\dim_{\mathbb{C}} S} f_*\underline{\mathbb{Q}}_X)|_S.$$

Since f is proper, we see that $\mathcal{L}_S$ is a local system on S. The monodromy representation of $\mathcal{L}_S$ factors through the finite group of symmetries of the set of irreducible components $E_1, \cdots, E_l$ of maximal dimension $\frac{1}{2}(\dim_{\mathbb{C}} X - \dim_{\mathbb{C}} S)$ of the fiber $f^{-1}(s)$, e.g., see the discussion in [187, Section 7.1.1]. In particular, $\mathcal{L}_S$ is semi-simple (by Maschke's theorem in finite group theory). This in turn implies that $IC_{\overline{S}}(\mathcal{L}_S)$ is a semi-simple perverse sheaf.

We can now formulate the decomposition theorem for semi-small maps (see [48, Theorem 3.4.1]):

Theorem 9.3.20 *Let $f : X \to Y$ be a proper surjective semi-small map with X nonsingular of pure complex dimension n. Let $\mathcal{Y}$ be a Whitney stratification of Y with respect to which f is stratified, and denote by $\mathcal{Y}_{rel} \subset \mathcal{Y}$ the set of relevant strata of f. For each $S \in \mathcal{Y}_{rel}$, let $\mathcal{L}_S$ be the corresponding local system with finite monodromy defined above. There is a canonical isomorphism in $\mathrm{Perv}(Y)$:*

$$Rf_*\underline{\mathbb{Q}}_X[n] \simeq {}^p\mathcal{H}^0(Rf_*\underline{\mathbb{Q}}_X[n]) \simeq \bigoplus_{S \in \mathcal{Y}_{rel}} IC_{\overline{S}}(\mathcal{L}_S). \qquad (9.9)$$

Remark 9.3.21 The semi-small situation is particularly nice since the decomposition is canonical and very explicit. As we will see later on, for general maps the decomposition is not canonical, and it is difficult to say a priori which summands occur in the direct image.

To prove Theorem 9.3.20, one proceeds one stratum at a time. Higher dimensional strata are dealt with inductively, by cutting transversally with a generic hyperplane section D on Y, so that one is reduced to a semi-small map $f : f^{-1}(D) \to D$, where the dimension of a positive dimensional stratum in Y has decreased by 1 on D. It then remains to deal with the case of a zero-dimensional relevant stratum S, i.e., the most singular points of f (where the dimension of the fiber of f equals half of the dimension of X). Next, it can be shown (by using the splitting criterion of Theorem 8.5.1) that the decomposition (9.9) is equivalent to the non-degeneracy of the refined intersection form (see below for its definition)

$$I_s : H_n(f^{-1}(s); \mathbb{Q}) \times H_n(f^{-1}(s); \mathbb{Q}) \longrightarrow \mathbb{Q} \qquad (9.10)$$

for $s \in S$, which in turn is a consequence of classical mixed Hodge theory.

Example 9.3.22 Let $f : X \to Y$ be a resolution of singularities of a singular surface Y (so, in particular, f is semi-small). Assume that Y has a single singular point $y \in Y$ with fiber $f^{-1}(y) = E$ a finite union of curves on X. As X is nonsingular, $IC_X = \mathbb{Q}_X[2]$, and we have an isomorphism

$$Rf_*\underline{\mathbb{Q}}_X[2] \simeq IC_Y \oplus T,$$

where T is a skyscraper sheaf at y with stalk $T = H^2(E; \mathbb{Q})$.

We conclude our discussion by rephrasing the statement of the decomposition theorem for a semi-small map $f : X \to Y$ entirely in terms of basic intersection theory on X (see [48]). Let $S \subseteq Y$ be a relevant stratum, and $s \in S$. Set $d := \dim_{\mathbb{C}} f^{-1}(s)$. Let N be a normal slice to S at s. The restriction $f|_{f^{-1}(N)} : f^{-1}(N) \to N$ is still semi-small and $\dim_{\mathbb{C}} f^{-1}(N) = 2d$. By composing the chain of maps

$$H_{2d}(f^{-1}(s); \mathbb{Q}) = H_{2d}^{BM}(f^{-1}(s); \mathbb{Q}) \to H_{2d}^{BM}(f^{-1}(N); \mathbb{Q})$$

$$\cong H^{2d}(f^{-1}(N); \mathbb{Q}) \to H^{2d}(f^{-1}(s); \mathbb{Q}),$$

one gets the *refined intersection pairing*

$$I_S : H_{2d}(f^{-1}(s); \mathbb{Q}) \times H_{2d}(f^{-1}(s); \mathbb{Q}) \longrightarrow \mathbb{Q}$$

associated to the relevant stratum S. A basis of $H_{2d}(f^{-1}(s); \mathbb{Q})$ is given by the classes of the complex d-dimensional irreducible components of $f^{-1}(s)$. The intersection pairing I_S is then represented by the intersection matrix of these components, computed in the manifold $f^{-1}(N)$. (If the stratum S is not relevant, then $H_{2d}(f^{-1}(s)) = 0$ by dimension count, so the above intersection form is trivial; it is non-degenerate in the sense that the corresponding linear map is an isomorphism of trivial vector spaces.) It can be shown (by using the splitting criterion of Theorem 8.5.1) that the splitting of $Rf_* \underline{\mathbb{Q}}_X[n]$ is governed by the non-degeneracy of the forms I_S, which in turn can be proved by classical Hodge theory. More precisely, one has the following:

Theorem 9.3.23 *Let $f : X \to Y$ be a proper semi-small map with X nonsingular. Then the statement of the decomposition theorem (Theorem 9.3.20) is equivalent to the non-degeneracy of the intersection forms I_S associated to relevant strata of f. These forms are non-degenerate, and if the typical fiber of f above a relevant stratum S is of complex dimension d, then the form $(-1)^d I_S$ is positive-definite.*

Remark 9.3.24 The fact that I_S in the above theorem has a precise sign is a generalization of a famous result of Grauert for surfaces, see [89].

The BBDG Decomposition Theorem for Arbitrary Maps

As already mentioned, if f is not a smooth proper map, or if X and/or Y are singular, Deligne's decomposition theorem (Theorem 9.3.3) fails. But if one replaces sheaf cohomology by perverse cohomology, then the following holds:

Theorem 9.3.25 (BBDG Decomposition Theorem [12]) *Let $f : X \to Y$ be a proper map of complex algebraic varieties, with X pure-dimensional. Then:*

(i) (Decomposition) There is a (non-canonical) isomorphism in $D_c^b(Y)$:

$$Rf_* IC_X \simeq \bigoplus_i {}^p\mathcal{H}^i(Rf_* IC_X)[-i]. \tag{9.11}$$

(ii) (Semi-simplicity) Each ${}^p\mathcal{H}^i(Rf_* IC_X)$ is a semi-simple object in $Perv(Y)$, i.e., if $\mathcal{Y}$ is the set of connected components of strata of Y in a stratification of f, there is a canonical isomorphism in $Perv(Y)$:

$$^p\mathcal{H}^i(f_*IC_X) \simeq \bigoplus_{S \in \mathcal{Y}} IC_{\overline{S}}(\mathcal{L}_{i,S}) \tag{9.12}$$

where the local systems $\mathcal{L}_{i,S}$ are semi-simple.

Remark 9.3.26 The direct summands appearing in the decomposition

$$Rf_*IC_X \simeq \bigoplus_{i \in \mathbb{Z}} \bigoplus_{S \in \mathcal{Y}} IC_{\overline{S}}(\mathcal{L}_{i,S})[-i] \tag{9.13}$$

are uniquely determined.

Exercise 9.3.27 Show that if U is a non-empty, nonsingular, and pure-dimensional open subset of a complex algebraic variety Y on which all the cohomology sheaves $\mathcal{H}^i(\mathcal{F}^\bullet)$ of $\mathcal{F}^\bullet \in D_c^b(Y)$ are local systems, then the restrictions to U of $^p\mathcal{H}^j(\mathcal{F}^\bullet)$ and $\mathcal{H}^{j-\dim_{\mathbb{C}} Y}(\mathcal{F}^\bullet)[\dim_{\mathbb{C}} Y]$ coincide.

Exercise 9.3.28 Show that if in Theorem 9.3.25 we assume that f is a projective submersion of nonsingular complex algebraic varieties, then one recovers Deligne's decomposition theorem (Theorem 9.3.3). (Hint: use Exercise 9.3.27.)

Standard facts in algebraic geometry reduce the proof of the BBDG decomposition theorem to the case of a projective morphism $f : X \to Y$, with X nonsingular (this is the situation considered in [49]). Specifically, one can derive the decomposition theorem in the general case of a proper map of complex algebraic varieties from the above-mentioned special case as follows: by resolution of singularities one can drop the assumption that X is nonsingular, while Chow's lemma allows us to replace "f projective" by "f proper."

Remark 9.3.29 If the morphism f is projective, then (9.11) is a formal consequence of the relative Hard Lefschetz theorem (cf. [57]). So the heart of the decomposition theorem (Theorem 9.3.25) consists of the semi-simplicity statement.

The proof of Theorem 9.3.25, once reduced to the case of a projective map with X nonsingular, is done by induction on the pair of indices $(\dim_{\mathbb{C}} Y, r(f))$, where $r(f)$ is degree of semi-smallness of f. The induction hypothesis (which applies simultaneously also to the relative Hard Lefschetz theorem) is that, if we fix f we may assume that Theorem 9.3.25 is known for every projective map $g : X' \to Y'$ with $r(g) < r(f)$, or $r(g) = r(f)$ and $\dim_{\mathbb{C}} g(X') < \dim_{\mathbb{C}} f(X)$. The induction starts with the case when f is the projection to a point, when the statement is trivial (or follows from classical Hodge theory in the case of relative Hard Lefschetz). This reduces the problem to proving the semi-simplicity of $^p\mathcal{H}^0(f_*\mathbb{Q}_X[n])$, and this is handled as in the semi-small case via non-degeneracy of a certain refined intersection pairing associated to the fibers over the most singular points of f.

Exercise 9.3.30 Work out the exact shape of the decomposition theorem for a projection $f : X = Y \times F \to Y$. Deduce the classical Künneth formula.

Exercise 9.3.31 Let $Y \subset \mathbb{C}P^{n+1}$ be the cone with vertex y over a nonsingular variety $V \subset \mathbb{C}P^n$. Work out the summands in the decomposition theorem for the map $f : X \to Y$ obtained by blowing up Y at the cone vertex y.

Exercise 9.3.32 Work out the exact shape of the decomposition theorem for the map obtained by blowing up a nonsingular subvariety of a complex algebraic manifold.

Exercise 9.3.33 Let X be a nonsingular surface and let Y be a nonsingular curve. Work out the decomposition theorem for a projective map $f : X \to Y$ with connected fibers.

First Applications of the Decomposition Theorem

In this section, we collect some of the direct applications of the decomposition theorem. For more results in this direction, the reader is advised to also consult, e.g., [47], [82], or [51].

One of the first consequences of the decomposition (9.13) is that it gives a splitting of $IH^*(X; \mathbb{Q})$ in terms of twisted intersection cohomology groups of closures of strata in Y, i.e.,

Corollary 9.3.34 *Under the assumptions and notations of Theorem 9.3.25, we have a splitting*

$$IH^j(X; \mathbb{Q}) \cong \bigoplus_{i \in \mathbb{Z}} \bigoplus_{S \in \mathcal{Y}} IH^{j-\dim_{\mathbb{C}} X + \dim_{\mathbb{C}} S - i}(\overline{S}; \mathcal{L}_{i,S}), \qquad (9.14)$$

for every $j \in \mathbb{Z}$.

Remark 9.3.35 If X is singular, there is no direct sum decomposition for $H^*(X; \mathbb{Q})$ analogous to (9.14), except for the case when X is a rational homology manifold (in which case one uses (9.14), together with (6.26)). Intersection cohomology is the relevant topological invariant designed precisely to deal with singular varieties and maps. Intersection cohomology is needed even when X and Y are both nonsingular, but the map $f : X \to Y$ is not a submersion.

An important application of the decomposition statement (9.11) for $f : X \to Y$ is the E_2-degeneration of the corresponding perverse Leray spectral sequence:

$$E_2^{i,j} = \mathbb{H}^i(Y; {}^p\mathcal{H}^j(Rf_* IC_X)) \Longrightarrow \mathbb{H}^{i+j}(Y; Rf_* IC_X) = IH^{i+j+\dim_{\mathbb{C}} X}(X; \mathbb{Q}).$$

Before discussing another application of Theorem 9.3.25, let us note that if $f : X \to Y$ is a proper algebraic map then $Rf_* = Rf_!$, and hence

$$\mathcal{D}(Rf_* IC_X) \simeq Rf_*(\mathcal{D} IC_X) \simeq Rf_* IC_X,$$

that is, Rf_*IC_X is Verdier self-dual. In particular, (8.2) yields that

$$^p\mathcal{H}^i(Rf_*IC_X) \simeq \mathcal{D}(^p\mathcal{H}^{-i}(Rf_*IC_X)).$$

(When coupled with the relative Hard Lefschetz theorem (Theorem 9.2.1), this further implies that each perverse cohomology sheaf $^p\mathcal{H}^i(Rf_*IC_X)$ is Verdier self-dual.) It then follows from (9.11) and (9.12) that the direct image Rf_*IC_X is *palindromic* (i.e., it reads the same, up to shifts and dualities, from right to left and left to right). More precisely, in the notations of Theorem 9.3.25, we get:

$$Rf_*IC_X \simeq \bigoplus_{i \in \mathbb{Z},\, S \in \mathcal{Y}} IC_{\overline{S}}(\mathcal{L}_{i,S})[-i]$$

$$\simeq \left(\bigoplus_{i<0,\, S \in \mathcal{Y}} IC_{\overline{S}}(\mathcal{L}_{i,S})[-i] \right) \oplus \left(\bigoplus_{S \in \mathcal{Y}} IC_{\overline{S}}(\mathcal{L}_{0,S}) \right)$$

$$\oplus \left(\bigoplus_{i<0,\, S \in \mathcal{Y}} IC_{\overline{S}}(\mathcal{L}_{i,S}^{\vee})[i] \right),$$

with $\mathcal{L}_{i,S}^{\vee}$ denoting the dual of the local system $\mathcal{L}_{i,S}$. This symmetry of the decomposition theorem has the following important consequence that was used by Ngô in the proof of the *support theorem* (a key step in his proof of the fundamental lemma in the *Langlands' program*), see [187, Section 7.3].

Theorem 9.3.36 *Let* $f : X \to Y$ *be a proper surjective map of complex algebraic varieties, with* X *nonsingular. Assume* f *has pure relative dimension* d *(i.e., all fibers of* f *have pure complex dimension* d*). Let* S *be a subvariety of* Y *appearing in the decomposition of* $Rf_*\mathbb{Q}_X[\dim_{\mathbb{C}} X]$*. Then* $\mathrm{codim}_Y(S) \leq d$*.*

Proof There is a maximum integer i_S^+ for which a term $IC_{\overline{S}}(\mathcal{L}_{i_S^+,S})[-i_S^+]$ appears in the decomposition of $Rf_*\mathbb{Q}_X[\dim_{\mathbb{C}} X]$. By the palindromicity of the decomposition, one may assume that $i_S^+ \geq 0$.

The local system $\mathcal{L}_{i_S^+,S}$ is defined on the stratum S. Let $U \subseteq Y$ be an open so that $U \cap \overline{S} = S$, i.e., S is a closed stratum in U. By proper base change, the decomposition is preserved by restriction to U, so one may replace Y by U. Let $X_U := f^{-1}(U)$. On U, $IC_{\overline{S}}(\mathcal{L}_{i_S^+,S})|_U = \mathcal{L}_{i_S^+,S}[\dim_{\mathbb{C}} S]$, so $\mathcal{L}_{i_S^+,S}[\dim_{\mathbb{C}} S][-i_S^+]$ is a direct summand of $Rf_*\mathbb{Q}_{X_U}[\dim_{\mathbb{C}} X]$. In particular, $\mathcal{L}_{i_S^+,S}$ is a direct summand of $R^{\dim_{\mathbb{C}} X - \dim_{\mathbb{C}} S + i_S^+} f_*\mathbb{Q}_{X_U}$ (which is therefore non-zero). Since f is proper, for every $s \in S$ one has

$$(R^{\dim_{\mathbb{C}} X - \dim_{\mathbb{C}} S + i_S^+} f_*\mathbb{Q}_{X_U})_s \cong H^{\dim_{\mathbb{C}} X - \dim_{\mathbb{C}} S + i_S^+}(f^{-1}(s); \mathbb{Q}).$$

On the other hand, as $\dim_{\mathbb{C}} f^{-1}(s) = d$ and $i_s^+ \geq 0$, it follows that

$$\dim_{\mathbb{C}} X - \dim_{\mathbb{C}} S \leq \dim_{\mathbb{C}} X - \dim_{\mathbb{C}} S + i_s^+ \leq 2d.$$

Since $\dim_{\mathbb{C}} X = \dim_{\mathbb{C}} Y + d$, the conclusion follows.   □

Another important application of the BBDG decomposition theorem is provided by the following result (e.g., see [47, Section 4.5]):

Theorem 9.3.37 *Let $f : X \to Y$ be a proper map of complex algebraic varieties with X irreducible, and let $Y' := f(X)$ be the image of f. Denote by $d = \dim_{\mathbb{C}} X - \dim_{\mathbb{C}} Y'$ the relative dimension of f. Then $IC_{Y'}[d]$ is a direct summand of Rf_*IC_X. In particular, $IH^j(Y'; \mathbb{Q})$ is a direct summand of $IH^j(X; \mathbb{Q})$ for every integer j.*

Proof By replacing Y by $Y' = f(X)$, one may assume without any loss of generality that f is surjective. Then one needs to show that $IC_Y[\dim_{\mathbb{C}} X - \dim_{\mathbb{C}} Y]$ is a direct summand of Rf_*IC_X.

The proof relies on the following principles:

(a) (*IC localization principle*) Let S be a stratum of Y with closure $\overline{S}$, let $\mathcal{L}$ be a local system on S, and consider $IC_{\overline{S}}(\mathcal{L}) \in Perv(Y)$. Let $U \subseteq Y$ be an open subset with $\overline{S} \cap U \neq \emptyset$. Then

$$IC_{\overline{S}}(\mathcal{L})|_U \simeq IC_{\overline{S} \cap U}(\mathcal{L}|_{S \cap U}).$$

(b) (*IC normalization principle*) If $\nu : \widetilde{X} \to X$ is the normalization morphism (e.g., see Section 2.4), then

$$R\nu_* IC_{\widetilde{X}} \simeq \nu_* IC_{\widetilde{X}} \simeq IC_X.$$

(c) (*DT localization principle*) A summand $IC_{\overline{S}}(\mathcal{L})$ appears in the decomposition theorem on Y if and only if there is an open $U \subseteq Y$ meeting S such that the restriction $IC_{\overline{S}}(\mathcal{L})|_U$ appears in the decomposition theorem on U.

Principle (*a*) is a direct consequence of the definition of an IC-complex, as the (co)support conditions are preserved by restriction to an open subset. Principle (*b*) follows again from the conditions of (co)support for IC-complexes by restricting the morphism to an open dense set where it is an isomorphism; indeed, conditions of support are preserved under a finite birational map. (See also the discussion in Section 2.4, as well as Exercise 8.3.9). Principle (*c*) follows from the validity of the decomposition theorem on Y and on every open subset $U \subseteq Y$, and from the fact that the summands of the decomposition theorem over U are uniquely determined (Remark 9.3.26).

Back to the proof of the theorem, by the two localization principles above, one can replace Y with any of its Zariski-dense open subsets. We may thus assume that there are no contributions from the proper subvarieties of Y to the decomposition theorem. In other words, after this reduction, the decomposition theorem becomes

$$Rf_*IC_X \simeq \bigoplus_i IC_Y(\mathcal{L}_i)[-i],$$

with $\mathcal{L}_i := \mathcal{L}_{i,Y}$.

By constructibility, and by further shrinking Y if necessary, one may also assume that $IC_Y = \underline{\mathbb{Q}}_Y[\dim_{\mathbb{C}} Y]$ and $IC_Y(\mathcal{L}_i) = \mathcal{L}_i[\dim_{\mathbb{C}} Y]$. So then

$$Rf_*IC_X \simeq \bigoplus_i \mathcal{L}_i[\dim_{\mathbb{C}} Y - i],$$

and it remains to show that $\underline{\mathbb{Q}}_Y$ is a direct summand of $\mathcal{H}^{-\dim_{\mathbb{C}} X}(Rf_*IC_X)$.

Without loss of generality, one may further assume that X is normal. Indeed, if $v : \widetilde{X} \to X$ is the normalization morphism and $\widetilde{f} := f \circ v$, then the IC normalization principle yields that

$$R\widetilde{f}_*IC_{\widetilde{X}} \simeq Rf_*Rv_*IC_{\widetilde{X}} \simeq Rf_*IC_X.$$

Assuming X is normal, there is an isomorphism

$$\underline{\mathbb{Q}}_X \simeq \mathcal{H}^{-\dim_{\mathbb{C}} X}(IC_X).$$

Since there is always a map $\underline{\mathbb{Q}}_X \to IC_X[-\dim_{\mathbb{C}} X]$ (cf. Exercise 6.7.1), inducing the above-mentioned isomorphism for X normal, there exists a distinguished triangle

$$\underline{\mathbb{Q}}_X \longrightarrow IC_X[-\dim_{\mathbb{C}} X] \longrightarrow \tau_{\geq 1}IC_X[-\dim_{\mathbb{C}} X] \longrightarrow$$

After applying Rf_* to it, one gets an isomorphism

$$R^0f_*\underline{\mathbb{Q}}_X \simeq \mathcal{H}^{-\dim_{\mathbb{C}} X}(Rf_*IC_X).$$

Therefore, after all these reductions, it remains to show that $\underline{\mathbb{Q}}_Y$ is a direct summand of $R^0f_*\underline{\mathbb{Q}}_X$.

Stein factorization ([224], [97, III, Corollary 11.5]) decomposes $f : X \to Y$ as

$$X \xrightarrow{g} Z \xrightarrow{h} Y,$$

where h is finite and g has connected fibers. Because of connected fibers, the stalk of $R^0g_*\underline{\mathbb{Q}}_X$ at a point $z \in Z$ is computed by $(R^0g_*\underline{\mathbb{Q}}_X)_z \cong H^0(g^{-1}(z); \mathbb{Q}) \cong \mathbb{Q}$, hence $R^0g_*\underline{\mathbb{Q}}_X \simeq \underline{\mathbb{Q}}_Z$. We are thus reduced to show that $\underline{\mathbb{Q}}_Y$ is a direct summand of $R^0h_*\underline{\mathbb{Q}}_Z$. Since $h : Z \to Y$ a finite map, by shrinking the target if necessary, one may assume that h is a covering map of finite degree $\ell \geq 1$. Then the adjunction map

$$adj_* : \underline{\mathbb{Q}}_Y \to Rh_*h^*\underline{\mathbb{Q}}_Y \simeq Rh_*\underline{\mathbb{Q}}_Z \simeq R^0h_*\underline{\mathbb{Q}}_Z$$

realizes $\underline{\mathbb{Q}}_Y$ as a direct summand of $R^0h_*\underline{\mathbb{Q}}_Z$ since the composition $\frac{1}{\ell} \cdot adj_! \circ adj_*$ is the identity of $\underline{\mathbb{Q}}_Y$, with the adjunction (or trace) map

$$adj_! : R^0h_*\underline{\mathbb{Q}}_Z \simeq Rh_*\underline{\mathbb{Q}}_Z \simeq Rh_!h^!\underline{\mathbb{Q}}_Y \to \underline{\mathbb{Q}}_Y$$

coming from $Rh_* = Rh_!$ and $h^* = h^!$ for the finite (unramified and oriented) covering h.   □

As a special case of Theorem 9.3.37 if X is irreducible, or of Theorem 9.3.25 if X is pure-dimensional, if $f : X \to Y$ is a resolution of singularities, then

$$Rf_*(\underline{\mathbb{Q}}_X[n]) \simeq IC_Y \oplus (\text{contribution from singularities of } Y).$$

Therefore,

Corollary 9.3.38 *The intersection cohomology of a pure-dimensional complex algebraic variety is a direct summand of the cohomology of a resolution of singularities.*

A nice application of Theorem 9.3.37 deals with the question of functoriality of intersection cohomology for algebraic maps. Specifically, given a proper surjective map $f : X \to Y$ of pure-dimensional complex algebraic varieties, it is natural to ask if there exists an induced homomorphism $IH^*(Y; \mathbb{Q}) \xrightarrow{?} IH^*(X; \mathbb{Q})$ in intersection cohomology that is compatible with the map f^* induced by f in cohomology, i.e., making the following diagram commutative:

$$
\begin{array}{ccc}
IH^*(Y; \mathbb{Q}) & \xrightarrow{\ ?\ } & IH^*(X; \mathbb{Q}) \\
\alpha_Y \uparrow & & \uparrow \alpha_X \\
H^*(Y; \mathbb{Q}) & \xrightarrow{\ f^*\ } & H^*(X; \mathbb{Q})
\end{array}
$$

(Here α_X, α_Y are the natural maps of Exercise 6.7.1.) A positive answer to this question was given in [7], by reduction to finite characteristic. In [238], Weber gave a simplified proof of this fact, as a direct consequence of Theorem 9.3.37. It suffices, of course, to prove the following sheaf-theoretic statement, which we formulate here as an exercise for the interested reader:

Exercise 9.3.39 Let $f : X \to Y$ be a proper surjective map of pure-dimensional complex algebraic varieties. Show that there exists a morphism

$\lambda_f : IC_Y[-\dim_\mathbb{C} Y] \longrightarrow Rf_*IC_X[-\dim_\mathbb{C} X]$ such that the following diagram commutes

$$
\begin{array}{ccc}
IC_Y[-\dim_\mathbb{C} Y] & \xrightarrow{\lambda_f} & Rf_*IC_X[-\dim_\mathbb{C} X] \\
\alpha_Y \Big\uparrow & & \Big\uparrow \alpha_X \\
\mathbb{Q}_Y & \xrightarrow{adj} & Rf_*\mathbb{Q}_X
\end{array}
$$

where α_X, α_Y are the (shifted) natural maps discussed in Exercise 6.7.1, and adj : $\mathbb{Q}_Y \to Rf_*f^*\mathbb{Q}_Y = Rf_*\mathbb{Q}_X$ is the adjunction morphism.

As an application of the BBDG decomposition theorem and the E_2-degeneration of the perverse Leray spectral sequence, one can also prove the singular version of Theorem 9.3.6 as well as a local version (see [12, Corollary 6.2.8, Corollary 6.2.9]):

Theorem 9.3.40 (Global and Local Invariant Cycle Theorems) *Let $f : X \to Y$ be a proper map of complex algebraic varieties, with X pure-dimensional. Let $U \subseteq Y$ be a Zariski-open subset on which the sheaf $R^i f_*IC_X$ is locally constant (i.e., a local system). Then the following assertions hold:*

(a) (Global) The natural restriction map

$$IH^i(X; \mathbb{Q}) \longrightarrow H^0(U; R^i f_*IC_X)$$

is surjective.

(b) (Local) Let $u \in U$ and $B_u \subset U$ be the intersection with a sufficiently small Euclidean ball (chosen with respect to a local embedding of (Y, u) into a manifold) centered at u. Then the natural restriction/retraction map

$$H^i(f^{-1}(u); IC_X) \cong H^i(f^{-1}(B_u); IC_X) \longrightarrow H^0(B_u; R^i f_*IC_X)$$

is surjective.

Set of Supports of an Algebraic Map

We conclude this section with a brief discussion on the *set of supports* of a proper algebraic map. In view of the decomposition (9.13), one can make the following:

Definition 9.3.41 Let $f : X \to Y$ be a proper algebraic map of complex algebraic varieties with X pure-dimensional. The *set of supports of f* is the collection of subvarieties $\overline{S}$ appearing in the decomposition (9.13) with some associated non-zero local system $\mathcal{L}_{i,S}$.

Example 9.3.42 It follows from Theorem 9.3.37 that if X is irreducible then $f(X)$ is always a support. Moreover, the BBDG decomposition theorem (Theorem 9.3.25) (see also Remark 9.3.26) shows that a support is a closure of a stratum of f.

Example 9.3.43 If $f : X \to Y$ is a semi-small map (Definition 9.3.7), it follows from Theorem 9.3.20 that the set of supports of f consists of the closures of the relevant strata of f. If, moreover, f is a small map, then Proposition 9.3.13 shows that the only support of f is Y.

Exercise 9.3.44 Let $f : X \to Y$ be a proper surjective map from a nonsingular surface X to a curve Y. Show that if all fibers of f are irreducible, then the only support of f is Y.

In general, it is very difficult to determine the set of supports of a map. Given the introductory level of these notes, we will not give any details here (besides the above discussion), but provide instead a list of available references so that the interested reader can delve further into this beautiful story.

The first breakthrough on this problem is the *Ngô support theorem* [187], which gives a sharp condition for the absence of supports in the case of abelian fibrations. The theorem was later extended to the relative Hilbert scheme map associated to families of irreducible curves, in [174] and [157]. The paper [53] gives a complete characterization of the supports of toric maps between toric varieties in terms of the combinatorics of the fans associated to the domain and target varieties. A more general approach to support-type theorems has been recently proposed by L. Migliorini and V. Schende in [175], by making use of "higher discriminants" of a map.

9.4 Applications of the Decomposition Package

In this section, we sample several of the numerous applications of the "decomposition package" (i.e., the (relative) Hard Lefschetz theorem, the BBDG decomposition theorem, and their various Hodge-theoretic aspects). For other important applications (including to representation theory), the reader may also consult, e.g., [47, 51, 222], and the references therein.

We begin with a discussion on the computation of topological invariants of Hilbert schemes of points on a nonsingular surface, then move to combinatorial applications and overview Stanley's proof of McMullen's conjecture (about a complete characterization of face vectors of simplicial polytopes) as well as Huh–Wang's recent resolution of the Dowling–Wilson top-heavy conjecture (on the enumeration of subspaces of a projective space generated by a finite set of points).

Topology of Hilbert Schemes of Points on a Smooth Complex Surface

Let X be a complex quasi-projective variety of pure complex dimension d. The n-th *symmetric product* of X is defined as

$$S^n X := X^{\times n} / S_n,$$

i.e., the quotient of the product of n copies of X by the natural action of the symmetric group on n elements, S_n. The n-th symmetric product of X parameterizes effective 0-cycles of degree n on X, i.e., formal linear combinations $\sum_{i=1}^{\ell} v_i[x_i]$ of points x_i in X with non-negative integer coefficients v_i satisfying $\sum_{i=1}^{\ell} v_i = n$. In short,

$$S^n X = \left\{ \sum_{i=1}^{\ell} v_i[x_i] \mid x_i \in X, v_i \in \mathbb{Z}_{\geq 0}, \sum_{i=1}^{\ell} v_i = n \right\}.$$

$S^n X$ has a natural stratification with strata defined in terms of the partitions of n. We denote by $P(n)$ the set of partitions of the natural number n. Then, to a partition $v = (v_1, \cdots, v_\ell)$ of n one associates a sequence $\underline{k} := (k_1, \cdots, k_n)$, with k_i denoting the number of times i appears in v. The length of such a partition is defined by $\ell(v) := \ell = \sum_{i=1}^{n} k_i$, and we have that $n = \sum_{i=1}^{n} i k_i$. In the above notations, the symmetric product $S^n X$ admits a stratification with strata $S_v^n X$ in one-to-one correspondence with such partitions $v = (v_1, \cdots, v_\ell) \in P(n)$ of n, defined by

$$S_v^n X := \left\{ \sum_{i=1}^{\ell} v_i[x_i] \mid x_i \neq x_j, \text{if } i \neq j \right\},$$

or, in terms of the sequence $\underline{k}$ associated with the given partition $v \in P(n)$,

$$S_v^n X \cong \left((\prod_{i=1}^{n} X^{k_i}) \setminus \Delta \right) / \prod_{i=1}^{n} S_{k_i},$$

with Δ denoting the large diagonal in $X^{\sum k_i}$.

The *Hilbert scheme* Hilb_X^n of a nonsingular complex quasi-projective variety X of pure complex dimension d describes collections of n (not necessarily distinct) points on X. It is the moduli space of zero-dimensional subschemes of X of length n. (Here, Hilb_X^n already denotes the reduced scheme structure, which suffices for our applications.) It comes equipped with a natural proper morphism

$$\pi_n : \text{Hilb}_X^n \longrightarrow S^n X, \quad Z \mapsto \sum_{x \in Z} \text{length}(Z_x) \cdot [x]$$

to the n-th symmetric product of X, the *Hilbert–Chow morphism*, taking a zero-dimensional subscheme to its associated zero-cycle. This morphism is birational for X of dimension at most two, but otherwise for large n the Hilbert scheme is in general reducible and has components of dimension much larger than that of the symmetric product. The subvariety

$$\mathrm{Hilb}^n_{X,x} := \pi_n^{-1}(n[x])_{red}$$

of subschemes supported at $x \in X$ is called the *punctual Hilbert scheme of length n at x*. (Here, *red* denotes the underlying reduced scheme structure.)

Symmetric products and Hilbert schemes of points are intimately connected via the Hilbert–Chow morphism. For example, if $d = 1$, then $\mathrm{Hilb}^n_X \cong S^n X$ is nonsingular. If $d = 2$, then Hilb^n_X is smooth and π_n is a resolution of singularities; we will get back to this particular case below. If $d \geq 3$, Hilb^n_X is singular for $n \geq 4$ and much less is known about its topology.

Let us now focus on the case $d = 2$, i.e., X is a nonsingular complex surface, see [185] for a nice reference. The Hilbert scheme Hilb^n_X is in this case nonsingular (and irreducible if X is irreducible) of complex dimension $2n$. Recall also that the n-th symmetric product $S^n X$ is stratified by strata $S^n_\nu X$ associated to partitions $\nu \in P(n)$. Each stratum $S^n_\nu X$ is nonsingular of dimension $2\ell(\nu)$. It can be shown that for each $\nu = (\nu_1, \cdots, \nu_\ell) \in P(n)$, the restriction $\pi_n^{-1}(S^n_\nu X) \longrightarrow S^n_\nu X$ is a locally trivial topological fibration with fiber isomorphic to the product $\prod_{i=1}^\ell \mathrm{Hilb}^{\nu_i}_{\mathbb{C}^2,0}$. In particular, the strata $S^n_\nu X$ provide a stratification for the Hilbert–Chow morphism π_n, and the fiber $\pi_n^{-1}(x_\nu)$ of π_n over a point $x_\nu \in S^n_\nu X$ is irreducible of dimension $n - \ell(\nu)$. It follows that the Hilbert–Chow morphism π_n is a *semi-small* map, and all strata are relevant. Note also that since the fibers of π_n are all irreducible, the local systems appearing in the decomposition of Theorem 9.3.20 are all constant of rank one. In particular, the decomposition theorem for π_n takes in this case the form

$$R\pi_{n*}(\underline{\mathbb{Q}}_{\mathrm{Hilb}^n_X}[2n]) \simeq \bigoplus_{\nu \in P(n)} IC_{\overline{S^n_\nu X}}. \tag{9.15}$$

Furthermore, the closures $\overline{S^n_\nu X}$ of strata of π_n, and their desingularizations can be explicitly determined as follows. First, we have a disjoint set decomposition

$$\overline{S^n_\nu X} = \bigsqcup_{\mu \leq \nu} S^n_\mu X,$$

where for $\mu, \nu \in P(n)$ we write $\mu \leq \nu$ if there exists a decomposition $I_1, \cdots, I_{\ell(\mu)}$ of the set $\{1, \cdots, \ell(\nu)\}$ so that $\mu_1 = \sum_{i \in I_1} \nu_i, \cdots, \mu_{\ell(\mu)} = \sum_{i \in I_{\ell(\mu)}} \nu_i$. This decomposition reflects the fact that a cycle $\sum_i \nu_i[x_i] \in S^n_\nu X$ may degenerate to a cycle in which some of the x_i's come together. Secondly, if $\underline{k} := (k_1, \cdots, k_n)$ is the sequence associated to the given partition ν (i.e., k_i is the number of times i appears in ν), let us set

$$S^\nu X := \prod_{i=1}^{n} S^{k_i} X.$$

The variety $S^\nu X$ has complex dimension $2\ell(\nu)$, and there is a natural finite map

$$\pi_n^\nu : S^\nu X \longrightarrow \overline{S_\nu^n X}$$

that is an isomorphism when restricted to $(\pi_n^\nu)^{-1}(S_\nu^n X)$. Since $S^\nu X$ has only finite quotient singularities, it is a normal variety, so π_n^ν is the normalization map for $\overline{S_\nu^n X}$. By the normalization principle for IC-complexes, we then have an isomorphism:

$$IC_{\overline{S_\nu^n X}} \simeq \pi_{n*}^\nu IC_{S^\nu X} \simeq \pi_{n*}^\nu \underline{\mathbb{Q}}_{S^\nu X}[2\ell(\nu)], \tag{9.16}$$

where the last identification follows since $S^\nu X$ is a rational homology manifold (see Theorem 6.6.3). Altogether, (9.15) and (9.16) yield the decomposition:

$$R\pi_{n*}(\underline{\mathbb{Q}}_{\mathrm{Hilb}_X^n}[2n]) \simeq \bigoplus_{\nu \in P(n)} \pi_{n*}^\nu \underline{\mathbb{Q}}_{S^\nu X}[2\ell(\nu)]. \tag{9.17}$$

This is the form obtained by Göttsche and Soergel [88]. Taking hypercohomology in (9.17), one obtains the following computation for the cohomology of Hilb_X^n:

Theorem 9.4.1 *Let X be a nonsingular complex surface. In the above notations, for every $i \geq 0$ one has:*

$$H^i(\mathrm{Hilb}_X^n; \mathbb{Q}) \cong \bigoplus_{\nu \in P(n)} H^{i+2\ell(\nu)-2n}(S^\nu X; \mathbb{Q}). \tag{9.18}$$

The Betti numbers of symmetric products of a variety were computed by Macdonald in [147]. Together with (9.18), this yields a computation of the Betti numbers of Hilbert schemes of points on a nonsingular complex surface. The resulting formulae are more convenient to state in the form of generating series. For simplicity, we state here only the corresponding Euler characteristic identity (see [87]):

$$\sum_{n \geq 0} \chi(\mathrm{Hilb}_X^n) \cdot t^n = \left(\prod_{k=1}^{\infty} \frac{1}{1-t^k} \right)^{\chi(X)}. \tag{9.19}$$

The original proof of this result was based on the Weil conjectures and counting subschemes over finite fields.

For higher dimensional generalizations of such results and for more applications, see, e.g., [37] and the references therein.

Stanley's Proof of McMullen's Conjecture

There are already several instances when properties of intersection cohomology groups of complex projective varieties have been successfully translated into combinatorics for solving long-standing conjectures. One of the most relevant examples of the interplay between combinatorics and geometry is Stanley's proof [221, 222] of (the necessity part of) McMullen's conjecture [171], giving an if and only if condition for the existence of a simplicial polytope with a given number f_i of i-dimensional faces. The sufficiency part of McMullen's conjecture was verified by Billera–Lee [14].

Stanley's idea of proof is beautiful in its simplicity, and it can be roughly summarized as follows: to a simplicial polytope P one associates a projective (toric) variety X_P so that McMullen's combinatorial conditions for P are translated into properties of the Betti numbers of X_P. The latter properties would then follow if we knew that Poincaré duality and the Hard Lefschetz theorem hold for the rational cohomology of X_P. The trick is to notice that, while X_P is in general singular, its singularities are rather mild (finite quotient singularities), making X_P into a rational homology manifold. Hence $H^*(X_P; \mathbb{Q}) \cong IH^*(X_P; \mathbb{Q})$, and the assertions follow now from the Poincaré duality and the Hard Lefschetz theorem for intersection cohomology.

In this section, we give an overview of Stanley's approach to McMullen's conjecture (see also [20], [51, Section 4.1] and [77, Section 5.6] for nice accounts of this story), while referring the reader to [45] and [77] for comprehensive references on polytopes and toric varieties.

A convex polytope is the convex hull of a finite set in the real Euclidean space. For a d-dimensional convex polytope P, denote by $f_i = f_i(P)$ the number of its i-dimensional faces, $0 \le i \le d - 1$. These numbers are collected into a string

$$f(P) := (f_0, \cdots, f_{d-1}),$$

called the *face vector* of P. One of the most natural problems in the study of f-vectors of polytopes is to obtain a complete characterization, i.e., a set of conditions by which one can recognize if a given string of natural numbers is the f-vector of a convex polytope or not. There are obvious obstructions for such a realization problem. For example, a simple-minded topological argument shows that the face vector of a d-dimensional convex polytope satisfies the *generalized Euler formula*:

$$f_0 - f_1 + f_2 - \cdots + (-1)^{d-1} f_{d-1} = 1 + (-1)^{d-1}.$$

In 1971, McMullen [171] gave a conjectural description of the f-vectors of *simplicial* polytopes. A convex polytope is said to be *simplicial* if all its faces are simplices. The class of simplicial polytopes is special in some sense, but nevertheless very important in polytope theory. For instance, if one seeks to maximize the number of i-faces of a d-dimensional polytope with n vertices, the maximum is obtained simultaneously for all i by certain simplicial polytopes. Before stating McMullen's conjecture, we need to introduce an important auxiliary concept. The *h-vector*

$$h(P) := (h_0, \cdots, h_d)$$

of a d-dimensional simplicial polytope P with face vector $f(P)$ is defined by the coefficients of the *h-polynomial*

$$h(P,t) = \sum_{i=0}^{d} h_i t^i := (t-1)^d + f_0(t-1)^{d-1} + \cdots + f_{d-1},$$

that is,

$$h_i = \sum_{j=0}^{i} \binom{d-j}{d-i} (-1)^{i-j} f_{j-1},$$

where $f_{-1} := 1$. (Occasionally, we write $h_i(P)$ for h_i to indicate the dependence of P if such P is given.) With the above notations, the McMullen conjecture can be (re)formulated as follows (see [221]):

Conjecture 9.4.2 (McMullen) *A vector $f = (f_0, \cdots, f_{d-1}) \in \mathbb{N}^d$ is a face vector $f(P)$ for some d-dimensional simplicial polytope P if and only if the following conditions are satisfied:*

(1) (Dehn–Sommerville) $h_i = h_{d-i}$ for all $0 \leq i \leq d$;

(2) there is a graded commutative $\mathbb{Q}$-algebra $R = \bigoplus_{i \geq 0} R_i$, with $R_0 = \mathbb{Q}$, generated by R_1, and with $\dim_{\mathbb{Q}} R_i = h_i - h_{i-1}$ for $1 \leq i \leq \lfloor d/2 \rfloor$. (In particular, $h_{i-1} \leq h_i$ for $1 \leq i \leq \lfloor d/2 \rfloor$.)

As already mentioned, Stanley's proof of the necessity part of McMullen's conjecture makes use of the theory of toric varieties. A d-dimensional *toric variety* X is an irreducible normal complex algebraic variety on which the complex d-torus $\mathbb{T} = (\mathbb{C}^*)^d$ acts with an open orbit. More precisely, one has an embedding $\mathbb{T} \hookrightarrow X$, and the action of $\mathbb{T}$ on itself extends to an algebraic action of $\mathbb{T}$ on all of X. Toric varieties arise from combinatorial objects called *fans*, which are finite collections of cones in a lattice. We denote by X_Σ the d-dimensional toric variety associated to a fan Σ in a d-dimensional lattice of $\mathbb{R}^d$. The toric variety X_Σ is said to be *simplicial* if the fan Σ is simplicial, i.e., each cone of Σ is spanned by linearly independent elements of the lattice. Simplicial toric varieties are rational homology manifolds.

Let us now assume that a d-dimensional simplicial convex polytope P is given. Since the polytope is simplicial, moving all of its vertices slightly yields a polytope with the same face vector. By such a perturbation, we can assume that P is *rational*, i.e., its vertices are in the rational points of the given lattice. Furthermore, we may translate P so that the origin is in its interior. We next define the fan $\Sigma(P)$ consisting of the cones (with vertex at the origin) over the proper faces of P, and let $X_P :=$ $X_{\Sigma(P)}$ be the associated toric variety. It turns out that X_P is projective and simplicial. In particular, X_P is a rational homology manifold. Moreover, the following holds (e.g., see [77, Section 5.2]): for every $0 \le i \le d$,

$$H^{2i+1}(X_P; \mathbb{Q}) = 0 \quad \text{and} \quad \dim_{\mathbb{Q}} H^{2i}(X_P; \mathbb{Q}) = h_i(P).$$

Note also that, since X_P is a rational homology manifold, we have an isomorphism $H^*(X_P; \mathbb{Q}) \cong IH^*(X_P; \mathbb{Q})$. Therefore, Poincaré duality and the Hard Lefschetz theorem also hold for the usual rational cohomology $H^*(X_P; \mathbb{Q})$ of X_P. As a consequence, the Dehn–Sommerville relation (1) in McMullen's conjecture follows from Poincaré duality, while the unimodality of the h_i's in (2) follows from the Hard Lefschetz theorem. By the general theory of toric varieties, the cohomology ring $H^*(X_P; \mathbb{Q})$ is generated by elements of degree 2 (classes of divisors). Set

$$R^i := H^{2i}(X_P; \mathbb{Q})/L \cdot H^{2i-2}(X_P; \mathbb{Q}),$$

where $L \in H^2(X_P; \mathbb{Q})$ is the class appearing in the statement of the Hard Lefschetz theorem. Then the Hard Lefschetz theorem shows that

$$R^* = H^*(X_P; \mathbb{Q})/(L)$$

satisfies the requirements of item (2) of McMullen's conjecture.

Remark 9.4.3 When the polytope P is non-simplicial, the h-polynomial $h(P,t)$ may have negative coefficients and the Dehn–Sommerville relations do not hold. Moreover, assuming P is rational, the cohomology of the associated toric variety X_P can exist in odd degrees, and the corresponding Betti numbers are not invariants of the combinatorics of faces. The correct way to generalize the above discussion to the rational non-simplicial context is to replace the rational cohomology of X_P with the intersection cohomology groups $IH^*(X_P; \mathbb{Q})$. The "generalized" h-polynomial is then defined as $h(P,t) = \sum_{i=0}^{d} h_i(P)t^i$, with

$$h_i(P) := \dim_{\mathbb{Q}} IH^{2i}(X_P; \mathbb{Q}),$$

and its coefficients satisfy the Dehn–Sommerville relations and unimodality by Poincaré duality and, respectively, the Hard Lefschetz theorem for intersection cohomology groups. Furthermore, the generalized h-polynomial turns out to be a combinatorial invariant, i.e., it can be defined only in terms of the partially ordered set of faces of P (see, e.g., [220, 72, 20]).

Remark 9.4.4 Addressing the above problems for more general (i.e., non-rational) convex polytopes is far more challenging, since the toric description is missing in this context. However, there has been success in abstracting toric computations polyhedrally, without constructing any sort of toric space at all, and such questions have been settled purely combinatorially by defining groups that can substitute for intersection cohomology. Moreover, the decomposition theorem and the Hard Lefschetz theorem now have proofs in this combinatorial setting. For more details, the interested reader is advised to consult, e.g., [8, 24, 116].

Huh–Wang's Proof of Dowling–Wilson's Conjecture

One of the earliest results in enumerative combinatorial geometry is the following theorem of de Bruijn and Erdös [46]:

Theorem 9.4.5 (de Bruijn–Erdös) *Every set of points E in a projective plane determines at least $|E|$ lines, unless all the points are contained in a line. (Here, $|E|$ denotes the number of elements of the set E.)*

Motzkin [184] and others extended the above result to higher dimensions, showing that every set of points E in a projective space determines at least $|E|$ hyperplanes, unless E is contained in a hyperplane.

Let $E = \{v_1, \cdots, v_n\}$ be a spanning subset of a d-dimensional vector space V, and let $w_i(E)$ be the number of i-dimensional subspaces spanned by subsets of E.

Example 9.4.6 If E is the set of 4 general vectors in $\mathbb{R}^3$ (that is, 4 general points in the projective plane), then one can readily check that: $w_0(E) = 1$, $w_1(E) = 4$, $w_2(E) = 6$, $w_3(E) = 1$.

In 1974, Dowling and Wilson [65, 66] proposed the following conjecture:

Conjecture 9.4.7 (Dowling–Wilson Top-Heavy Conjecture) *For all $i < d/2$ one has:*

$$w_i(E) \leq w_{d-i}(E). \tag{9.20}$$

Remark 9.4.8 The de Bruijn–Erdös theorem (Theorem 9.4.5) is a special case of Conjecture 9.4.7, as it can be reformulated by saying that in the case $d = 3$ one has $w_1(E) \leq w_2(E)$. More generally, Motzkin's result shows that $w_1(E) \leq w_{d-1}(E)$.

Another conjecture concerning the numbers $w_i(E)$ was proposed by Rota [202, 203] in 1971. In the above notations, it can be formulated as follows:

Conjecture 9.4.9 (Rota's Unimodal Conjecture) *There is some j so that*

$$w_0(E) \leq \cdots \leq w_{j-1}(E) \leq w_j(E) \geq w_{j+1}(E) \geq \cdots \geq w_d(E). \tag{9.21}$$

In this section, we give a brief account of the recent proof by Huh–Wang of the Dowling–Wilson top-heavy conjecture, and of the unimodality of the "lower half" of the sequence $\{w_i(E)\}$, as also conjectured by Dowling–Wilson in [66]; see [109] for complete details.[6] In the above notations, the main result of Huh–Wang can be stated as follows:

Theorem 9.4.10 (Huh–Wang) *For all* $i < d/2$, *the following properties hold:*

(a) (top heavy) $w_i(E) \leq w_{d-i}(E)$.
(b) (unimodality) $w_i(E) \leq w_{i+1}(E)$.

Proof For simplicity, we assume that the vector space V is defined over the field $\mathbb{C}$ of complex numbers. Then the proof of Theorem 9.4.10 rests on the following two key steps:

(1) There exists a complex d-dimensional projective variety Y such that for every $0 \leq i \leq d$ one has:

$$H^{2i+1}(Y; \mathbb{Q}) = 0 \text{ and } \dim_{\mathbb{Q}} H^{2i}(Y; \mathbb{Q}) = w_i(E).$$

(2) There exists a resolution of singularities $\pi : X \to Y$ of Y such that the induced cohomology map

$$\pi^* : H^*(Y; \mathbb{Q}) \longrightarrow H^*(X; \mathbb{Q})$$

is injective in each degree.

To define the variety Y of Step (1), one first uses $E = \{v_1, \cdots, v_n\}$ to construct a map $i_E : V^{\vee} \to \mathbb{C}^n$ by regarding each $v_i \in E$ as a linear map on the dual vector space $V^{\vee}$. Precomposing i_E with the open inclusion $\mathbb{C}^n \hookrightarrow (\mathbb{C}P^1)^n$, one then gets a map

$$f : V^{\vee} \to (\mathbb{C}P^1)^n.$$

Finally, set

$$Y := \overline{\text{Image } (f)} \subset (\mathbb{C}P^1)^n.$$

By work of Ardila–Boocher [3], the variety Y has an algebraic cell decomposition (i.e., it is paved by complex affine spaces), and the number of $\mathbb{C}^i$'s appearing in the decomposition of Y is exactly $w_i(E)$. Having defined Y, the resolution X is a sequence of blowups (a *wonderful model*) associated to a certain canonical stratification of Y. The cohomology rings of both Y and X are well-understood combinatorially and Step (2) can be checked directly.

The proof of Theorem 9.4.10 follows the pattern of Stanley's proof of McMullen's conjecture. However, the space Y is in this case highly singular, so

---

[6]The original formulations of the Dowling–Wilson and Rota conjectures concern *matroids*. The proof by Huh–Wang is applicable only for matroids realizable over some field.

its rational cohomology does not satisfy the Kähler package. Instead, one needs to use the corresponding intersection cohomology results. More precisely, assuming (1) and (2), one proceeds as follows. First, by Exercise 9.3.39, the map π^* factorizes through intersection cohomology, i.e.,

$$\pi^* : H^*(Y; \mathbb{Q}) \overset{\alpha}{\to} IH^*(Y; \mathbb{Q}) \overset{\beta}{\hookrightarrow} H^*(X; \mathbb{Q}),$$

where the fact that β is injective follows from Corollary 9.3.38. Since π^* is injective by Step (2), it follows that $\alpha : H^*(Y; \mathbb{Q}) \to IH^*(Y; \mathbb{Q})$ is injective as well. For $i < d/2$, consider the following commutative diagram:

$$
\begin{array}{ccc}
H^{2i}(Y; \mathbb{Q}) & \overset{\alpha}{\hookrightarrow} & IH^{2i}(Y; \mathbb{Q}) \\
{\scriptstyle \cup L^{d-2i}} \downarrow & & \cong \downarrow {\scriptstyle \cup L^{d-2i}} \\
H^{2d-2i}(Y; \mathbb{Q}) & \overset{\alpha}{\hookrightarrow} & IH^{2d-2i}(Y; \mathbb{Q})
\end{array}
$$

where the right-hand vertical arrow is the Hard Lefschetz isomorphism for the intersection cohomology groups of Y (see Corollary 9.2.3). Since the maps labelled by α are injective, it follows that

$$H^{2i}(Y; \mathbb{Q}) \overset{\cup L^{d-2i}}{\longrightarrow} H^{2d-2i}(Y; \mathbb{Q})$$

is injective as well. In particular,

$$w_i(E) = \dim_{\mathbb{Q}} H^{2i}(Y; \mathbb{Q}) \leq \dim_{\mathbb{Q}} H^{2d-2i}(Y; \mathbb{Q}) = w_{d-i}(E)$$

for every $i < d/2$, thus proving part (a) of the theorem.

Part (b) follows similarly, by using the unimodality of the intersection cohomology Betti numbers (as discussed in Section 9.2). $\qquad\square$

Chapter 10
Hypersurface Singularities. Nearby and Vanishing Cycles

Some of the early applications of the theory of perverse sheaves appear in Singularity Theory, for the study of complex hypersurface singularities. In Section 10.1 we give a brief overview of the local topological structure of hypersurface singularities, as originally described in Milnor's seminal book [179]. In Section 10.2 we investigate the global topology of complex hypersurface complements by means of invariants inspired by knot theory. The nearby and vanishing cycle functors are introduced in Section 10.3, where we also discuss their relation with perverse sheaves. Nearby and vanishing cycles are used to glue the local topological information around hypersurface singularities. Concrete applications of nearby and vanishing cycles are presented in Section 10.4 (to the computation of Euler characteristics of complex projective hypersurfaces), Section 10.5 (for obtaining generalized Riemann–Hurwitz-type formulae), and in Section 10.6 (for deriving homological connectivity statements for the local topology of complex singularities). The chapter concludes with the introduction in Section 10.7 of several concepts of fundamental importance for the construction of Saito's mixed Hodge modules of Chapter 11. The interested reader may also consult [61, 154, Chapter 6] and the references therein for other interesting applications of perverse sheaves and of nearby and vanishing cycles in Singularity Theory.

10.1 Brief Overview of Complex Hypersurface Singularities

Let $f : \mathbb{C}^{n+1} \to \mathbb{C}$ be a regular (or analytic) map with $0 \in \mathbb{C}$ a critical value. Let $X_0 = f^{-1}(0)$ be the special (singular) fiber of f, and $X_s = f^{-1}(s), s \neq 0$ (s small enough) the generic (smooth) fiber. Pick $x \in X_0$, and choose a small enough ϵ-ball $B_{\epsilon,x}^{2n+2}$ in $\mathbb{C}^{n+1}$ centered at x, with boundary the $(2n+1)$-sphere $S_{\epsilon,x}^{2n+1}$. The topology of the hypersurface singularity germ (X_0, x) is described by the following

© Springer Nature Switzerland AG 2019
L. G. Maxim, *Intersection Homology & Perverse Sheaves*, Graduate Texts in Mathematics 281, https://doi.org/10.1007/978-3-030-27644-7_10

fundamental result in Singularity Theory (see [179], and also [60, Chapter 3] and the references therein).

Theorem 10.1.1 *In the above notations, the following hold:*

1) $B_{\epsilon,x}^{2n+2} \cap X_0$ *is contractible, and it is homeomorphic to the cone on* $K_x :=$ $S_{\epsilon,x}^{2n+1} \cap X_0$, *the (real) link of x in* X_0.
2) *The link* K_x *is* $(n-2)$-*connected.*
3) $\frac{f}{|f|} : S_{\epsilon,x}^{2n+1} - K_x \longrightarrow S^1$ *is a topologically locally trivial fibration, called the Milnor fibration of the hypersurface singularity germ* (X_0, x).
4) *Let* F_x *be the fiber of the Milnor fibration, that is, the Milnor fiber of f at x. If the complex dimension of the germ of the critical set of* X_0 *at x is r, then* F_x *is* $(n-r-1)$-*connected. In particular, if x is an isolated singularity, then* F_x *is* $(n-1)$-*connected. (Here we use the convention that* $\dim_{\mathbb{C}} \emptyset = -1$.)
5) *The Milnor fiber* F_x *has the homotopy type of a finite CW complex of real dimension n.*

Remark 10.1.2 Regarding the connectivity of the Milnor fiber in item (4), the case $r = 0$ (i.e., that of an isolated hypersurface singularity) was treated by Milnor [179, Lemma 6.4], while the general case is due to Kato–Matsumoto [123].

The following statements describe the homotopy type of the Milnor fiber in very simple situations:

Proposition 10.1.3 ([179, Lemma 2.13]) *If* (X_0, x) *is a nonsingular hypersurface singularity germ, then* F_x *is contractible.*

Proposition 10.1.4 ([179, Theorem 6.5, Theorem 7.2]) *If* (X_0, x) *is an isolated hypersurface singularity germ, then the Milnor fiber* F_x *has the homotopy type of a bouquet of* μ_x *n-spheres,*

$$F_x \simeq \bigvee_{\mu_x} S^n,$$

where

$$\mu_x = \dim_{\mathbb{C}} \mathbb{C}\{x_1, \ldots, x_n\}/(\frac{\partial f}{\partial x_1}, \ldots, \frac{\partial f}{\partial x_n}) \qquad (10.1)$$

is the Milnor number of f at x. Here, $\mathbb{C}\{x_1, \ldots, x_n\}$ *is the* $\mathbb{C}$-*algebra of analytic function germs at* $0 \in \mathbb{C}^{n+1}$. *(The n-spheres in the above bouquet are called the "vanishing cycles" at x.)*

Remark 10.1.5 Different proofs of formula (10.1) for the middle Betti number of the Milnor fiber of an isolated hypersurface singularity can be found, e.g., in [26] and [134].

Example 10.1.6

(i) If $A_1 = \{x^2 + y^2 = 0\} \subset (\mathbb{C}^2, \underline{0})$, then the origin $\underline{0} = (0,0) \in A_1$ is the only singular point of A_1, and the corresponding Milnor number and Milnor fiber at $\underline{0}$ are: $\mu_{\underline{0}} = 1$ and $F_{\underline{0}} \simeq S^1$.

(ii) If $A_2 = \{x^3 + y^2 = 0\} \subset (\mathbb{C}^2, \underline{0})$, the origin $\underline{0} \in A_2$ is the only singular point of A_2, and the corresponding Milnor number and Milnor fiber at $\underline{0}$ are: $\mu_{\underline{0}} = \dim_{\mathbb{C}} \mathbb{C}\{x, y\}/(x^2, y) = 2$ and $F_{\underline{0}} \simeq S^1 \vee S^1$. The link $K_{\underline{0}}$ of the singular point $\underline{0} \in A_2$ is the famous *trefoil knot*, i.e., the $(2, 3)$-torus knot.

Remark 10.1.7 Due to their high connectivity, links of isolated hypersurface singularities are the main source of *exotic spheres*. This was in fact Milnor's motivation for studying complex hypersurface singularities. Indeed, by using the generalized Poincaré hypothesis of Smale–Stallings, it can be shown that if $n \neq 2$ the link K of an isolated hypersurface singularity is homeomorphic to the sphere S^{2n-1} if, and only if, K is a $\mathbb{Z}$-homology sphere (i.e., $H_*(K; \mathbb{Z}) \cong H_*(S^{2n-1}; \mathbb{Z})$), see [179, Lemma 8.1]. The integral homology of such a link K can be studied by using the *monodromy* and the Wang sequence associated to the Milnor fibration, see Exercise 10.1.9. For nice accounts on exotic spheres arising from Singularity Theory, see [179, 126, 104, Sections 8,9], as well as [60, Chapter 3].

Definition 10.1.8 The *monodromy* homeomorphism of f at x is the homeomorphism $h_x : F_x \to F_x$ induced on the fiber of the Milnor fibration at x by circling the base of the fibration once. (When the point x is clear from the context, we simply write h for h_x; occasionally, we may also use the notation $h_{f,x}$ (or h_f) to emphasize the map f.)

Exercise 10.1.9 Let $f = 0$ be an isolated hypersurface singularity at the origin of $\mathbb{C}^{n+1}$, $n \geq 2$, and let K, F, and h denote the corresponding link, Milnor fiber, and monodromy homeomorphism, respectively. Then K is a $(n-2)$-connected, closed, oriented, $(2n-1)$-dimensional manifold, and hence, by Poincaré duality, the only interesting integer (co)homology of K appears in degrees $n-1$ and n. Moreover, the Milnor fiber F has the homotopy type of a bouquet of n-spheres. Let $\Delta(t)$ denote the local Alexander polynomial at the origin, i.e.,

$$\Delta(t) = \det(t \cdot I - h_* : H_n(F; \mathbb{Z}) \to H_n(F; \mathbb{Z})).$$

Use the Wang long exact sequence associated to the Milnor fibration to prove the following statements:

(a) K is a $\mathbb{Q}$-homology sphere (i.e., it has the $\mathbb{Q}$-homology of S^{2n-1}) if, and only if, $\Delta(1) \neq 0$ (i.e., $t = 1$ is not an eigenvalue of the algebraic monodromy operator $h_* : H_n(F; \mathbb{Z}) \to H_n(F; \mathbb{Z})$.)

(b) K is a $\mathbb{Z}$-homology sphere if, and only if, $\Delta(1) = \pm 1$.

In particular, as already mentioned in Remark 10.1.7, it follows that if $n \geq 3$ and $\Delta(1) = \pm 1$ then K is homeomorphic to S^{2n-1}. Moreover, the embedding $K \subset$

S_ε^{2n+1} is not equivalent to the trivial equatorial embedding $S_\varepsilon^{2n-1} \subset S_\varepsilon^{2n+1}$ (i.e., K is an exotic $(2n-1)$-sphere) except for the smooth case $df(0) \neq 0$.

Milnor conjectured, and it was proved by Grothendieck [92] and Landman [132], that the monodromy homeomorphism induces a *quasi-unipotent* operator on the Milnor fiber (co)homology. The following result came to be known as the *monodromy theorem* and it has been subsequently reproved and sharpened by several quite different approaches, e.g., see [44, 74, 124, 137, 212]:

Theorem 10.1.10 (Monodromy Theorem) *All eigenvalues of the algebraic monodromy $h_x^* : H^i(F_x; \mathbb{C}) \to H^i(F_x; \mathbb{C})$ are roots of unity. In fact, there are positive integers p and q such that*

$$\big((h_x^*)^p - id\big)^q = 0.$$

Moreover, one can take $q = i + 1$.

For future reference, we record here the following definition (already used in Exercise 10.1.9), which is inspired by knot theory.

Definition 10.1.11 The characteristic polynomial

$$\Delta_x^i(t) := \det(t \cdot I - h_x^* : H^i(F_x; \mathbb{C}) \to H^i(F_x; \mathbb{C}))$$

of the algebraic monodromy is called the *i-th (local) cohomological Alexander polynomial* associated with the hypersurface singularity germ at the point x. Similarly, one defines local homological Alexander polynomials $\Delta_{i,x}(t)$ by using the operators induced by monodromy on the homology of the Milnor fiber.

Remark 10.1.12 Theorem 10.1.10 asserts that all zeros of the local Alexander polynomials $\Delta_x^i(t)$ and $\Delta_{i,x}(t)$ are roots of unity, or equivalently, these local Alexander polynomials are products of cyclotomic polynomials.

It is also important to note that the algebraic monodromy can already detect singularities. Specifically, A'Campo [1] proved the following:

Theorem 10.1.13 (A'Campo) *Let*

$$L(h_x) := \sum_i (-1)^i trace \left(h_x^* : H^i(F_x; \mathbb{C}) \to H^i(F_x; \mathbb{C}) \right)$$

be the Lefschetz number of the monodromy homeomorphism h_x. Then $L(h_x) = 0$ if x is a singular point for f (i.e., if $df(x) = 0$).

As a consequence of the above statement, one has the following:

Corollary 10.1.14 *If x is a singular point for f, then the associated Milnor fiber F_x cannot be homologically contractible, i.e., $H^*(F_x; \mathbb{C}) \neq H^*(point; \mathbb{C})$.*

The Milnor fibration construction can be *globalized* if the hypersurface is defined by a *weighted homogeneous polynomial*. Let $f \in \mathbb{C}[x_0, \cdots, x_n]$ be a weighted homogeneous polynomial of degree d with respect to the weights $wt(x_i) = w_i$, where w_i is a positive integer, for all $i = 0, \ldots, n$. This means that

$$f(t^{w_0} x_0, \ldots, t^{w_n} x_n) = t^d \cdot f(x_0, \ldots, x_n).$$

There is a natural $\mathbb{C}^*$-action on $\mathbb{C}^{n+1}$ associated to these weights, given by

$$t \cdot x = (t^{w_0} x_0, \ldots, t^{w_n} x_n)$$

for all $t \in \mathbb{C}^*$ and $x = (x_0, \ldots, x_n) \in \mathbb{C}^{n+1}$, which can be used to show that the restriction of the polynomial mapping f given by

$$f : \mathbb{C}^{n+1} - f^{-1}(0) \longrightarrow \mathbb{C}^*$$

is a locally trivial fibration. This fibration is usually referred to as the *global (affine) Milnor fibration*, and its fiber $F = f^{-1}(1)$ is called the *global (affine) Milnor fiber* of f. It is easy to see that F is homotopy equivalent to the Milnor fiber associated to the germ of f at the origin. The monodromy homeomorphism $h : F \to F$ is in this case particularly simple: it is given by multiplication by a primitive d-th root of unity, that is,

$$h(x) = \exp \frac{2\pi i}{d} \cdot x$$

(e.g., see [60, Example 3.1.19]). In particular, $h^d = id$, and hence the complex algebraic monodromy operator

$$h^* : H^*(F; \mathbb{C}) \longrightarrow H^*(F; \mathbb{C})$$

is diagonalizable (semi-simple) and has as eigenvalues only d-th roots of unity. Finally, if the above weighted homogeneous polynomial f has an isolated singularity at the origin, the corresponding Milnor number of f at $0 \in \mathbb{C}^{n+1}$ is computed by the formula (see, e.g., [59, Proposition 7.27]):

$$\mu_0 = \prod_{i=0}^{n} \frac{d - w_i}{w_i}. \tag{10.2}$$

Exercise 10.1.15 Let $f : \mathbb{C}^{n+1} \to \mathbb{C}$ be given by $f = x_0 x_1 \cdots x_n$. Show that the Milnor fiber of the singularity of f at the origin is homotopy equivalent to $(S^1)^n$. In particular, the homology groups $H_i(F; \mathbb{Z})$ are non-zero in all dimensions $0 \le i \le n$, which shows that the connectivity statement of Theorem 10.1.1(4) is sharp.

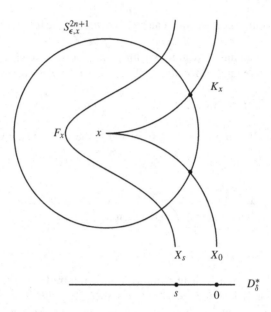

Fig. 10.1 Milnor fibration

We also mention here another version of the Milnor fibration, as developed by Lê [136], but see also [179, Theorem 5.11]. If $\mathring{D}_\delta^*$ is the open punctured disc (at the origin) of radius δ in $\mathbb{C}$, then the above Milnor fibration is fiber diffeomorphic equivalent to the smooth locally trivial fibration (Fig. 10.1)

$$\mathring{B}_{\epsilon,x}^{2n+2} \cap f^{-1}(\mathring{D}_\delta^*) \longrightarrow \mathring{D}_\delta^*, \ 0 < \delta \ll \epsilon \ll 1,$$

which is usually referred to as the *Milnor–Lê fibration* . In particular, the Milnor fiber $F_x \cong \mathring{B}_{\epsilon,x}^{2n+2} \cap f^{-1}(s)$ (for $0 < |s| \ll \delta \ll \epsilon$) can be viewed as a *local smoothing* of X_0 near x. Since most of our calculations in this Chapter are (co)homological, in what follows we do not make any distinction between the two types of fibrations.

Remark 10.1.16 The Milnor fibration associated to a complex hypersurface singularity germ does not depend on the choice of a local equation for that germ, see [134] for details. Moreover, for reduced hypersurfaces, the homotopy-type of the Milnor fiber is an invariant of the local, ambient topological-type of the hypersurface; see [134, 135] for a precise formulation.

At this end, it is important to note that the concepts of Milnor fibration, Milnor fiber, and monodromy operator have been also extended to the more general situation when $f : \mathbb{C}^{n+1} \to \mathbb{C}$ is replaced by a non-constant regular or analytic function $f : X \to \mathbb{C}$, with X a complex algebraic or analytic variety, see [136]. In this case, the open ball $\mathring{B}_{\epsilon,x}$ of radius ϵ about $x \in X$ is defined by using an embedding of the germ (X, x) in an affine space $\mathbb{C}^N$. Then $F_x = \mathring{B}_{\epsilon,x} \cap X_s$, for $0 < |s| \ll \delta \ll \epsilon$, is the (local) Milnor fiber of the function f at the point x.

One of the most versatile tools for studying the homotopy type of the Milnor fiber is the *Thom–Sebastiani theorem*, see [217, 191, 211, 186]. Results of the Thom–Sebastiani type consist of exhibiting topological or analytical properties of a function $f(x_0, \ldots, x_n) + g(y_0, \ldots, y_m)$ with separated variables from analogous properties of the components f and g. Topologically, these correspond to the well-known join construction that we now recall.

Definition 10.1.17 Given two topological spaces X and Y, the *join* of X and Y, denoted $X * Y$, is the space obtained from the product $X \times [0, 1] \times Y$ by making the following identifications:

(i) $(x, 0, y) \sim (x', 0, y)$ for all $x, x' \in X$, $y \in Y$;
(ii) $(x, 1, y) \sim (x, 1, y')$ for all $x \in X$, $y, y' \in Y$.

Informally, $X * Y$ is the union of all segments joining points $x \in X$ to points $y \in Y$. For example, if X is a point, then $X * Y$ is just the cone cY on Y. If $X = S^0$, then $X * Y$ is the suspension ΣY of Y.

For future reference, we denote by $[x, t, y]$ the equivalence class in $X * Y$ of $(x, t, y) \in X \times [0, 1] \times Y$.

The homology of a join $X * Y$ was computed by Milnor [176] in terms of homology groups of the factors X and Y as follows:

Lemma 10.1.18 *Let X, Y be topological spaces with self-maps $a : X \to X$ and $b : Y \to Y$. Define a self-map $a * b : X * Y \to X * Y$ by setting*

$$(a * b)([x, t, y]) := [a(x), t, b(y)].$$

Then there is an isomorphism (with integer coefficients)

$$\widetilde{H}_{r+1}(X * Y) \cong \bigoplus_{i+j=r} \left(\widetilde{H}_i(X) \otimes \widetilde{H}_j(Y) \right) \oplus \bigoplus_{i+j=r-1} \mathrm{Tor}(\widetilde{H}_i(X), \widetilde{H}_j(Y)),$$

which is compatible with the homomorphisms induced by $a * b$, a, and b, respectively, at the homology level.

Let $f : (\mathbb{C}^{n+1}, 0) \to (\mathbb{C}, 0)$ and $g : (\mathbb{C}^{m+1}, 0) \to (\mathbb{C}, 0)$ be two hypersurface singularity germs, and consider their sum

$$f + g : (\mathbb{C}^{n+m+2}, 0) \to (\mathbb{C}, 0), \quad (f + g)(x, y) = f(x) + g(y)$$

for $x = (x_0, \ldots, x_n) \in \mathbb{C}^{n+1}$, $y = (y_0, \ldots, y_m) \in \mathbb{C}^{m+1}$. Let F_f, F_g, F_{f+g} be the corresponding Milnor fibers, and h_f, h_g, h_{f+g} the associated monodromy homeomorphisms. (If f and g are weighted homogeneous polynomials, then $f + g$ is also weighted homogeneous, and in this case we can consider the global (affine) Milnor objects as well.) In these notations, one has the following result (proved in [217] in the case of isolated singularities, and extended to arbitrary singularities in [211, 191]):

Theorem 10.1.19 *There is a homotopy equivalence*

$$j : F_f * F_g \longrightarrow F_{f+g}$$

so that the diagram

$$
\begin{array}{ccc}
F_f * F_g & \xrightarrow{\ j\ } & F_{f+g} \\
{\scriptstyle h_f * h_g} \downarrow & & \downarrow {\scriptstyle h_{f+g}} \\
F_f * F_g & \xrightarrow{\ j\ } & F_{f+g}
\end{array}
$$

is commutative up to homotopy.

As a consequence, one gets by Lemma 10.1.18 the following:

Corollary 10.1.20 (Thom–Sebastiani) *Assume that both f and g are isolated hypersurface singularity germs. Then $f + g$ is also an isolated hypersurface singularity and the following diagram is commutative:*

$$
\begin{array}{ccc}
\widetilde{H}_n(F_f; \mathbb{Z}) \otimes \widetilde{H}_m(F_g; \mathbb{Z}) & \longrightarrow & \widetilde{H}_{n+m+1}(F_{f+g}; \mathbb{Z}) \\
{\scriptstyle (h_f)_* \otimes (h_g)_*} \downarrow & & \downarrow {\scriptstyle (h_{f+g})_*} \\
\widetilde{H}_n(F_f; \mathbb{Z}) \otimes \widetilde{H}_m(F_g; \mathbb{Z}) & \longrightarrow & \widetilde{H}_{n+m+1}(F_{f+g}; \mathbb{Z})
\end{array}
$$

Example 10.1.21 (Whitney Umbrella) Let $f(x, y, z) = z^2 - xy^2$ be the Whitney umbrella, and denote by F its Milnor fiber at the singular point at the origin. Since f is a sum of two polynomials in different sets of variables and the Milnor fiber of $\{z^2 = 0\}$ at 0 is just two points, one can apply the Thom–Sebastiani theorem (Theorem 10.1.19) to deduce that F is the suspension on the Milnor fiber G of $g(x, y) = xy^2$ at the origin. Since g is homogeneous, its Milnor fiber G is defined by $xy^2 = 1$, and hence G is homotopy equivalent to a circle S^1. Therefore, the Milnor fiber F of the Whitney umbrella at the origin is homotopy equivalent to a 2-sphere S^2.

Exercise 10.1.22 Show that the Milnor fiber at the origin of the hypersurface defined by $y^2z + x^2 + v^3 = 0$ is homotopy equivalent to $S^3 \vee S^3$.

We conclude this section with a discussion on the important class of examples provided by the *Brieskorn–Pham singularities*, see [25], [179, Section 9], or [60, Chapter 3, Section 4]. Consider the isolated singularity at the origin defined by the weighted homogeneous polynomial

$$f_{\mathbf{a}} = x_0^{a_0} + \cdots + x_n^{a_n},$$

where $n \geq 2$, $a_i \geq 2$ are integers, and $\mathbf{a} = (a_0, \ldots, a_n)$. Let $K(\mathbf{a})$, $F(\mathbf{a})$, $\mu(\mathbf{a})$, $h(\mathbf{a})$ denote the corresponding link, Milnor fiber, Milnor number, and monodromy homeomorphism, respectively. The Thom–Sebastiani theorem (Theorem 10.1.19) can be used to deduce the following result:

Theorem 10.1.23 (Brieskorn–Pham) *The eigenvalues of the algebraic monodromy operator*

$$h(\mathbf{a})_* : H_n(F(\mathbf{a}); \mathbb{Z}) \longrightarrow H_n(F(\mathbf{a}); \mathbb{Z})$$

are the products $\lambda_0 \lambda_1 \cdots \lambda_n$, where each λ_j ranges over all a_j-th roots of unity other than 1. In particular, the corresponding Milnor number is

$$\mu(\mathbf{a}) = (a_0 - 1)(a_1 - 1) \cdots (a_n - 1),$$

and the Alexander polynomial at the origin is given by

$$\Delta(t) = \prod (t - \lambda_0 \lambda_1 \cdots \lambda_n).$$

By combining Exercise 10.1.9 and Theorem 10.1.23, one can now obtain examples of exotic spheres of type $K(\mathbf{a})$, i.e., which are links of Brieskorn–Pham singularities.

Example 10.1.24 ([25]) Let $f : \mathbb{C}^5 \to \mathbb{C}$ be given by

$$f(x, y, z, t, u) = x^2 + y^2 + z^2 + t^3 + u^{6k-1}$$

Then, for $1 \leq k \leq 28$, the link of the singularity at the origin of $f^{-1}(0)$ is a topological 7-sphere. Furthermore, these give the 28 different types of exotic 7-spheres from [126].

10.2 Global Aspects of Hypersurface Singularities

In this section, we show how the theory of perverse sheaves and the local topological information at singular points can be combined to derive global topological statements about complex hypersurface complements. For similar applications and various approaches, the reader may consult [62, 141, 142, 160, 163], and also [61, Section 6.4].

We begin with the following simple application of Ehreshmann's fibration theorem (e.g., see [60, Corollary (1.3.4)]):

Proposition 10.2.1 *All nonsingular complex projective hypersurfaces in $\mathbb{C}P^{n+1}$ of a fixed degree d are diffeomorphic to each other.*

Let V be a (globally defined) degree d reduced hypersurface in $\mathbb{C}P^{n+1}$ ($n \geq 1$) and let H be a hyperplane in $\mathbb{C}P^{n+1}$, which we refer to as the "hyperplane at infinity." Consider the complement

$$\mathcal{U} := \mathbb{C}P^{n+1} - (V \cup H) = \mathbb{C}^{n+1} - V^a,$$

where $V^a \subset \mathbb{C}^{n+1} = \mathbb{C}P^{n+1} - H$ denotes the affine part of V. Alternatively, one can start with a degree d polynomial $f(z_1, \ldots, z_{n+1}) \colon \mathbb{C}^{n+1} \to \mathbb{C}$, and take $V^a = \{f = 0\}$, with $V \subset \mathbb{C}P^{n+1}$ the projectivization of V^a, and H given by $z_0 = 0$. (Here, $z_0, z_1, \ldots, z_{n+1}$ denote the homogeneous coordinates on $\mathbb{C}P^{n+1}$.)

Denote by $V_1, \ldots, V_r$ the irreducible components of V, with $d_i = \deg(V_i)$ for $i = 1, \ldots, r$.

Exercise 10.2.2 In the above notations, show that

$$H_1(\mathbb{C}P^{n+1} - V; \mathbb{Z}) \cong \mathbb{Z}^{r-1} \oplus \mathbb{Z}/\gcd(d_1, \ldots, d_r), \tag{10.3}$$

generated by the homology classes v_i of meridians γ_i about the irreducible components V_i of V. These generators satisfy the relation (cf. [60, Proposition 4.1.3]):

$$\sum_{i=1}^{r} d_i v_i = 0. \tag{10.4}$$

By applying the statement of Exercise 10.2.2 to the complement $\mathcal{U} := \mathbb{C}P^{n+1} - (V \cup H)$, one gets that

$$H_1(\mathcal{U}; \mathbb{Z}) \cong \mathbb{Z}^r, \tag{10.5}$$

generated by the homology classes v_i of meridians γ_i about the irreducible components V_i of V. Moreover, if γ_∞ denotes the meridian loop in $\mathcal{U}$ about the hyperplane H at infinity, with homology class v_∞, then the following relation holds in $H_1(\mathcal{U}; \mathbb{Z})$:

$$v_\infty + \sum_{i=1}^{r} d_i v_i = 0. \tag{10.6}$$

Consider the epimorphism

$$f_\# \colon \pi_1(\mathcal{U}) \longrightarrow \pi_1(\mathbb{C}^*) \cong \mathbb{Z}$$

induced by the restriction of f to $\mathcal{U}$. It can be seen that $f_\#$ coincides with the *total linking number* homomorphism (cf. [60, pp. 76–77]); in particular, $f_\#(\gamma_i) = 1$, for $i = 1, \ldots, r$, and hence $f_\#(\gamma_\infty) = -d$ by (10.6).

Denote by $\mathcal{U}^c$ the (infinite cyclic) covering of $\mathcal{U}$ defined by Ker $f_{\#}$. The group of covering transformations of $\mathcal{U}^c$ is isomorphic to $\mathbb{Z}$, so the chain groups $C_i(\mathcal{U}^c; \mathbb{C})$ and homology groups $H_i(\mathcal{U}^c; \mathbb{C})$ become modules over the group ring $\mathbb{C}[\mathbb{Z}] \cong \mathbb{C}[t, t^{-1}]$.

Definition 10.2.3 The $\mathbb{C}[t, t^{-1}]$-module $H_i(\mathcal{U}^c; \mathbb{C})$ is called the (global) *i-th Alexander module* of the hypersurface complement $\mathcal{U}$.

The goal of this section is to investigate the Alexander modules $H_i(\mathcal{U}^c; \mathbb{C})$. This is the global analogue of the problem of studying the topology of the Milnor fiber of a hypersurface singularity germ, as the Milnor fiber can itself be regarded as an infinite cyclic cover of the local germ complement.

First note that, since $\mathcal{U}$ is the complement of a complex affine hypersurface in $\mathbb{C}^{n+1}$, it is an $(n+1)$-dimensional affine variety, hence it has the homotopy type of a finite CW-complex of real dimension $n+1$ (e.g., see [178], or [61, (1.6.7), (1.6.8)]). Therefore, $H_i(\mathcal{U}^c; \mathbb{C}) = 0$ for $i \geq n+1$, $H_{n+1}(\mathcal{U}^c; \mathbb{C})$ is a free $\mathbb{C}[t, t^{-1}]$-module, and the $\mathbb{C}[t, t^{-1}]$-modules $H_i(\mathcal{U}^c; \mathbb{C})$ are of finite type for $0 \leq i \leq n$.

Definition 10.2.4 The hypersurface $V \subset \mathbb{C}P^{n+1}$ is said to be *in general position (with respect to the hyperplane H) at infinity* if H is transversal in the stratified sense to V, i.e., if $\mathcal{S}$ is a Whitney stratification of V then H is transversal to every stratum $S \in \mathcal{S}$.

Exercise 10.2.5 Show that if $V \subset \mathbb{C}P^{n+1}$ is a nonsingular hypersurface of degree d, then $\pi_1(\mathbb{C}P^{n+1} - V) \cong \mathbb{Z}/d$ and $\pi_i(\mathbb{C}P^{n+1} - V) = 0$ for $1 < i \leq n$.

Exercise 10.2.6 Show that if $V \subset \mathbb{C}P^{n+1}$ is a nonsingular hypersurface in general position (with respect to the hyperplane H) at infinity, and $\mathcal{U} := \mathbb{C}P^{n+1} - (V \cup H)$, then $\pi_1(\mathcal{U}) \cong \mathbb{Z}$ and $\pi_i(\mathcal{U}) = 0$ for $1 < i \leq n$. In particular, $\widetilde{H}_i(\mathcal{U}^c; \mathbb{C}) = 0$ for $i < n+1$.

More generally, the statements of Exercises 10.2.5 and 10.2.6, coupled with a Lefschetz-type result for homotopy groups (see [95], and also the formulation in [60, Theorem (1.6.5)]), yield the following result proved in [140, Lemma 1.5]:

Proposition 10.2.7 *Let $V \subset \mathbb{C}P^{n+1}$ be a projective hypersurface in general position (with respect to the hyperplane H) at infinity, so that V has no codimension one singularities. Let k denote the complex dimension of the singular locus of V. Then:*

(i) $\pi_1(\mathbb{C}P^{n+1} - V) \cong \mathbb{Z}/d$ and $\pi_1(\mathbb{C}P^{n+1} - (V \cup H)) \cong \mathbb{Z}$.

(ii) $\pi_i(\mathbb{C}P^{n+1} - V) = \pi_i(\mathbb{C}P^{n+1} - (V \cup H)) = 0$ for $1 < i < n - k$.

In particular, if a hypersurface V as in the above proposition has only isolated singularities, the only interesting Alexander module is $H_n(\mathcal{U}^c; \mathbb{C})$. This is the situation considered in [140] (where also isolated singularities "at infinity" are allowed). In what follows, no restrictions will be imposed on the singularities of V, provided that V is in general position at infinity.

The following result can be regarded as a global version of the monodromy theorem (Theorem 10.1.10). It was initially obtained in [160], and then reproved in [62] in a more general setup. (See also [30] for further generalizations of the monodromy theorem.)

Theorem 10.2.8 *Assume that the degree d hypersurface $V \subset \mathbb{C}P^{n+1}$ is in general position at infinity. Then, for $i \leq n$, the Alexander module $H_i(\mathcal{U}^c; \mathbb{C})$ is a torsion semi-simple $\mathbb{C}[t, t^{-1}]$-module that is annihilated by $t^d - 1$.*

By analogy with Definition 10.1.11, for every $i \leq n$ we let $\Delta_i(t)$ denote the *i-th (global) Alexander polynomial* of the hypersurface complement $\mathcal{U}$, that is, the characteristic polynomial of the action induced by a generating covering homomorphism on $H_i(\mathcal{U}^c; \mathbb{C})$. Theorem 10.2.8 asserts that all zeros of the global Alexander polynomial $\Delta_i(t)$, $i \leq n$, are roots of unity of order $d = \deg(V)$.

As a consequence of Theorem 10.2.8, one may easily calculate the rank of the free $\mathbb{C}[t, t^{-1}]$-module $H_{n+1}(\mathcal{U}^c; \mathbb{C})$ in terms of the Euler characteristic of the complement as follows:

Corollary 10.2.9 *Under the assumptions and notations of the previous theorem,*

$$rank_{\mathbb{C}[t,t^{-1}]} H_{n+1}(\mathcal{U}^c; \mathbb{C}) = (-1)^{n+1} \chi(\mathcal{U}).$$

In order to prove Theorem 10.2.8, we follow here the approach from [62] and first note that the Alexander modules $H_i(\mathcal{U}^c; \mathbb{C})$ are related to the (co)homology of rank-one $\mathbb{C}$-local systems defined on the complement $\mathcal{U}$ as follows. For $\lambda \in \mathbb{C}^*$, consider the *Milnor long exact sequence* (e.g., see [64, Theorem 4.2]):

$$\cdots \to H_i(\mathcal{U}^c; \mathbb{C}) \xrightarrow{t-\lambda} H_i(\mathcal{U}^c; \mathbb{C}) \longrightarrow H_i(\mathcal{U}; \mathcal{L}_\lambda) \longrightarrow H_{i-1}(\mathcal{U}^c; \mathbb{C}) \to \cdots$$

where $\mathcal{L}_\lambda$ denotes the rank-one $\mathbb{C}$-local system on $\mathcal{U}$ corresponding to the representation

$$\rho_\lambda : \pi_1(\mathcal{U}) \to H_1(\mathcal{U}) \to \mathbb{C}^*$$

obtained by composing the abelianization morphism $\pi_1(\mathcal{U}) \to H_1(\mathcal{U})$ with the homomorphism $H_1(\mathcal{U}) \to \mathbb{C}^*$ defined by mapping v_i ($i = 1, \ldots, r$) to λ and v_∞ to λ^{-d}. Then Theorem 10.2.8 follows from the following *generic vanishing* statement:

Proposition 10.2.10 *Let $\lambda \in \mathbb{C}^*$ be such that $\lambda^d \neq 1$, and denote by $\mathcal{L}_\lambda$ the corresponding rank-one $\mathbb{C}$-local system on $\mathcal{U}$. Then $H_i(\mathcal{U}; \mathcal{L}_\lambda) = 0$ for all $i \neq n + 1$.*

Proof By using (4.1), it is of course sufficient to show that $H^i(\mathcal{U}; \mathcal{L}_\lambda) = 0$ for all $i \neq n + 1$.

Let $\mathbb{C}^{n+1} = \mathbb{C}P^{n+1} - H$, and denote by $u: \mathcal{U} \hookrightarrow \mathbb{C}^{n+1}$ and $v: \mathbb{C}^{n+1} \hookrightarrow \mathbb{C}P^{n+1}$ the two inclusion maps. Since $\mathcal{U}$ is nonsingular of complex dimension $n + 1$, and

$\mathcal{L}_\lambda$ is a local system on $\mathcal{U}$, it follows by Example 10.29 that $\mathcal{L}_\lambda[n+1] \in Perv(\mathcal{U})$. Moreover, since u is a quasi-finite affine morphism, Theorem 8.6.3 yields that

$$\mathcal{F}^\bullet := Ru_*(\mathcal{L}_\lambda[n+1]) \in Perv(\mathbb{C}^{n+1}).$$

But $\mathbb{C}^{n+1}$ is an $(n+1)$-dimensional affine variety, so by Artin's vanishing theorem for perverse sheaves (Corollary 8.6.6), one has that:

$$\mathbb{H}^i(\mathbb{C}^{n+1}; \mathcal{F}^\bullet) = 0, \text{ for all } i > 0, \tag{10.7}$$

and

$$\mathbb{H}^i_c(\mathbb{C}^{n+1}; \mathcal{F}^\bullet) = 0, \text{ for all } i < 0. \tag{10.8}$$

Let $a: \mathbb{C}P^{n+1} \to pt$ be the constant map to a point space. Then:

$$\mathbb{H}^i(\mathbb{C}^{n+1}; \mathcal{F}^\bullet) \cong H^{i+n+1}(\mathcal{U}; \mathcal{L}_\lambda) \cong H^i(Ra_* Rv_* \mathcal{F}^\bullet). \tag{10.9}$$

Similarly,

$$\mathbb{H}^i_c(\mathbb{C}^{n+1}; \mathcal{F}^\bullet) \cong H^i(Ra_! Rv_! \mathcal{F}^\bullet), \tag{10.10}$$

where the last equality follows since a is a proper map, hence $Ra_! = Ra_*$.

Consider now the canonical morphism $Rv_! \mathcal{F}^\bullet \to Rv_* \mathcal{F}^\bullet$, and extend it to the distinguished triangle:

$$Rv_! \mathcal{F}^\bullet \longrightarrow Rv_* \mathcal{F}^\bullet \longrightarrow \mathcal{G}^\bullet \overset{[1]}{\longrightarrow} \tag{10.11}$$

in $D^b_c(\mathbb{C}P^{n+1})$. Since $v^* Rv_! \simeq id \simeq v^* Rv_*$, after applying v^* to the above triangle it follows that $v^*\mathcal{G} \simeq 0$, or equivalently, $\mathcal{G}$ is supported on H. Next, apply $Ra_! = Ra_*$ to the distinguished triangle (10.11) to obtain a new triangle in $D^b_c(pt)$:

$$Ra_! Rv_! \mathcal{F}^\bullet \longrightarrow Ra_* Rv_* \mathcal{F}^\bullet \longrightarrow Ra_* \mathcal{G}^\bullet \overset{[1]}{\longrightarrow} . \tag{10.12}$$

Upon applying the cohomology functor to the distinguished triangle (10.12), one gets the following long exact sequence of complex vector spaces

$$\to \mathbb{H}^i_c(\mathbb{C}^{n+1}; \mathcal{F}^\bullet) \to \mathbb{H}^i(\mathbb{C}^{n+1}; \mathcal{F}^\bullet) \to \mathbb{H}^i(\mathbb{C}P^{n+1}; \mathcal{G}^\bullet)$$
$$\to \mathbb{H}^{i+1}_c(\mathbb{C}^{n+1}; \mathcal{F}^\bullet) \to$$

Using the vanishing from (10.7) and (10.8), together with the identifications (10.9) and (10.10), one then has that:

$$H^{i+n+1}(\mathcal{U}; \mathcal{L}_\lambda) \cong \mathbb{H}^i(\mathbb{C}P^{n+1}; \mathcal{G}^\bullet) \cong \mathbb{H}^i(H; \mathcal{G}^\bullet) \text{ for } i < -1,$$

and $H^n(\mathcal{U}; \mathcal{L}_\lambda)$ is a subspace of the $\mathbb{C}$-vector space $\mathbb{H}^{-1}(H; \mathcal{G}^\bullet)$. So in order to complete the proof, it suffices to show that $\mathbb{H}^i(H; \mathcal{G}^\bullet) = 0$ for all $i \leq -1$. The following arguments will in fact prove the stronger vanishing $\mathcal{G}^\bullet \simeq 0$, which in turn is equivalent to the vanishing of all stalk cohomology groups $\mathcal{H}^q(\mathcal{G}^\bullet)_x$ at points $x \in H$.

For $x \in H$, denote by $\mathcal{U}_x = \mathcal{U} \cap B_x$ the local complement at x, for B_x a small ball in $\mathbb{C}P^{n+1}$ centered at x. Then one has the following identification:

$$\mathcal{H}^q(\mathcal{G}^\bullet)_x \cong \mathcal{H}^q(Rv_*\mathcal{F}^\bullet)_x$$
$$\cong \mathcal{H}^{q+n+1}(Rv_*Ru_*\mathcal{L}_\lambda)_x$$
$$\cong \mathbb{H}^{q+n+1}(B_x; R(v \circ u)_*\mathcal{L})$$
$$\cong H^{q+n+1}(\mathcal{U}_x; \mathcal{L}_x),$$

with $\mathcal{L}_x$ denoting the restriction of $\mathcal{L}_\lambda$ to $\mathcal{U}_x$.

If $x \in H - H \cap V$, then $\mathcal{U}_x$ is homotopy equivalent to S^1, and the corresponding local system $\mathcal{L}_x$ is defined by the action of γ_∞, i.e., by multiplication by $v = \lambda^{-d} \neq 1$. The desired vanishing follows in this case by Exercise 4.2.13.

If $x \in H \cap V$, then the transversality assumption implies that the local complement $\mathcal{U}_x$ is homotopy equivalent to a product $(B_x' - (V \cap B_x')) \times S^1$, with B_x' a small open ball centered at x in H, and the S^1-factor corresponding to a meridian loop about H. It follows that the corresponding local system $\mathcal{L}_x$ is an external tensor product, the second factor being defined by the action of γ_∞ as in the previous case. The desired vanishing follows then from the Künneth formula of Theorem 7.1.8.

$\square$

Exercise 10.2.11 Let $h := f_d$ be the degree d homogeneous polynomial given by the top-degree part of the polynomial $f : \mathbb{C}^{n+1} \to \mathbb{C}$ defining V^a. Let $F_h = \{h = 1\}$ be the global Milnor fiber of h. Show that there are $\mathbb{C}[t, t^{-1}]$-module isomorphisms $H_i(F_h; \mathbb{C}) \to H_i(\mathcal{U}^c; \mathbb{C})$ for $i < n$, and a $\mathbb{C}[t, t^{-1}]$-module epimorphism $H_n(F_h; \mathbb{C}) \to H_n(\mathcal{U}^c; \mathbb{C})$. Deduce an alternative proof of Theorem 10.2.8.

Exercise 10.2.12 Assume that the degree d hypersurface $V \subset \mathbb{C}P^{n+1}$ is in general position at infinity, has no codimension one singularities, and is a rational homology manifold. Show that for all $1 \leq i \leq n$, $\Delta_i(1) \neq 0$.

Perverse sheaves can also be used to estimate the zeros of the global Alexander polynomials in terms of those of the local Alexander polynomials at (affine) points along some irreducible component of V. More precisely, one has the following result of [160] (see also [62] and [141]), which in the case of hypersurfaces with only isolated singularities was proved by Libgober in [140]:

Theorem 10.2.13 Let $\lambda \in \mathbb{C}^*$ be such that $\lambda^d = 1$, and let σ be a non-negative integer. If λ is not a root of the i-th local Alexander polynomial $\Delta_{i,x}(t)$ for $i < n + 1 - \sigma$ and any (affine) point $x \in V_1 - H$ in the irreducible component V_1 of V, then λ is not a root of the global Alexander polynomial $\Delta_i(t)$ for $i < n + 1 - \sigma$.

Proof As in the proof of Theorem 10.2.8, after replacing $\mathbb{C}^{n+1}$ by $\mathcal{U}_1 = \mathbb{C}P^{n+1} - V_1$, it follows that for $i \leq -1$, $H^{i+n+1}(\mathcal{U}; \mathcal{L}_\lambda)$ is a subspace of $\mathbb{H}^i(\mathbb{C}P^{n+1}; \mathcal{G}^\bullet)$, where $\mathcal{G}^\bullet$ is now a complex of $\mathbb{C}$-sheaves supported on V_1. It thus suffices to show that $\mathbb{H}^i(\mathbb{C}P^{n+1}; \mathcal{G}^\bullet) = 0$, for $i < -\sigma$.

The cohomology stalks of $\mathcal{G}^\bullet$ at a point $x \in V_1$ are given as above by

$$\mathcal{H}^q(\mathcal{G}^\bullet)_x \cong H^{q+n+1}(\mathcal{U}_x, \mathcal{L}_x).$$

Therefore, for a fixed $x \in V_1$ the fact that λ is not a root of $\Delta_{i,x}(t)$ for $i < n + 1 - \sigma$ is equivalent to the assertion that $\mathcal{H}^q(\mathcal{G}^\bullet)_x = 0$ for all $q < -\sigma$. The desired vanishing follows now by using the hypercohomology spectral sequence with E_2-term defined by $E_2^{p,q} = H^p(V_1; \mathcal{H}^q(\mathcal{G}^\bullet))$, which computes the groups $\mathbb{H}^i(V_1; \mathcal{G}^\bullet) \cong \mathbb{H}^i(\mathbb{C}P^{n+1}; \mathcal{G}^\bullet)$.   □

Example 10.2.14 Suppose that V is a degree d reduced projective hypersurface that is also a rational homology manifold, has no codimension one singularities, and is in general position at infinity. Assume that the local monodromies at points of strata contained in some irreducible component V_1 of V have orders that are relatively prime to d (e.g., the transversal singularities along strata of V_1 are Brieskorn–Pham type singularities, having all exponents relatively prime to d). Then, by Theorem 10.2.8, Theorem 10.2.13, and Exercise 10.2.12, it follows that $\Delta_i(t) = 1$, for all $1 \leq i \leq n$.

Exercise 10.2.15 Let V be the trifold in $\mathbb{C}P^4 = \{(x : y : z : t : v)\}$, defined by the polynomial: $y^2z + x^3 + tx^2 + v^3 = 0$. Let $H := \{t = 0\}$ be the hyperplane at infinity. Show that V is in general position at infinity (with respect to H), and that the corresponding global Alexander polynomials of the complement $\mathcal{U} = \mathbb{C}P^4 - (V \cup H)$ are computed by:

$$\Delta_0(t) = t - 1, \ \Delta_1(t) = 1, \ \Delta_2(t) = 1, \ \Delta_3(t) = 1.$$

We conclude this section with a discussion on the intersection homology realization of the Alexander modules, see [160] for complete details.

One can think of an n-dimensional projective hypersurface V as the singular locus of $\mathbb{C}P^{n+1}$, which is now regarded as a filtered space stratified by V and its singularities. This yields a Whitney stratification of the pair $(\mathbb{C}P^{n+1}, V)$. By the assumption on transversality at infinity (with respect to the hyperplane H), one may also consider the induced stratification for the pair $(\mathbb{C}P^{n+1}, V \cup H)$. Let $\mathcal{L}$ be the local system on $\mathcal{U}$ with stalk $\mathbb{C}[t, t^{-1}]$ and action by an element $\alpha \in \pi_1(\mathcal{U})$ determined by multiplication by $t^{f_\#(\alpha)}$. Then, for every perversity $\overline{p}$, the intersection homology complex $IC_{\overline{p}}^\bullet(\mathcal{L}) \in D_c^b(\mathbb{C}P^{n+1})$ is defined by the axiomatic construction as in Chapter 6. The following result, left here as an exercise for the interested reader, was proved in [160, Lemma 3.1]:

Lemma 10.2.16 *If* $i : V \cup H \hookrightarrow \mathbb{C}P^{n+1}$ *is the inclusion, then* $i^* IC_{\overline{m}}^{\bullet}(\mathcal{L})$ *is quasi-isomorphic to the zero complex, i.e., the cohomology stalks of the complex* $IC_{\overline{m}}^{\bullet}(\mathcal{L})$ *vanish at points in* $V \cup H$.

As an immediate consequence, one obtains the following intersection homology realization of the Alexander modules of the complement:

Corollary 10.2.17 *There is an isomorphism of* $\mathbb{C}[t, t^{-1}]$-*modules:*

$$IH_i^{\overline{m}}(\mathbb{C}P^{n+1}; \mathcal{L}) \cong H_i(\mathcal{U}; \mathcal{L}) \cong H_i(\mathcal{U}^c; \mathbb{C}),$$

for every integer i.

Exercise 10.2.18 Reprove the results of Theorems 10.2.8 and 10.2.13 by combining Corollary 10.2.17 with the Cappell–Shaneson superduality isomorphism of Theorem 6.4.5. (Hint: you will also need the Artin-type vanishing results for perverse sheaves over the ring $\mathbb{C}[t, t^{-1}]$, as referenced in Remark 8.6.12; see [160] for complete details.)

10.3 Nearby and Vanishing Cycles

In this section, we will address the following *question*: how can one piece together, in a consistent way, the (local) Milnor information at various points along a singular fiber of a regular (or analytic) map? One possible answer is provided by the *nearby and vanishing cycles* of f, which provide a sheafification of this local information.

In order to better motivate the constructions in this section, let us consider a family $\{X_s\}_{s \in D^*}$ of nonsingular complex hypersufaces degenerating to a singular hypersurface X_0, where D^* is a small punctured disc about $0 \in \mathbb{C}$. The goal is then to derive topological information about X_0 from the monodromy of the family and the (local) smoothing(s) of X_0.

For example, if the projection map of the family is proper, then there exists a *specialization map*

$$sp : X_s \to X_0$$

($s \in D^*$) that collapses the vanishing cycles to the singularities of X_0. For a construction of the specialization map, see [85, Part II, Section 6.13]; an overview of the construction is also given in [150, Section 5.8].

On the cohomology level, the specialization map

$$sp^* : H^*(X_0; \mathbb{Z}) \to H^*(X_s; \mathbb{Z})$$

is constructed as follows. If the above family of complex hypersurfaces is given by a (proper) map $f : X \to S$ on a complex manifold X, so that $X_s = f^{-1}(s), s \neq 0$, is the generic fiber, and $X_0 = f^{-1}(0)$ is the special fiber, then for a small enough disc D_δ about $0 \in \mathbb{C}$ and for $s \in D_\delta^*$, we have:

$$X_s \stackrel{i_s}{\hookrightarrow} f^{-1}(D_\delta) \simeq X_0,$$

which induces cohomology maps:

$$H^*(X_0; \mathbb{Z}) \cong H^*(f^{-1}(D_\delta); \mathbb{Z}) \stackrel{i_s^*}{\longrightarrow} H^*(X_s; \mathbb{Z}).$$

The specialization homomorphism sp^* corresponds to the composition of the above homomorphisms in cohomology.

Construction

In this section, we follow Deligne's approach [93, Exposés 13 et 14] to build such a specialization homomorphism by using sheaf theory.

We assume that the base ring A is commutative and noetherian, of finite global dimension, and we work with constructible complexes of sheaves of A-modules.

Let $f : X \to D \subset \mathbb{C}$ be a holomorphic map from a reduced complex space X to a disc $D \subset \mathbb{C}$. Denote by $X_0 = f^{-1}(0)$ the fiber over the center of the disc, with $i : X_0 \hookrightarrow X$ the inclusion map. Let $X^* := X - X_0$ with induced map $f : X^* \to D^*$ to the punctured disc. Consider the following cartesian diagram:

$$
\begin{array}{ccccccc}
X_0 & \stackrel{i}{\hookrightarrow} & X & \stackrel{j}{\longleftarrow} & X^* & \stackrel{\widehat{\pi}}{\longleftarrow} & \widetilde{X}^* \\
\downarrow & & {\scriptstyle f}\downarrow & & {\scriptstyle f^*}\downarrow & & \downarrow \\
\{0\} & \stackrel{}{\hookrightarrow} & D & \longleftarrow & D^* & \underset{\pi}{\longleftarrow} & \widetilde{D}^*
\end{array}
$$

Here $\pi : \widetilde{D}^* \to D^*$ is the infinite cyclic (and universal) cover of D^* defined by the map $z \mapsto \exp(2\pi i z)$, so that $\widehat{\pi} : \widetilde{X}^* \to X^*$ is an infinite cyclic cover with deck group $\mathbb{Z}$. The space $\widetilde{X}^*$ is identified with the *canonical fiber of* f (and it is homotopy equivalent to the generic fiber X_s), and the map $j \circ \widehat{\pi}$ is a canonical model (i.e., independent of the choice of the specific fiber) for the inclusion of the generic fiber X_s in X^*. One can then make the following definition.

Definition 10.3.1 (Nearby Cycle Functor)   The *nearby cycle functor of* f assigns to $\mathcal{F}^\bullet \in D_c^b(X)$ the complex on X_0 defined by

$$\psi_f \mathcal{F}^\bullet := i^* R(j \circ \widehat{\pi})_*(j \circ \widehat{\pi})^* \mathcal{F}^\bullet \in D^b(X_0). \tag{10.13}$$

Moreover, $\psi_f \mathcal{F}^\bullet$ is a constructible complex, i.e., $\psi_f \mathcal{F}^\bullet \in D_c^b(X_0)$ (e.g., see [122, p. 352] if X is nonsingular, and also [214, Theorem 4.0.2, Lemma 4.2.1]), so we get a functor

$$\psi_f : D_c^b(X) \longrightarrow D_c^b(X_0).$$

Remark 10.3.2 Roughly speaking, to define $\psi_f \mathcal{F}^\bullet$ we pull back $\mathcal{F}^\bullet$ to the "generic fiber" of f and then retract onto the "special fiber" X_0. In particular, $\psi_f \mathcal{F}^\bullet$ contains more information about the behavior of $\mathcal{F}^\bullet$ near X_0 then the naive restriction $i^* \mathcal{F}^\bullet$. It is also worth noting that $\psi_f \mathcal{F}^\bullet$ depends only on the restriction of $\mathcal{F}^\bullet$ to X^*. Also, since the definition of $\psi_f \mathcal{F}^\bullet$ involves non-algebraic maps, its constructibility is not clear a priori.

It follows directly from the above definition that the stalk cohomology at a point of X_0 computes the (hyper)cohomology of the corresponding Milnor fiber. Indeed, for $x \in X_0$, let $\mathring{B}_{\epsilon,x}$ be an open ball of radius ϵ in X, centered at x. If X is singular, such a ball is defined by using an embedding of the germ (X, x) in a complex affine space. Then, as in Section 10.1, for $|s|$ non-zero and sufficiently small, $F_x = \mathring{B}_{\epsilon,x} \cap X_s$ is the (local) Milnor fiber of f at x. In these notations, one has the following:

Corollary 10.3.3 *For every $x \in X_0$ there is an A-module isomorphism:*

$$\mathcal{H}^k(\psi_f \mathcal{F}^\bullet)_x \cong \mathbb{H}^k(\mathring{B}_{\epsilon,x} \cap X_s; \mathcal{F}^\bullet|_{X_s}) = \mathbb{H}^k(F_x; \mathcal{F}^\bullet), \qquad (10.14)$$

for all $k \in \mathbb{Z}$. In particular, if $\mathcal{F}^\bullet = \underline{A}_X$ is the constant sheaf on X, then

$$\mathcal{H}^k(\psi_f \underline{A}_X)_x \cong H^k(F_x; A). \qquad (10.15)$$

When f is *proper*, the nearby cycle functor provides a sheaf-theoretic interpretation of the specialization map

$$sp : X_s \longrightarrow X_0$$

mentioned above, which collapses the "topology of a local smoothing" to the singular points. Indeed, in this setup one has the following (see [85, Part II, Section 6.13], and also [51, Remark 5.5.1]):

Theorem 10.3.4

$$\psi_f \mathcal{F}^\bullet \simeq Rsp_*(\mathcal{F}^\bullet|_{X_s}) \in D_c^b(X_0). \qquad (10.16)$$

As an immediate consequence, one then has the identification:

Corollary 10.3.5

$$\mathbb{H}^k(X_0; \psi_f \mathcal{F}^\bullet) \cong \mathbb{H}^k(X_s; \mathcal{F}^\bullet|_{X_s}) \qquad (10.17)$$

for every $k \in \mathbb{Z}$ *and* $s \in D^*$. *In particular, if* $\mathcal{F}^\bullet = \underline{A}_X$ *is the constant sheaf on* X, *then*

$$\mathbb{H}^k(X_0; \psi_f \underline{A}_X) \cong H^k(X_s; A). \tag{10.18}$$

Remark 10.3.6 The deck group action on $\widetilde{D^*}$ in Definition 10.3.1 induces a monodromy transformation $h = h_f$ on ψ_f, which is compatible with the monodromy of the family $\{X_s\}_{s \in D^*}$ via (10.18), and, resp., with the Milnor monodromy via (10.15).

Definition 10.3.7 The sheaf complex $\psi_f \underline{A}_X$ is called the *nearby cycle complex of* f *with* A-coefficients.

Consider the adjunction morphism

$$\mathcal{F}^\bullet \to R(j \circ \widehat{\pi})_*(j \circ \widehat{\pi})^* \mathcal{F}^\bullet$$

and apply i^* to obtain the *specialization morphism*

$$sp : i^* \mathcal{F}^\bullet \to \psi_f \mathcal{F}^\bullet.$$

By taking the cone of sp, one gets a unique distinguished triangle

$$i^* \mathcal{F}^\bullet \xrightarrow{sp} \psi_f \mathcal{F}^\bullet \xrightarrow{can} \varphi_f \mathcal{F}^\bullet \xrightarrow{[1]} \tag{10.19}$$

in $D^b_c(X_0)$, where $\varphi_f \mathcal{F}^\bullet$ is, by definition, the *vanishing cycles* of $\mathcal{F}^\bullet$. One gets a functor

$$\varphi_f : D^b_c(X) \to D^b_c(X_0)$$

called the *vanishing cycle functor of* f. Note that cones are not functorial, so the above construction is not enough to get φ_f as a functor; for complete details see, e.g., [122, Chapter 8] or [214, pages 25–26].

The vanishing cycle functor also comes equipped with a monodromy automorphism, which (in order to avoid cumbersome notations later on) shall still be denoted by h.

Definition 10.3.8 The sheaf complex $\varphi_f \underline{A}_X$ is called the *vanishing cycle complex of* f *with* A-coefficients.

For the computation of the stalk cohomology $\mathcal{H}^k(\varphi_f \mathcal{F}^\bullet)_x$ of the vanishing cycles at $x \in X_0$, one can use the long exact sequence associated to the triangle (10.19), that is,

$$\cdots \longrightarrow \mathcal{H}^k(i^* \mathcal{F}^\bullet)_x \longrightarrow \mathcal{H}^k(\psi_f \mathcal{F}^\bullet)_x \longrightarrow \mathcal{H}^k(\varphi_f \mathcal{F}^\bullet)_x \longrightarrow \cdots,$$

together with the A-module isomorphisms

$$\mathbb{H}^k(\mathring{B}_{\epsilon,x} \cap X_0; \mathcal{F}^\bullet) \cong \mathcal{H}^k(i^* \mathcal{F}^\bullet)_x \cong \mathcal{H}^k(\mathcal{F}^\bullet)_x \cong \mathbb{H}^k(\mathring{B}_{\epsilon,x}; \mathcal{F}^\bullet)$$

and

$$\mathcal{H}^k(\psi_f \mathcal{F}^\bullet)_x \cong \mathbb{H}^k(\mathring{B}_{\epsilon,x} \cap X_s; \mathcal{F}^\bullet)$$

for $s \in D^*$, to obtain the identification

$$\mathcal{H}^k(\varphi_f \mathcal{F}^\bullet)_x \cong \mathbb{H}^{k+1}(\mathring{B}_{\epsilon,x}, \mathring{B}_{\epsilon,x} \cap X_s; \mathcal{F}^\bullet). \tag{10.20}$$

Example 10.3.9 As a particular case of (10.20), assume that X is nonsingular and $\mathcal{F}^\bullet = \underline{A}_X$ is the constant sheaf on X. Then, since $\mathring{B}_{\epsilon,x} \cap X_0$ is contractible, one gets (for $s \in D^*$)

$$\mathcal{H}^k(\varphi_f \underline{A}_X)_x \cong H^{k+1}(\mathring{B}_{\epsilon,x}, \mathring{B}_{\epsilon,x} \cap X_s; A)$$
$$\cong \widetilde{H}^k(\mathring{B}_{\epsilon,x} \cap X_s; A)$$
$$\cong \widetilde{H}^k(F_x; A),$$

with F_x the Milnor fiber of f at x. Recall from Proposition 10.1.3 that if x is a nonsingular point of X_0, then F_x is contractible, so, in view of the above stalk calculation, one gets that $\mathcal{H}^k(\varphi_f \underline{A}_X)_x = 0$ at such a nonsingular point. It then follows that in this case one has the inclusion:

$$\mathrm{supp}(\varphi_f \underline{A}_X) \subseteq \mathrm{Sing}(X_0).$$

In fact, by using Corollary 10.1.14, it follows readily that these sets coincide if A is a field (see, e.g., [61, Corollary 6.1.18]).

Example 10.3.10 Assume that X is nonsingular of complex dimension $n + 1$, and $X_0 = f^{-1}(0)$ has a singular locus Σ that is nonsingular as a variety. For simplicity, assume that Σ is connected. Suppose that $\varphi_f \underline{A}_X$ is constructible with respect to the stratification of X_0 given by the strata Σ and $X_0 - \Sigma$ (e.g., this is the case if the filtration $\Sigma \subset X_0$ corresponds to a Whitney stratification of X_0). If $r = \dim_{\mathbb{C}} \Sigma < n = \dim_{\mathbb{C}} X_0$, by the local product structure of neighborhoods of points in Σ, the Milnor fiber F_x at a point $x \in \Sigma$ has the homotopy type of an $(n - r)$-dimensional CW complex, which moreover is $(n - r - 1)$-connected. If $i : \Sigma \hookrightarrow X_0$ denotes the inclusion map, it follows from the stalk calculation of Example 10.3.9 that

$$\varphi_f \underline{A}_X \simeq i_! \mathcal{L}_\Sigma[r - n],$$

where

$$\mathcal{L}_\Sigma \simeq \mathcal{H}^{n-r}(\varphi_f \underline{A}_X)|_\Sigma$$

is the local system on Σ whose stalk at $x \in \Sigma$ is $H^{n-r}(F_x; A)$.

For a more concrete example, let $f : \mathbb{C}^{n+1} \to \mathbb{C}$ be a polynomial function that depends only on the first $n - r + 1$ coordinates of $\mathbb{C}^{n+1}$. Furthermore, suppose that f has an isolated singularity at $0 \in \mathbb{C}^{n-r+1}$ when regarded as a polynomial function on $\mathbb{C}^{n-r+1}$. If $X_0 = f^{-1}(0) \subset \mathbb{C}^{n+1}$, then the singular locus Σ of X_0 is the affine space $\mathbb{C}^r$ in the remaining coordinates of $\mathbb{C}^{n+1}$, and the filtration $\Sigma \subset X_0$ induces a Whitney stratification of X_0. Since Σ is in this case nonsingular and simply connected, the local system $\mathcal{H}^{n-r}(\varphi_f \underline{A}_{\mathbb{C}^{n+1}})|_\Sigma$ is the constant sheaf $\underline{M}_\Sigma$ with stalk $H^{n-r}(F_0; A)$. Therefore,

$$\varphi_f \underline{A}_{\mathbb{C}^{n+1}} \simeq i_! \underline{M}_\Sigma[r - n].$$

A more general estimation of the support of vanishing cycles is provided by the following result, see, e.g., [152] or [61, Proposition 4.2.8].

Proposition 10.3.11 *Let X be a complex analytic variety with a given Whitney stratification $\mathcal{X}$, and let $f : X \to \mathbb{C}$ be an analytic function. For every $\mathcal{X}$-constructible complex $\mathcal{F}^\bullet$ on X and every integer k, one has the inclusion*

$$\mathrm{supp}\mathcal{H}^k(\varphi_f \mathcal{F}^\bullet) \subseteq X_0 \cap \mathrm{Sing}_{\mathcal{X}}(f), \tag{10.21}$$

where

$$\mathrm{Sing}_{\mathcal{X}}(f) := \bigcup_{V \in \mathcal{X}} \mathrm{Sing}(f|_V)$$

is the stratified singular set of f with respect to the stratification $\mathcal{X}$.

The above construction of the vanishing and nearby cycle functors can also be performed in the following global context (for details, see [61, Section 4.2]). Let X be a complex algebraic (resp., analytic) variety, and let $f : X \to \mathbb{C}$ be a non-constant regular (resp., analytic) function. Then, for every $s \in \mathbb{C}$, one has functors

$$\mathcal{F}^\bullet \in D^b_c(X) \longmapsto \psi_{f-s}\mathcal{F}^\bullet, \ \varphi_{f-s}\mathcal{F}^\bullet \in D^b_c(X_s),$$

where $X_s = f^{-1}(s)$ is assumed to be a non-empty hypersurface, by simply repeating the above considerations for the function $f - s$ restricted to a tube $T(X_s) := f^{-1}(\Delta_s)$ around the fiber X_s (here Δ_s is a small disc centered at s). By stratification theory, in the algebraic context (or in the analytic context for a proper holomorphic map f) one can find a small disc Δ_s centered at $s \in \mathbb{C}$ such that $f : f^{-1}(\Delta_s^*) \to \Delta_s^*$ is a (stratified) locally trivial fibration, with $\Delta_s^* := \Delta_s - \{s\}$ the punctured disc at s.

Exercise 10.3.12 Let $Y \xrightarrow{\pi} X \xrightarrow{f} \mathbb{C}$ be two complex analytic morphisms with π proper. Set $g = f \circ \pi$. Show that for $\mathcal{F}^\bullet \in D^b_c(Y)$ the following *base change property* holds:

$$R\widehat{\pi}_*(\psi_g \mathcal{F}^\bullet) \simeq \psi_f(R\pi_* \mathcal{F}^\bullet), \tag{10.22}$$

and

$$R\widehat{\pi}_*(\varphi_g \mathcal{F}^\bullet) \simeq \varphi_f(R\pi_* \mathcal{F}^\bullet), \tag{10.23}$$

where $\widehat{\pi} : g^{-1}(0) \to f^{-1}(0)$ is induced by π. (Hint: use the proper base change formula of Theorem 5.1.7(b) for the cartesian squares induced by the map π; see, e.g., [61, Proposition 4.2.11].)

Relation with Perverse Sheaves

Let $f : X \to \mathbb{C}$ be a non-constant regular (or complex analytic) function, and we work with a base ring that is commutative, noetherian, of finite dimension. The interplay between perverse sheaves, on the one hand, and nearby and vanishing cycle functors, on the other hand, is reflected by the following result, see [84], [122, Corollary 10.3.13], [214, Theorem 6.0.2], [155, Theorem 3.1, Corollary 3.2]:

Theorem 10.3.13 *The shifted functors*

$$\psi_f[-1], \ \varphi_f[-1] : D_c^b(X) \longrightarrow D_c^b(X_0)$$

are t-exact. In particular, there are induced functors

$$\psi_f[-1], \ \varphi_f[-1] : Perv(X) \longrightarrow Perv(X_0).$$

Moreover, the functors $\psi_f[-1]$ and $\varphi_f[-1]$ commute with the Verdier duality functor $\mathcal{D}$ up to natural isomorphisms.

To simplify the notation, it is customary to define the *perverse nearby and perverse vanishing cycle functors* by

$$^P\psi_f := \psi_f[-1] \quad \text{and} \quad {}^P\varphi_f := \varphi_f[-1],$$

respectively.

Example 10.3.14 If X is a pure $(n + 1)$-dimensional locally complete intersection (e.g., X is nonsingular), then by using Theorem 8.3.12 one gets that $\psi_f \underline{A}_X[n]$ and $\varphi_f \underline{A}_X[n]$ are perverse sheaves on X_0.

Exercise 10.3.15 Let $f : \mathbb{C}^{n+1} \to \mathbb{C}$ be a polynomial so that $Z = f^{-1}(0)$ has only isolated singularities. Let Z_{reg} denote the nonsingular locus of Z, and set

$$\mathcal{G}^\bullet := {}^P\psi_f(\underline{\mathbb{Q}}_{\mathbb{C}^{n+1}}[n + 1]) = \psi_f(\underline{\mathbb{Q}}_{\mathbb{C}^{n+1}}[n]).$$

Then $\mathcal{G}^\bullet$, $\underline{\mathbb{Q}}_Z[n]$ and IC_Z are Verdier self-dual perverse sheaves on the hypersurface Z so that:

$$\mathcal{G}^\bullet|_{Z_{\text{reg}}} \simeq \underline{\mathbb{Q}}_{Z_{\text{reg}}}[n] \simeq IC_Z|_{Z_{\text{reg}}}.$$

Show that $\mathcal{G}^\bullet \simeq IC_Z$ if, and only if, the hypersurface Z is nonsingular. (Hint: use Corollary 10.1.14.)

Thom–Sebastiani for Vanishing Cycles

In this section, we follow [153] (see also [214, Corollary 1.3.4]) to discuss a corresponding Thom–Sebastiani result for vanishing cycles, generalizing Corollary 10.1.20 to functions defined on singular ambient spaces, with arbitrary critical loci, and with arbitrary sheaf coefficients.

Let $f : X \to \mathbb{C}$ and $g : Y \to \mathbb{C}$ be complex analytic functions. Let pr_1 and pr_2 denote the projections of $X \times Y$ onto X and Y, respectively. Consider the function

$$f \boxtimes g := f \circ pr_1 + g \circ pr_2 : X \times Y \to \mathbb{C}.$$

We work over a regular noetherian base ring of finite dimension (e.g., $\mathbb{Z}$, $\mathbb{Q}$, or $\mathbb{C}$). The goal is to express the vanishing cycle functor $\varphi_{f \boxtimes g}$ in terms of φ_f and φ_g. It is more convenient here to use the corresponding perverse vanishing cycles, as introduced in the previous section.

We let $V(f) = \{f = 0\}$, and similarly for $V(g)$ and $V(f \boxtimes g)$. Denote by k the inclusion of $V(f) \times V(g)$ into $V(f \boxtimes g)$. With these notations, the following result holds (see [153, Theorem on p.354]):

Theorem 10.3.16 *For $\mathcal{F}^\bullet \in D_c^b(X)$ and $\mathcal{G}^\bullet \in D_c^b(Y)$, there is a natural isomorphism*

$$k^{*\,P}\varphi_{f \boxtimes g}(\mathcal{F}^\bullet \overset{L}{\boxtimes} \mathcal{G}^\bullet) \simeq {}^P\varphi_f \mathcal{F}^\bullet \overset{L}{\boxtimes} {}^P\varphi_g \mathcal{G}^\bullet \qquad (10.24)$$

commuting with the corresponding monodromies.

Moreover, if $p = (x, y) \in X \times Y$ is such that $f(x) = 0$ and $g(y) = 0$, then, in an open neighborhood of p, the complex ${}^P\varphi_{f \boxtimes g}(\mathcal{F}^\bullet \overset{L}{\boxtimes} \mathcal{G}^\bullet)$ has support contained in $V(f) \times V(g)$, and, in every open set in which such a containment holds, there are natural isomorphisms

$$ {}^P\varphi_{f \boxtimes g}(\mathcal{F}^\bullet \overset{L}{\boxtimes} \mathcal{G}^\bullet) \simeq k_!({}^P\varphi_f \mathcal{F}^\bullet \overset{L}{\boxtimes} {}^P\varphi_g \mathcal{G}^\bullet) \simeq k_*({}^P\varphi_f \mathcal{F}^\bullet \overset{L}{\boxtimes} {}^P\varphi_g \mathcal{G}^\bullet). \qquad (10.25)$$

An immediate consequence of Theorem 10.3.16 and of the algebraic Künneth formula is the following generalization of Corollary 10.1.20 to functions defined on singular spaces, and with arbitrary critical loci.

Corollary 10.3.17 *In the notations of the above theorem and with integer coefficients, there is an isomorphism*

$$\widetilde{H}^{i-1}(F_{f\boxtimes g,p}) \cong \bigoplus_{a+b=i} \left(\widetilde{H}^{a-1}(F_{f,pr_1(p)}) \otimes \widetilde{H}^{b-1}(F_{g,pr_2(p)}) \right)$$

$$\oplus \bigoplus_{c+d=i+1} \mathrm{Tor}\left(\widetilde{H}^{c-1}(F_{f,pr_1(p)}), \widetilde{H}^{d-1}(F_{g,pr_2(p)}) \right), \qquad (10.26)$$

where $F_{f,x}$ denotes as usual the Milnor fiber of a function f at x, and similarly for $F_{g,y}$.

Example 10.3.18 (Brieskorn Singularities and Intersection Cohomology) Let us now indicate how Theorem 10.3.16 applies in the context of Brieskorn–Pham singularities, with twisted intersection cohomology coefficients; see [153, Section 2.4] for complete details.

Recall that a rank r local system $\mathcal{L}$ of complex vector spaces on $\mathbb{C}^*$ is determined up to isomorphism by a monodromy automorphism $h_{\mathcal{L}} : \mathbb{C}^r \to \mathbb{C}^r$. By Exercise 8.4.8, the intersection cohomology complex $IC_{\mathbb{C}}(\mathcal{L})$ on $\mathbb{C}$ agrees with $\mathcal{L}[1]$ on $\mathbb{C}^*$ and has stalk cohomology at the origin concentrated in degree -1, where it is isomorphic to Ker $(id - h_{\mathcal{L}})$.

Next, consider a collection of complex local systems $\mathcal{L}_i$ of rank r_i on $\mathbb{C}^*$, with monodromy automorphisms h_i, and denote the corresponding intersection cohomology complexes on $\mathbb{C}$ by $IC_{\mathbb{C}}(\mathcal{L}_i)$. For positive integers a_i, consider the functions $f_i(x) = x^{a_i}$ on $\mathbb{C}$. The complex ${}^p\varphi_{f_i}IC_{\mathbb{C}}(\mathcal{L}_i)$ is a perverse sheaf supported only at 0; therefore, ${}^p\varphi_{f_i}IC_{\mathbb{C}}(\mathcal{L}_i)$ is non-zero only in degree zero, where it has dimension $a_i r_i - \dim \mathrm{Ker}\, (id - h_i)$. On the other hand, it is a simple exercise to see that the external product of intersection cohomology complexes is an intersection cohomology complex (see Exercise 6.3.14), and hence

$$IC_{\mathbb{C}}(\mathcal{L}_1) \overset{L}{\boxtimes} \cdots \overset{L}{\boxtimes} IC_{\mathbb{C}}(\mathcal{L}_n) \simeq IC_{\mathbb{C}^n}(\mathcal{L}_1 \boxtimes \cdots \boxtimes \mathcal{L}_n),$$

where $\mathcal{L}_1 \boxtimes \cdots \boxtimes \mathcal{L}_n$ is the $\mathbb{C}$-local system on $(\mathbb{C}^*)^n$ with monodromy automorphism

$$h : \mathbb{Z}^n \cong \pi_1((\mathbb{C}^*)^n) \to Aut(\mathbb{C}^{r_1} \times \cdots \times \mathbb{C}^{r_n})$$

defined by

$$h(t_1,\ldots,t_n)(v_1,\ldots,v_n) := (h_1(v_1),\ldots,h_n(v_n)).$$

It then follows by iterating the Thom–Sebastiani isomorphism that

$$^P \varphi_{x_1^{a_1} + \cdots + x_n^{a_n}} \, IC_{\mathbb{C}^n} (\mathcal{L}_1 \boxtimes \cdots \boxtimes \mathcal{L}_n)$$

is a perverse sheaf supported only at the origin, and hence concentrated only in degree zero, where it has dimension equal to

$$\prod_i (a_i r_i - \dim \operatorname{Ker} (id - h_i)) \, .$$

In the particular situation when $r_i = 1$ and $h_i = id_{\mathbb{C}}$ for all i, one is back in the case of the constant sheaf coefficients, and the above calculation recovers the Brieskorn–Pham result of Theorem 10.1.23 that the dimension of the vanishing cycles in degree $n - 1$ (i.e., the Milnor number of the isolated singularity at the origin of $x_1^{a_1} + \cdots + x_n^{a_n} = 0$) is $\prod_i (a_i - 1)$.

On Euler Characteristic Computations

Several results in this chapter concern the calculation of Euler characteristics of hypersurfaces. As it will become clear from the considerations below, nearby and vanishing cycles provide an ideal tool for such computations. For simplicity, in this section we assume that the base ring A is a field.

Let $f : X \to D \subset \mathbb{C}$ be a proper holomorphic map defined on a complex analytic variety X. Then one has a distinguished triangle

$$i^* \underline{A}_X = \underline{A}_{X_0} \longrightarrow \psi_f \underline{A}_X \longrightarrow \varphi_f \underline{A}_X \xrightarrow{[1]}$$

and, by considering the associated long exact sequence in hypercohomology, one obtains by (10.18) the following long exact sequence of A-vector spaces:

$$\cdots \longrightarrow H^k(X_0; A) \longrightarrow H^k(X_s; A) \longrightarrow \mathbb{H}^k(X_0; \varphi_f \underline{A}_X) \longrightarrow \cdots \qquad (10.27)$$

for $s \in D^*$. Moreover, since the fibers of f are compact, it follows from Corollary 7.3.1 that the corresponding Euler characteristics are well defined, and one gets

$$\chi(X_s) = \chi(X_0) + \chi(X_0, \varphi_f \underline{A}_X), \qquad (10.28)$$

with

$$\chi(X_0, \varphi_f \underline{A}_X) := \chi \left(\mathbb{H}^*(X_0; \varphi_f \underline{A}_X) \right) .$$

Let us next assume that the fibers of f are complex algebraic varieties, like in the situations considered below. Then $\chi(X_0, \varphi_f \underline{A}_X)$ can be computed in terms of a stratification of X_0, by using the additivity of Euler characteristic for constructible complexes as in Chapter 7. More precisely, if X is nonsingular and $\mathcal{S}$ is a stratification of X_0 such that $\varphi_f \underline{A}_X$ is $\mathcal{S}$-constructible, then Theorem 7.3.6 and Example 10.3.9 yield the following:

Lemma 10.3.19

$$\chi(X_0, \varphi_f \underline{A}_X) = \sum_{S \in \mathcal{S}} \chi(S) \cdot \mu_S, \qquad (10.29)$$

where

$$\mu_S := \chi\left(\mathcal{H}^*(\varphi_f \underline{A}_X)_{x_S}\right) = \chi\left(\widetilde{H}^*(F_{x_S}; A)\right)$$

is the Euler characteristic of the reduced cohomology of the Milnor fiber F_{x_S} of f at some point $x_S \in S$.

Example 10.3.20 (Specialization Sequence) In the above notations, let us moreover assume that X is nonsingular and the singular fiber X_0 has *only isolated singularities.*

Assume that $\dim_{\mathbb{C}} X = n+1$, and hence $\dim_{\mathbb{C}} X_0 = n$. Then, for $x \in \mathrm{Sing}(X_0)$, the corresponding Milnor fiber $F_x \simeq \bigvee_{\mu_x} S^n$ is up to homotopy a bouquet of n-spheres, and the stalk calculation of Example 10.3.9 yields:

$$\mathbb{H}^k(X_0; \varphi_f \underline{A}_X) \cong \bigoplus_{x \in \mathrm{Sing}(X_0)} \mathcal{H}^k(\varphi_f \underline{A}_X)_x$$

$$\cong \bigoplus_{x \in \mathrm{Sing}(X_0)} \widetilde{H}^k(F_x; A)$$

$$= \begin{cases} 0, & k \neq n, \\ \bigoplus_{x \in \mathrm{Sing}(X_0)} \widetilde{H}^n(F_x; A), & k = n. \end{cases}$$

Then the long exact sequence (10.27) becomes the following *specialization sequence*:

$$0 \longrightarrow H^n(X_0; A) \longrightarrow H^n(X_s; A) \longrightarrow \bigoplus_{x \in \mathrm{Sing}(X_0)} \widetilde{H}^n(F_x; A)$$

$$\longrightarrow H^{n+1}(X_0; A) \longrightarrow H^{n+1}(X_s; A) \longrightarrow 0,$$

for $s \in D^*$, together with isomorphisms

$$H^k(X_0; A) \cong H^k(X_s; A), \text{ for } k \neq n, n+1.$$

(For a generalization of this result to the case when X_0 has an arbitrarily large singular locus, see [61, Corollary 6.2.2].) Taking Euler characteristics, one gets for $s \in D^*$ the identity:

$$\chi(X_s) = \chi(X_0) + \sum_{x \in \mathrm{Sing}(X_0)} \chi(\widetilde{H}^*(F_x; A))$$

$$= \chi(X_0) + (-1)^n \sum_{x \in \mathrm{Sing}(X_0)} \mu_x$$

or, equivalently,

$$\chi(X_0) = \chi(X_s) + (-1)^{n+1} \sum_{x \in \mathrm{Sing}(X_0)} \mu_x. \qquad (10.30)$$

10.4   Euler Characteristics of Complex Projective Hypersurfaces

We begin our calculation of Euler characteristics of complex hypersurfaces with the following well-known result:

Proposition 10.4.1 *Let $Y \subset \mathbb{C}P^{n+1}$ be a degree d nonsingular complex projective hypersurface defined by the homogeneous polynomial $g : \mathbb{C}^{n+2} \to \mathbb{C}$. Then the Euler characteristic of Y depends only on its degree d and complex dimension n, and it is given by the formula:*

$$\chi(Y) = (n+2) - \frac{1}{d}\{1 + (-1)^{n+1}(d-1)^{n+2}\}. \qquad (10.31)$$

Proof The affine cone $\widehat{Y} = \{g = 0\} \subset \mathbb{C}^{n+2}$ on Y has an isolated singularity at the cone point $0 \in \mathbb{C}^{n+2}$. Since g is homogeneous, the local Milnor fibration of g at the origin in $\mathbb{C}^{n+2}$ is fiber homotopic equivalent to the *global* Milnor fibration

$$F = \{g = 1\} \hookrightarrow \mathbb{C}^{n+2} - \widehat{Y} \xrightarrow{g} \mathbb{C}^*.$$

The associated Milnor fiber F of g at $0 \in \mathbb{C}^{n+2}$ is in this case a d-fold cover of $\mathbb{C}P^{n+1} - Y$. So

$$\chi(F) = d \cdot \chi(\mathbb{C}P^{n+1} - Y) = d \cdot (\chi(\mathbb{C}P^{n+1}) - \chi(Y)). \qquad (10.32)$$

On the other hand, since the diffeomorphism type of a nonsingular complex projective hypersurface is determined only by its degree and dimension (see Proposition 10.2.1), one can assume that g is defined by the degree d homogeneous polynomial: $g = \sum_{i=1}^{n+2} x_i^d$. Since the Milnor number, hence also the Euler characteristic of the associated Milnor fiber F are very easy to compute in this case (see, e.g., (10.2)), one gets by (10.32) the desired expression for the Euler characteristic of Y as a function of n and d. □

Proposition 10.4.1, coupled with the theory of nearby and vanishing cycles, yields the following:

Proposition 10.4.2 *Let* $V = \{k = 0\} \subset \mathbb{C}P^{n+1}$ *be a degree* d *hypersurface with only isolated singularities. Then the Euler characteristic* $\chi(V)$ *of* V *is computed by the formula:*

$$\chi(V) = (n+2) - \frac{1}{d}\{1 + (-1)^{n+1}(d-1)^{n+2}\} + (-1)^{n+1} \sum_{x \in \mathrm{Sing}(V)} \mu_x. \quad (10.33)$$

Proof The strategy is to define a family of complex projective hypersurfaces with singular fiber V, and generic fiber Y a smooth degree d projective hypersurface as in Proposition 10.4.1, then make use of formula (10.30).

Let

$$k_s := k + s\ell^d,$$

where $\{\ell = 0\}$ is a generic hyperplane in $\mathbb{C}P^{n+1}$, and set $X_s = \{k_s = 0\}$. This defines a family

$$X := \bigcup_s \{s\} \times X_s \subset \mathbb{C}P^1 \times \mathbb{C}P^{n+1}.$$

Note that X is an $(n+1)$-dimensional complex manifold with a projection $f : X \to \mathbb{C}P^1$. Restricting f to a small disc near $0 \in \mathbb{C}$ yields a proper family f with algebraic fibers, with $X_0 = V$ and $X_s = f^{-1}(s) \cong Y$ for $s \neq 0$, where Y is a smooth degree d complex projective hypersurface. The desired formula follows now from (10.30), which computes $\chi(V)$ as a function of $\chi(Y)$ (as seen in Proposition 10.4.1, $\chi(Y)$ depends only on n and d) and the Milnor numbers $\{\mu_x\}_{x \in \mathrm{Sing}(V)}$ at the singular points of V. □

In the case of complex projective hypersurfaces with arbitrarily large singularities, the base locus of the pencil k_s of hypersurfaces defined in the proof of the above result may have singularities, so additional care is needed. One may proceed as follows, see [61, Example 6.2.6(iii)]. Let $\mathcal{S}_d$ be the set of homogeneous degree d polynomials in $x_0, x_1, \ldots, x_{n+1}$. Let

$$\pi : \mathcal{X}_d \to \mathcal{S}_d$$

be the universal family of degree d complex projective hypersurfaces in $\mathbb{C}P^{n+1}$, so the fiber over $f \in \mathcal{S}_d$ is the projective hypersurface $X_f := \{f = 0\} \subset \mathbb{C}P^{n+1}$.

Let $\mathcal{D}_d$ be the *discriminant* of π, i.e., the set of degree d homogeneous polynomials corresponding to singular hypersurfaces. The discriminant $\mathcal{D}_d$ is a hypersurface in $\mathcal{S}_d$. Let C be a nonsingular curve in $\mathcal{S}_d$ that meets $\mathcal{D}_d$ only at $f \in \mathcal{D}_d$. Restricting π over C yields a family X of nonsingular hypersurfaces degenerating to X_f, and the proper projection map $\pi : X \to C$ has algebraic fibers: for $f' \neq f$, $f' \in C$, the fiber of π over f' is the nonsingular degree d complex projective hypersurface $X_{f'}$, while X_f is the (singular) fiber of $\pi : X \to C$ over f. Moreover, the total space X of the deformation is nonsingular. Therefore, (10.28) yields:

$$\chi(X_{f'}) - \chi(X_f) = \chi(X_f, \varphi_\pi \underline{A}_X). \tag{10.34}$$

As seen in Proposition 10.4.1, the Euler characteristic $\chi(X_{f'})$ of the nonsingular hypersurface $X_{f'}$ can be expressed only in terms of its degree and dimension, whereas the term $\chi(X_f, \varphi_\pi \underline{A}_X)$ on the right-hand side of (10.34) can be computed in terms of a stratification of X_f and the Milnor fibers of $\pi : X \to C$, as in Lemma 10.3.19.

The disadvantage of the above computational method is that (except for the case when the hypersurface has at most isolated singularities), the Milnor fiber information is computed in the deformation space X.

We conclude this section with a more general approach to computing Euler characteristics of *very ample divisors* on complex projective manifolds, which in particular also applies to complex projective hypersurfaces in $\mathbb{C}P^{n+1}$ with arbitrary singularities.

Let Z be a nonsingular complex projective variety, and let L be a very ample line bundle on Z. Let $f \in H^0(Z; L)$ be a holomorphic section of L, with $V = \{f = 0\}$ the hypersurface in Z defined by f. In order to simplify the statement of Theorem 10.4.4 below, let $\chi(Z|L)$ denote the Euler characteristic of the zero set of a sufficiently general section of L. Let $g \in H^0(Z; L)$ be a section of L whose zero set $W = \{g = 0\}$ is a nonsingular hypersurface in Z that is transverse to the strata of a Whitney stratification of Z compatible with V. For $s \in \mathbb{C}P^1$, consider the pencil of hypersurfaces in Z defined by

$$k_s = f - s \cdot g,$$

with $X_s := \{k_s = 0\}$. In particular, $V = X_0$ and $W = X_\infty$. Consider the *incidence variety*

$$X := \{(s, z) \in \mathbb{C}P^1 \times Z \mid z \in X_s\},$$

and note that X is just the blowup of Z along the pencil base locus $V \cap W$. Let $\pi : X \to \mathbb{C}P^1$ be the projection map, hence $X_s = \pi^{-1}(s)$ for every $s \in \mathbb{C}P^1$. Let

$$h = f/g : Z - W \subset X \longrightarrow \mathbb{C},$$

with $h^{-1}(0) = V - W$. Note that $h = \pi^* s$, with s the affine coordinate on $\mathbb{C} \subset \mathbb{C}P^1$. If $z \notin V \cap W$ (hence $g(z) \neq 0$), then in a neighborhood of z one can describe X_s as

$$\{z \mid k_s(z) = 0\} = \{z \mid h(z) = s\},$$

hence in a neighborhood of z, X_s ($s \neq 0$) can be identified with the Milnor fiber of h at z. Note that h also defines V in a neighborhood of $z \notin V \cap W$. Since the Milnor fiber of a complex hypersurface singularity germ does not depend on the choice of a local equation (Remark 10.1.16), one can freely use h or a local representative of f when considering Milnor fibers at points $z \notin V \cap W$.

The key technical result in the above setup asserts that the projection map on the incidence variety does not acquire vanishing cycles along the base locus $V \cap W$, i.e., the following holds:

Lemma 10.4.3 *In the above notations,*

$$\varphi_{\pi^* s} \underline{\mathbb{Q}}_X |_{V \cap W} \simeq 0, \tag{10.35}$$

with s denoting the inhomogeneous coordinate of $\mathbb{C}P^1$. Equivalently,

$$\psi_{\pi^* s} \underline{\mathbb{Q}}_X |_{V \cap W} \simeq \underline{\mathbb{Q}}_{V \cap W}. \tag{10.36}$$

The vanishing property (10.35) follows from [194, Proposition 5.1], where the contractibility of the Milnor fiber at a point in $V \cap W$ is obtained by integrating a controlled vector field as in [179]. A sheaf-theoretic proof of Lemma 10.4.3 was given in [165, Proposition 4.1] by using an embedded resolution of V.

Let us denote by $j : V - W \hookrightarrow V$ the inclusion map. Lemma 10.4.3 and the discussion preceding it yield the following quasi-isomorphism:

$$\varphi_{\pi^* s} \underline{\mathbb{Q}}_X \simeq j_! \varphi_h \underline{\mathbb{Q}}_{Z-W}. \tag{10.37}$$

Applying (10.28) to the function $\pi^* s$, and using (10.37), one gets that (recall that $V = X_0$)

$$\begin{aligned}
\chi(X_s) &= \chi(X_0) + \chi(X_0, \varphi_{\pi^* s} \underline{\mathbb{Q}}_X) \\
&= \chi(X_0) + \chi(X_0, j_! \varphi_h \underline{\mathbb{Q}}_{Z-W}) \\
&= \chi(V) + \chi(V - W, \varphi_h \underline{\mathbb{Q}}_{Z-W}),
\end{aligned} \tag{10.38}$$

where the last identity uses Corollary 7.12. In particular, as in Lemma 10.3.19 and using the fact already mentioned above that the Milnor fibration of a hypersurface singularity germ does not depend on the choice of a local equation for the germ, one gets from (10.38) the following result (see also [193, Proposition 7]):

Theorem 10.4.4 *Let L be a very ample complex line bundle over the complex projective manifold Z. Assume that the hypersurface V in Z is the zero set of a holomorphic section $f \in H^0(Z; L)$, and let $g \in H^0(Z; L)$ be a section of L so that its zero set W is nonsingular and transverse to a Whitney stratification $\mathcal{S}$ of V. Then*

$$\chi(Z|L) - \chi(V) = \sum_{S \in \mathcal{S}} \chi(S - W) \cdot \mu_S, \tag{10.39}$$

where

$$\mu_S := \chi\left(\widetilde{H}^*(F_{x_S}; \mathbb{Q})\right)$$

is the Euler characteristic of the reduced cohomology of the Milnor fiber F_{x_S} of V at some point $x_S \in S$.

Remark 10.4.5 In practice, one may use W to compute the generic Euler characteristic $\chi(Z|L)$, as general sections of L have the same topology. In particular, in the notations of Theorem 10.4.4 formula (10.39) can be restated as follows:

$$\chi(V) = \chi(W) - \sum_{S \in \mathcal{S}} \chi(S - W) \cdot \mu_S. \tag{10.40}$$

Exercise 10.4.6 Apply Theorem 10.4.4 in the context of a complex projective hypersurface with only isolated singularities, and deduce formula (10.33) as a special case of (10.39).

Remark 10.4.7 Formula (10.39) has been generalized to the case of global complete intersections in [165], also in the context of Hodge-theoretic characteristic classes that encode the complexity of singularities. A new formula for the Euler characteristics of complex projective hypersurfaces has been recently obtained in [168], by using iterated hyperplane sections. For a recent application of formula (10.40) to the triangulation problem in computer vision, see [166]. For further applications to applied algebra and algebraic statistics, see [169].

10.5 Generalized Riemann–Hurwitz-Type Formulae

In this section, we use the nearby and vanishing cycle functors to discuss generalized Riemann–Hurwitz-type formulae. We assume that the base ring A is a field.

Recall that if X is a finite CW complex and $\pi : Y \to X$ is a d-fold covering map, then

$$\chi(Y) = d \cdot \chi(X). \tag{10.41}$$

Say now that X and Y are nonsingular complex projective curves, and $\pi : Y \to X$ is a *branched* covering, i.e., π is a covering away from a finite set of points (called the branching set), near which π looks like $\pi(z) = z^k$, where k is called the *branching index* at a branch point. Then one has the following (see [111] or [91, Pages 216–219]):

Theorem 10.5.1 (Riemann–Hurwitz Formula) *If $\pi : Y \to X$ is a degree d branched covering of nonsingular complex projective curves, then*

$$\chi(Y) = d \cdot \chi(X) - \sum_{y \in Y} (e_y - 1),$$

where e_y is the branching index of π at $y \in Y$.

Let us next consider fibrations in place of coverings. First, the following generalization of (10.41) holds:

Proposition 10.5.2 *If $\pi : Y \to X$ is a locally trivial fibration with fiber F, so that F and X are finite dimensional CW complexes, then*

$$\chi(Y) = \chi(X) \cdot \chi(F).$$

Proof First, recall from Exercise 4.2.15 that if A is a field and $\mathcal{L}$ is a rank r A-local system on a finite CW complex X, then

$$\chi(X, \mathcal{L}) := \sum_i (-1)^i \dim H^i(X; \mathcal{L}) = \chi(X) \cdot r.$$

Secondly, since π is a fibration, the cohomology of the total space Y is computed by the Leray spectral sequence for π, that is,

$$E_2^{p,q} = H^p(X; R^q \pi_* \underline{A}_Y) \implies H^{p+q}(Y; A),$$

with $R^q \pi_* \underline{A}_Y$ the local system on X with stalk $H^q(F; A)$. It follows that $H^j(Y; A)$ is finite dimensional for all j, so $\chi(Y)$ is well defined. Moreover, one has the following sequence of equalities:

$$\chi(Y) = \chi(E_\infty) = \ldots = \chi(E_2)$$

$$= \sum_{p,q} (-1)^{p+q} \dim E_2^{p,q}$$

$$= \sum_{p,q} (-1)^{p+q} \dim H^p(X; R^q \pi_* \underline{A}_Y)$$

$$= \sum_q (-1)^q \left(\sum_p (-1)^p \dim H^p(X; R^q \pi_* \underline{A}_Y) \right)$$

$$= \sum_q (-1)^q \, \chi(X, R^q \pi_* \underline{A}_Y)$$

$$= \sum_q (-1)^q \, \chi(X) \cdot b_q(F)$$

$$= \chi(X) \cdot \chi(F).$$

$\square$

The following result can be regarded as a generalization of the Riemann–Hurwitz formula (see [112] and [61, Corollary 6.2.5]):

Theorem 10.5.3 (Iversen's Formula) *Let $f : X \to C$ be a proper holomorphic map from an $(n+1)$-dimensional complex analytic space X onto a curve. Let $B \subset C$ be the finite bifurcation set of f, i.e., f is a topologically locally trivial fibration over $C^* := C - B$. Then for an arbitrary $c \in C^*$, the following formula holds:*

$$\chi(X) = \chi(C) \cdot \chi(X_c) - \sum_{b \in B} \chi(X_b, \varphi_{f-b} \underline{A}_X), \tag{10.42}$$

with X_c, X_b denoting the fibers of f over $c \in C^*$ and $b \in B$, respectively.

Proof From the fibration $X - f^{-1}(B) \to C^*$, one gets by Proposition 10.5.2 that

$$\chi(X) - \chi(f^{-1}(B)) = \chi(X - f^{-1}(B))$$
$$= \chi(C^*) \cdot \chi(X_c)$$
$$= (\chi(C) - \chi(B)) \cdot \chi(X_c),$$

where $c \in C^*$. Here, one uses again the fact that for complex algebraic/analytic variety, the Euler characteristic χ is additive, i.e., $\chi(X) = \chi(Z) + \chi(X \backslash Z)$, for Z a closed subvariety of X. Therefore,

$$\chi(X) = \chi(C) \cdot \chi(X_c) + \sum_{b \in B} \chi(f^{-1}(b)) - \chi(B) \cdot \chi(X_c)$$

$$= \chi(C) \cdot \chi(X_c) + \sum_{b \in B} \left(\chi(X_b) - \chi(X_c) \right)$$

$$= \chi(C) \cdot \chi(X_c) - \sum_{b \in B} \chi(X_b, \varphi_{f-b} \underline{A}_X),$$

where the last equality follows from (10.28). $\square$

Remark 10.5.4 If the fibers of f are algebraic varieties, then each term $\chi(X_b, \varphi_{f-b} \underline{A}_X)$ of (10.42) can be computed in terms of a stratification of X_b as in Lemma 10.3.19.

Example 10.5.5 (Isolated Singularities) In the setting of Theorem 10.5.3, assume moreover that X is nonsingular, and f has only isolated singularities (i.e., each singular fiber $X_{b \in B}$ has only isolated singularities). Then $\varphi_{f-b}\underline{A}_X$ has support on $\mathrm{Sing}(X_b)$, so

$$
\begin{aligned}
\chi(X_b, \varphi_{f-b}\underline{A}_X) &= \chi(\mathrm{Sing}(X_b), \varphi_{f-b}\underline{A}_X) \\
&= \sum_{x \in \mathrm{Sing}(X_b)} \chi\left(\mathcal{H}^*(\varphi_{f-b}\underline{A}_X)_x\right) \\
&= \sum_{x \in \mathrm{Sing}(X_b)} \chi\left(\widetilde{H}^*(F_x; A)\right) \\
&= (-1)^n \sum_{x \in \mathrm{Sing}(X_b)} \mu_x,
\end{aligned}
$$

where F_x and μ_x denote the corresponding Milnor fiber and, resp., Milnor number at the singular point x. Thus, in this case, Iversen's formula (10.42) becomes:

$$
\begin{aligned}
\chi(X) &= \chi(C) \cdot \chi(X_c) - (-1)^n \sum_{b \in B} \sum_{x \in \mathrm{Sing}(X_b)} \mu_x \\
&= \chi(C) \cdot \chi(X_c) + (-1)^{n+1} \sum_{x \in \mathrm{Sing}(f)} \mu_x.
\end{aligned}
$$

Remark 10.5.6 For other generalizations of Proposition 10.5.2 to the stratified context, the reader may also consult [32, 33, 34].

10.6 Homological Connectivity of Milnor Fiber and Link of Hypersurface Singularity Germs

In this section, we indicate some immediate applications of the perverse nearby and vanishing cycles to the study of local topology of singularity germs; see [61, Section 6.1] and [154] for more such results.

Consider first the classical case of the Milnor fiber of a non-constant analytic function germ $f : (\mathbb{C}^{n+1}, 0) \to (\mathbb{C}, 0)$. Denote the Milnor fiber of the singularity at the origin in $X_0 = f^{-1}(0)$ by F_0, and let K be the corresponding link. The following result is a homological version of some of the statements contained in Theorem 10.1.1:

Proposition 10.6.1

(i) If $r = \dim_{\mathbb{C}} \mathrm{Sing}(f)$, then

$$
\widetilde{H}^k(F_0; A) = 0
$$

for any base ring A and for $k \notin [n - r, n]$. (Here we use the convention that $\dim_\mathbb{C} \emptyset = -1$.)

(ii) *The link K is homologically $(n - 2)$-connected, i.e.,*

$$\tilde{H}_i(K; \mathbb{Z}) = 0$$

for every integer $i \leq n - 2$.

Proof For (i), note that since $\underline{A}_X[n + 1]$ is a perverse sheaf on $X = \mathbb{C}^{n+1}$, it follows that ${}^P\varphi_f(\underline{A}_X[n + 1])$ is a perverse sheaf on X_0. Furthermore, since $\mathrm{supp}({}^P\varphi_f(\underline{A}_X[n + 1])) \subseteq \mathrm{Sing}(f)$, Corollary 8.2.10 yields that ${}^P\varphi_f(\underline{A}_X[n + 1])|_{\mathrm{Sing}(f)}$ is a perverse sheaf on $\mathrm{Sing}(f)$. Since $r = \dim_\mathbb{C} \mathrm{Sing}(f)$, the support condition for perverse sheaves yields that

$$\mathcal{H}^q({}^P\varphi_f(\underline{A}_X[n + 1])|_{\mathrm{Sing}(f)})_0 = 0$$

for $q \notin [-r, 0]$ (cf. Exercise 8.3.5). But since

$$\mathcal{H}^q({}^P\varphi_f(\underline{A}_X[n + 1])|_{\mathrm{Sing}(f)})_0 = \mathcal{H}^q({}^P\varphi_f(\underline{A}_X[n + 1]))_0,$$

this implies that

$$\mathcal{H}^q({}^P\varphi_f(\underline{A}_X[n + 1]))_0 = 0$$

for $q \notin [-r, 0]$. The assertion follows now from the stalk identification of Example 10.3.9, namely:

$$\mathcal{H}^q({}^P\varphi_f(\underline{A}_X[n + 1]))_0 = \mathcal{H}^{q+n}(\varphi_f(\underline{A}_X))_0 = \tilde{H}^{q+n}(F_0; A).$$

For (ii), by using Universal Coefficients formulae, it suffices to prove the assertion for coefficients in a field A. Recall that the link K is represented as the intersection $X_0 \cap S$, where S is a small $(2n + 1)$-dimensional sphere centered at the origin in $\mathbb{C}^{n+1}$. Then Alexander duality (Proposition 5.5.3) yields the isomorphism:

$$H^i(K; A) \cong H^{2n+1-i}(S, S - K; A)^\vee. \tag{10.43}$$

On the other hand, by the local conical structure of analytic sets, we have that

$$H^j(S - K; A) = H^j(B - X_0; A) = 0 \tag{10.44}$$

for all $j > n + 1$, where B is the open ball bounded by S and for the vanishing one uses the fact that $B - X_0$ is a Stein manifold of complex dimension $n + 1$ (see the discussion from Remark 8.6.10 and the references therein). The assertion follows now from (10.43) and (10.44), by using the long exact sequence for the cohomology of a pair. $\qquad\qquad\square$

In fact, the proof of Proposition 10.6.1(i) yields the following more general result:

Proposition 10.6.2 *Let $(X, 0)$ be an $(n + 1)$-dimensional complex singularity germ such that $\underline{A}_X[n + 1]$ is a perverse sheaf on X (e.g., X is a local complete intersection), and let $\mathcal{X}$ be a Whitney stratification of X. Let $f : (X, 0) \to (\mathbb{C}, 0)$ be an analytic function germ with $r = \dim_0 \mathrm{Sing}_{\mathcal{X}}(f)$, the dimension at the origin of the stratified singular locus of f. Then*

$$\widetilde{H}^k(F_0; A) = 0$$

for any base ring A and for $k \notin [n - r, n]$, where F_0 denotes the Milnor fiber of f at the origin.

The *complex link* of a singularity plays an important role in the study of the topology of singular spaces, especially in the *stratified Morse theory* of Goresky–MacPherson [85]. If $(X, 0)$ is a singularity germ, and we choose an embedding of $(X, 0)$ into a smooth germ $(\mathbb{C}^N, 0)$, the complex link $L(X, 0)$ of $(X, 0)$ is the Milnor fiber of the restriction $\ell|_X : (X, 0) \to (\mathbb{C}, 0)$ of a generic linear form ℓ on $\mathbb{C}^N$. If $\mathcal{X}$ is a Whitney stratification of X, the generic choice of ℓ ensures that $r = \dim_0 \mathrm{Sing}_{\mathcal{X}}(\ell|_X) = 0$. The topological type of the complex link depends only on the stratum containing the singularity and is independent of all other choices (see [85, Section 2.3]). Then Proposition 10.6.2 yields the following:

Corollary 10.6.3 *If $(X, 0)$ is an $(n + 1)$-dimensional complex singularity germ such that $\underline{A}_X[n + 1]$ is a perverse sheaf on X (e.g., X is a local complete intersection), then*

$$\widetilde{H}^k(L(X, 0); A) = 0$$

for all $k \neq n$ and any base ring A.

10.7 Canonical and Variation Morphisms

In this section, we introduce some new terminology that plays an important role in Saito's theory of mixed Hodge modules of the next chapter. Here we assume that $A = \mathbb{Q}$.

Let f be a non-constant holomorphic function on a complex analytic space X, with corresponding nearby and vanishing cycle functors ψ_f, φ_f, respectively. Recall that these two functors come equipped with monodromy automorphisms, both of which are denoted here by h.[1] The morphism

---

[1] The use of the same symbol h for both monodromy automorphisms acting on the nearby and, resp., vanishing cycle functors is not optimal, but it seems to be widely accepted in standard references

$$can : \psi_f \mathcal{F}^\bullet \longrightarrow \varphi_f \mathcal{F}^\bullet$$

of (10.19) is called the *canonical morphism*, and it is compatible with monodromy. There is a similar distinguished triangle associated to the *variation morphism*, namely:

$$\varphi_f \mathcal{F}^\bullet \xrightarrow{var} \psi_f \mathcal{F}^\bullet \longrightarrow i^![2]\mathcal{F}^\bullet \xrightarrow{[1]}$$

The variation morphism

$$var : \varphi_f \mathcal{F}^\bullet \to \psi_f \mathcal{F}^\bullet$$

is heuristically defined by the cone of the pair of morphisms:

$$(0, h - 1) : [i^* \mathcal{F}^\bullet \to \psi_f \mathcal{F}^\bullet] \longrightarrow [0 \to \psi_f \mathcal{F}^\bullet].$$

(See [122, pp. 351–352] for a formal definition.) Moreover, in the above notations, the following important result holds:

Proposition 10.7.1

$$can \circ var = h - 1, \quad var \circ can = h - 1. \tag{10.45}$$

The monodromy automorphisms acting on the nearby and vanishing cycle functors have Jordan decompositions

$$h = h_u \circ h_s = h_s \circ h_u,$$

where h_s is semi-simple (and locally of finite order) and h_u is unipotent.

For $\lambda \in \mathbb{Q}$ and $\mathcal{F}^\bullet \in D_c^b(X)$ a (shift of a) perverse sheaf, define

$$\psi_{f,\lambda} \mathcal{F}^\bullet := \mathrm{Ker}\,(h_s - \lambda \cdot id)$$

and similarly for $\varphi_{f,\lambda} \mathcal{F}^\bullet$; these are well-defined (shifted) perverse sheaves since $Perv(X)$ is an abelian category. By the definition of vanishing cycles, the canonical morphism *can* induces morphisms

$$can : \psi_{f,\lambda} \mathcal{F}^\bullet \longrightarrow \varphi_{f,\lambda} \mathcal{F}^\bullet,$$

which are isomorphisms for $\lambda \neq 1$, and there is a distinguished triangle

such as [122] or [214]. Nevertheless, it should be clear from the context which operator one refers to.

$$i^* \mathcal{F}^\bullet \xrightarrow{\ sp\ } \psi_{f,1} \mathcal{F}^\bullet \xrightarrow{\ can\ } \varphi_{f,1} \mathcal{F}^\bullet \xrightarrow{\ [1]\ } \ . \tag{10.46}$$

If $A = \mathbb{C}$, there are decompositions

$$\psi_f \mathcal{F}^\bullet = \bigoplus_\lambda \psi_{f,\lambda} \mathcal{F}^\bullet, \ \ \varphi_f \mathcal{F}^\bullet = \bigoplus_\lambda \varphi_{f,\lambda} \mathcal{F}^\bullet.$$

To better illustrate the meaning of the above decompositions, let us also mention that if X is nonsingular then:

$$\mathcal{H}^k(\psi_{f,\lambda} \mathbb{C}_X)_x \cong H^k(F_x; \mathbb{C})_\lambda, \ \ \mathcal{H}^k(\varphi_{f,\lambda} \mathbb{C}_X)_x \cong \widetilde{H}^k(F_x; \mathbb{C})_\lambda,$$

where the right-hand side denotes the λ-eigenspace of the monodromy acting on the (reduced) Milnor fiber cohomology.

In general, there are decompositions

$$\psi_f = \psi_{f,1} \oplus \psi_{f,\neq 1} \ \text{ and } \ \varphi_f = \varphi_{f,1} \oplus \varphi_{f,\neq 1} \tag{10.47}$$

so that $h_s = 1$ on $\psi_{f,1}$ and $\varphi_{f,1}$, and h_s has no 1-eigenspace on $\psi_{f,\neq 1}$ and $\varphi_{f,\neq 1}$. Moreover, $can : \psi_{f,\neq 1} \to \varphi_{f,\neq 1}$ and $var : \varphi_{f,\neq 1} \to \psi_{f,\neq 1}$ are isomorphisms.

It is technically convenient to define a modification Var of the variation morphism var as follows. Let

$$N := \log(h_u),$$

and define the morphism

$$Var : \varphi_f \mathcal{F}^\bullet \longrightarrow \psi_f \mathcal{F}^\bullet \tag{10.48}$$

by the cone of the pair $(0, N)$, see [205]. Then one has that

$$can \circ Var = N, \ Var \circ can = N,$$

and there is a distinguished triangle:

$$\varphi_{f,1} \mathcal{F}^\bullet \xrightarrow{\ Var\ } \psi_{f,1} \mathcal{F}^\bullet \longrightarrow i^![2] \mathcal{F}^\bullet \xrightarrow{\ [1]\ } \ . \tag{10.49}$$

The morphism Var is used in the following *semi-simplicity criterion for perverse sheaves* that has been used by M. Saito in his proof of the decomposition theorem (see [205, Lemma 5.1.4], [208, (1.6)]):

Proposition 10.7.2 *Let X be a complex manifold and let $\mathcal{F}^\bullet$ be a perverse sheaf on X. Then the following conditions are equivalent:*

(a) *In the category Perv(X), one has a splitting*

$$^P\varphi_{g,1}(\mathcal{F}^\bullet) = \mathrm{Ker}\left(Var : {}^P\varphi_{g,1}(\mathcal{F}^\bullet) \to {}^P\psi_{g,1}(\mathcal{F}^\bullet)\right)$$

$$\oplus \mathrm{Image}\left(can : {}^P\psi_{g,1}(\mathcal{F}^\bullet) \to {}^P\varphi_{g,1}(\mathcal{F}^\bullet)\right)$$

for every locally defined holomorphic function g on X.

(b) $\mathcal{F}^\bullet$ *can be written canonically as a direct sum of twisted intersection cohomology complexes.*

Chapter 11
Overview of Saito's Mixed Hodge Modules, and Immediate Applications

In this chapter, we give a brief overview of Morihiko Saito's theory of mixed Hodge modules, and explain some immediate applications of this deep theory. Mixed Hodge modules are extensions in the singular context of variations of mixed Hodge structures, and can be regarded, informally, as sheaves of mixed Hodge structures.

For complete details on this theory, the reader is advised to consult [205, 207]; see also [206] and [210] for an introduction.

11.1 Classical Hodge Theory

We begin with a short review of concepts and results from classical mixed Hodge theory, due to Deligne [55, 56]; see also the book [195] for a comprehensive reference.

Definition 11.1.1 Let H be a finite dimensional $\mathbb{Q}$-vector space with complexification $H_{\mathbb{C}} := H \otimes \mathbb{C}$. A *pure Hodge structure of weight k* on H is a direct sum decomposition (called *Hodge decomposition*)

$$H_{\mathbb{C}} = \bigoplus_{p+q=k} H^{p,q}$$

such that $H^{q,p} = \overline{H^{p,q}}$. (Here, $\overline{}$ denotes the complex conjugation.) The numbers

$$h^{p,q}(H) := \dim_{\mathbb{C}} H^{p,q}$$

are called the *Hodge numbers* of the Hodge structure H.

Remark 11.1.2 To a weight k pure Hodge structure H, one associates a decreasing *Hodge filtration* $F^{\bullet}$ on $H_{\mathbb{C}}$ by setting

© Springer Nature Switzerland AG 2019

L. G. Maxim, *Intersection Homology & Perverse Sheaves*, Graduate Texts in Mathematics 281, https://doi.org/10.1007/978-3-030-27644-7_11

$$F^p = \bigoplus_{s \geq p} H^{s,k-s}.$$

Conversely, a decreasing filtration $F^\bullet$ on $H_{\mathbb{C}}$ with the property that $F^p \cap \overline{F^q} = 0$ if $p + q = k + 1$, defines a weight k pure Hodge structure on H by setting

$$H^{p,q} = F^p \cap \overline{F^q}.$$

Example 11.1.3 If X is a nonsingular complex projective variety, then $H^k(X; \mathbb{Q})$ has a pure Hodge structure of weight k, with

$$H^{p,q} \cong H^q(X; \Omega_X^p).$$

This fact imposes severe restrictions on the topology of X. For example, the odd Betti numbers of X are even: indeed, the Hodge numbers $h^{p,q}(X) :=$ $\dim_{\mathbb{C}} H^q(X; \Omega_X^p)$ of X are symmetric, i.e., $h^{p,q} = h^{q,p}$. Moreover, important topological invariants of X, such as the signature, can be entirely expressed in terms of the Hodge numbers of X. Indeed, in this context, the *Hirzebruch polynomial* $\chi_y(X)$ of Example 3.2.5 can be written as:

$$\chi_y(X) := \sum_{p,q} (-1)^q h^{p,q}(X) \cdot y^p,$$

and the important *Hodge index theorem* (see [106] and [102, Theorem 15.8.2]) shows that, if $\dim_{\mathbb{C}} X$ is even, the signature $\sigma(X)$ of X can be computed from the Hodge numbers of X as follows:

$$\sigma(X) = \chi_1(X) = \sum_{p,q} (-1)^q h^{p,q}(X). \tag{11.1}$$

Note also that $H_k(X; \mathbb{Q}) \cong H^k(X; \mathbb{Q})^\vee$ gets an induced Hodge structure of weight $-k$. Moreover, one gets a Hodge structure of weight $k + \ell$ on $H^k(X; \mathbb{Q}) \otimes H^\ell(X; \mathbb{Q})$ by declaring that $H^{p,q} \otimes H^{r,s}$ has type $(p + r, q + s)$. Similarly, one has a Hodge structure of weight $k - \ell$ on $H^k(X; \mathbb{Q}) \otimes H_\ell(X; \mathbb{Q})$. These play an important role in the context of cup and cap product of Example 11.1.8 below.

Definition 11.1.4 The r-th *Tate twist* $H(r)$ of a weight k pure Hodge structure H is the weight $k - 2r$ pure Hodge structure with $H(r) = H$ and with Hodge filtration $F^p H(r)_{\mathbb{C}} := F^{p+r} H_{\mathbb{C}}$. In particular,

$$H(r)^{p,q} = H^{p+r,q+r}.$$

Definition 11.1.5 If H and H' are pure Hodge structures of the same weight k, a linear map $h : H \to H'$ is called a *morphism of pure Hodge structures* if $h_{\mathbb{C}} = h \otimes id_{\mathbb{C}} : H_{\mathbb{C}} \to H'_{\mathbb{C}}$ satisfies

$$h_{\mathbb{C}}(F^p H_{\mathbb{C}}) \subset F^p H'_{\mathbb{C}}$$

for all p.

Remark 11.1.6 If $h : H \to H'$ is a morphism of pure Hodge structures, then the rational vector spaces Ker h, Image h, and Coker h have canonically induced pure Hodge structures of the same weight.

Example 11.1.7 If $f : X \to Y$ is an algebraic map between nonsingular complex projective varieties, the induced morphism

$$f^* : H^k(Y; \mathbb{Q}) \to H^k(X; \mathbb{Q})$$

is a morphism of pure Hodge structures of weight k.

Example 11.1.8 If X is a nonsingular complex projective variety, it follows from Example 11.1.3 that the cup product map

$$H^k(X; \mathbb{Q}) \otimes H^\ell(X; \mathbb{Q}) \longrightarrow H^{k+\ell}(X; \mathbb{Q})$$

is a morphism of pure Hodge structures of weight $k + \ell$. Similarly, the cap product map

$$H^k(X; \mathbb{Q}) \otimes H_\ell(X; \mathbb{Q}) \longrightarrow H_{\ell-k}(X; \mathbb{Q})$$

is a morphism of pure Hodge structures of weight $k - \ell$.

Proposition 11.1.9 *If X is a nonsingular irreducible complex projective variety of complex dimension n, the Poincaré duality isomorphism*

$$H^{2n-k}(X; \mathbb{Q})(n) \xrightarrow{\cong} H_k(X; \mathbb{Q})$$

is an isomorphism of pure Hodge structures of weight $-k$.

Proposition 11.1.10 *If $h : H \to H'$ is a morphism of pure Hodge structures, then $h_{\mathbb{C}}$ is strictly compatible with the Hodge filtrations, i.e.,*

$$h_{\mathbb{C}}(F^p H_{\mathbb{C}}) = F^p H'_{\mathbb{C}} \cap \text{Image } h_{\mathbb{C}}$$

for all p.

Proposition 11.1.11 *The category Hs of pure Hodge structures is an abelian category.*

If H is a pure Hodge structure, the *Weil operator* $C : H_{\mathbb{C}} \to H_{\mathbb{C}}$ is defined by $C(x) = i^{p-q} x$ for every $x \in H^{p,q}$.

Definition 11.1.12 A *polarization* of a weight k pure Hodge structure is a bilinear form

$$S : H \otimes H \to \mathbb{Q}$$

that is $(-1)^k$-symmetric and such that

(i) the orthogonal complement of F^p is F^{k-p+1},
(ii) the hermitian form on $H_{\mathbb{C}}$ given by $S(Cx, \bar{y})$ is positive-definite, i.e., for every $0 \neq x \in H^{p,q}$,

$$(-1)^k i^{p-q} S(x, \bar{x}) > 0.$$

A Hodge structure that admits a polarization is said to be *polarizable*.

Example 11.1.13 The cohomology groups of nonsingular complex projective varieties are endowed with polarizable Hodge structures.

Definition 11.1.14 A *mixed Hodge structure* is a finite dimensional $\mathbb{Q}$-vector space H endowed with an increasing *weight filtration* $W_{\bullet}$, and a decreasing *Hodge filtration* $F^{\bullet}$ on $H_{\mathbb{C}}$, so that $(gr_k^W H, F^{\bullet})$, with the induced Hodge filtration on $(gr_k^W H)_{\mathbb{C}}$ (denoted again by $F^{\bullet}$), is a pure Hodge structure of weight k, for every $k \in \mathbb{Z}$. To a mixed Hodge structure $(H, W_{\bullet}, F^{\bullet})$ one associates *(mixed) Hodge numbers* $h^{p,q}(H)$ by the formula

$$h^{p,q}(H) = \dim_{\mathbb{C}} gr_F^p gr_{p+q}^W H_{\mathbb{C}}.$$

A mixed Hodge structure is said to be *(graded) polarizable* if the pure Hodge structures $(gr_k^W H, F^{\bullet})$ are polarizable.

Definition 11.1.15 The r-th Tate twist $H(r)$ of a mixed Hodge structure $(H, W_{\bullet}, F^{\bullet})$ is defined by setting $H(r) = H$, $W_k H(r) = W_{k+2r} H$, and $F^p H(r)_{\mathbb{C}} = F^{p+r} H_{\mathbb{C}}$, for all integers k, p. In particular,

$$h^{p,q}(H(r)) = h^{p+r,q+r}(H).$$

Definition 11.1.16 A linear map $h : H \to H'$ between two mixed Hodge structures is called a *morphism of mixed Hodge structures* if h is compatible with both filtrations, i.e.,

$$h(W_k H) \subset W_k H' \text{ for all } k,$$

$$h_{\mathbb{C}}(F^p H_{\mathbb{C}}) \subset F^p H'_{\mathbb{C}} \text{ for all } p.$$

The following generalization of Proposition 11.1.10 holds:

Proposition 11.1.17 *If $h : H \to H'$ is a mixed Hodge structure morphism, then h is strictly compatible with both the weight and the Hodge filtration, that is,*

$$h(W_k H) = W_k H' \cap \text{Image } h$$

for all k, and

$$h_{\mathbb{C}}(F^p H_{\mathbb{C}}) = F^p H'_{\mathbb{C}} \cap \text{Image } h_{\mathbb{C}}$$

for all p.

Moreover, one has:

Proposition 11.1.18 *The category mHs of mixed Hodge structure is abelian. The functors gr_k^W and gr_F^p are exact.*

A fundamental result of Deligne shows that the rational cohomology groups of complex algebraic varieties carry canonical mixed Hodge structures. More precisely, one has the following:

Theorem 11.1.19 (Deligne) *Let X be a complex algebraic variety. Then there is a canonical mixed Hodge structure on $H^*(X; \mathbb{Q})$ such that the following properties hold for every integer $k \geq 0$:*

(i) *The weight filtration $W_{\bullet}$ on $H^k(X; \mathbb{Q})$ satisfies*

$$0 = W_{-1} \subseteq W_0 \subseteq \cdots \subseteq W_{2k} = H^k(X; \mathbb{Q}).$$

For $k \geq n = \dim_{\mathbb{C}} X$, one also has that $W_{2n} = \cdots = W_{2k}$.
(ii) *The Hodge filtration $F^{\bullet}$ on $H^k(X; \mathbb{C})$ satisfies*

$$H^k(X; \mathbb{C}) = F^0 \supseteq \cdots \supseteq F^{k+1} = 0.$$

For $k \geq n = \dim_{\mathbb{C}} X$, one also has that $F^{n+1} = 0$.
(iii) *If X is nonsingular, then*

$$W_{k-1} H^k(X; \mathbb{Q}) = 0,$$

i.e., all weights on $H^k(X; \mathbb{Q})$ are $\geq k$. Moreover,

$$W_k H^k(X; \mathbb{Q}) = j^* H^k(\overline{X}; \mathbb{Q})$$

for every compactification $j : X \hookrightarrow \overline{X}$ of X.
(iv) *If X is a projective variety, then*

$$W_k H^k(X; \mathbb{Q}) = H^k(X; \mathbb{Q}),$$

i.e., all weights on $H^k(X; \mathbb{Q})$ are $\leq k$. Moreover,

$$W_{k-1}H^k(X; \mathbb{Q}) = \text{Ker } \pi^*$$

for every proper map $\pi : \widetilde{X} \to X$ with $\widetilde{X}$ nonsingular.

(v) *The assignment $X \mapsto H^k(X; \mathbb{Q})$ is functorial, in the sense that every morphism $f : X \to Y$ of complex algebraic varieties induces a mixed Hodge structure morphism $f^* : H^k(Y; \mathbb{Q}) \to H^k(X; \mathbb{Q})$.*

Remark 11.1.20 If X is a nonsingular complex projective variety, then (*iii*) and (*iv*) yield that $H^k(X; \mathbb{Q})$ is a pure weight k Hodge structure, a fact already mentioned in Example 11.1.3. The assertion still holds if X is a rational homology manifold, by making use of (*iv*) and Poincaré duality (Proposition 11.1.9).

For completeness, we also include the following definition, which appears naturally in the study of families of algebraic varieties.

Definition 11.1.21 Let X be a nonsingular complex algebraic variety. A *variation $\mathcal{L}$ of Hodge structures of weight k* consists of the following data:

(a) a local system $\mathcal{L}_{\mathbb{Q}}$ of $\mathbb{Q}$-vector spaces on X.
(b) a finite decreasing filtration $\mathcal{F}^\bullet$ of the holomorphic vector bundle $\mathcal{V} := \mathcal{L} \otimes_{\mathbb{Q}_X} \mathcal{O}_X$ by holomorphic sub-bundles such that:

  (i) for all $x \in X$, the filtration $\{\mathcal{F}^p(x)\}_p \subset \mathcal{V}(x) := \mathcal{L}_{\mathbb{Q},x} \otimes_{\mathbb{Q}} \mathbb{C}$ defines a pure Hodge structure of weight k on $\mathcal{L}_{\mathbb{Q},x}$.
  (ii) the canonical connection $\nabla : \mathcal{V} \to \mathcal{V} \otimes_{\mathcal{O}_X} \Omega_X^1$ whose sheaf of horizontal sections is Ker $\nabla = \mathcal{L}_{\mathbb{C}}$ satisfies *Griffiths transversality*, i.e.,

$$\nabla(\mathcal{F}^p) \subset \mathcal{F}^{p-1} \otimes \Omega_X^1.$$

Definition 11.1.22 A weight k variation of Hodge structure $\mathcal{L}$ is *polarizable* if there is a duality pairing

$$S : \mathcal{L} \otimes \mathcal{L} \to \underline{\mathbb{Q}}_X(-k)$$

that on each fiber induces a polarization of the corresponding Hodge structure of weight k.

Example 11.1.23 Let $f : X \to Y$ be a smooth projective morphism between nonsingular complex algebraic varieties. By Ehreshmann's theorem, f is a smooth fibration, so the k-th higher direct image sheaf

$$\mathcal{L} := \mathcal{L}^k := R^k f_* \underline{\mathbb{Q}}_X$$

is a local system on Y. For $y \in Y$, let $X_y := f^{-1}(y)$ be the fiber of f over y, which is a nonsingular complex projective variety. Then $\left(R^k f_* \underline{\mathbb{Q}}_X \right)_y = H^k(X_y; \mathbb{Q})$ carries a Hodge structures of weight k, and $\mathcal{L}$ underlies a *geometric variation* of Hodge structures of weight k. Indeed, in this case, one gets

$$\mathcal{V} = \mathcal{L} \otimes_{\underline{\mathbb{Q}}_Y} \mathcal{O}_Y \simeq R^k f_*(\Omega^\bullet_{X|Y}),$$

with $\Omega^\bullet_{X|Y}$ the *relative holomorphic de Rham complex* (i.e., $\Omega^p_{X|Y} := \Lambda^p \Omega^1_{X|Y}$, for $\Omega^1_{X|Y} := \Omega^1_X / f^* \Omega^1_Y$). The trivial (a.k.a. "stupid") filtration on $\Omega^\bullet_{X|Y}$ determines a decreasing filtration $\mathcal{F}^p$ of $\mathcal{V}$ by holomorphic sub-bundles, with

$$gr^p_{\mathcal{F}} \left((R^{p+q} f_* \underline{\mathbb{Q}}_X) \otimes_{\underline{\mathbb{Q}}_Y} \mathcal{O}_Y \right) \simeq R^q f_*(\Omega^p_{X|Y})$$

inducing for all $y \in Y$ the Hodge filtration on $\mathcal{V}(y) \cong H^k(X_y; \mathbb{Q}) \otimes \mathbb{C}$. Moreover, $\mathcal{V}$ gets an induced integrable *Gauss–Manin connection* $\nabla : \mathcal{V} \to \mathcal{V} \otimes_{\mathcal{O}_Y} \Omega^1_Y$ with $\mathcal{L} \simeq \operatorname{Ker} \nabla$ and $\nabla \circ \nabla = 0$, satisfying the Griffiths transversality condition $\nabla(\mathcal{F}^p) \subset \mathcal{F}^{p-1} \otimes_{\mathcal{O}_Y} \Omega^1_Y$, for all p.

One can similarly define the notion of a *variation of mixed Hodge structures* $\mathcal{L}$ on a nonsingular complex algebraic variety, by asking in addition that the local system $\mathcal{L}_\mathbb{Q}$ has an ascending filtration $\mathcal{W}$, by locally constant subsheaves so that each $gr^\mathcal{W}_k \mathcal{L}_\mathbb{Q}$ underlies a polarizable variation of weight k on X. For example, if $f : X \to Y$ is a proper morphism of complex algebraic varieties, then the restrictions of $R^k f_* \underline{\mathbb{Q}}_X$ to the strata in Y corresponding to a stratification of f are (geometric) variations of mixed Hodge structures.

11.2 Mixed Hodge Modules

In the remainder of this chapter, the base ring A for sheaves is assumed to be the rationals $\mathbb{Q}$.

We recall that for a complex algebraic variety X, the derived category of bounded constructible complexes of sheaves of $\mathbb{Q}$-vector spaces on X is denoted by $D^b_c(X)$, and it contains as a full subcategory the category $Perv(X)$ of perverse $\mathbb{Q}$-complexes. The Verdier duality functor $\mathcal{D}_X$ is an involution on $D^b_c(X)$ preserving $Perv(X)$. Associated to a morphism $f : X \to Y$ of complex algebraic varieties, there are pairs of adjoint functors (f^*, Rf_*) and $(Rf_!, f^!)$ between the respective categories of constructible complexes, which are interchanged by Verdier duality.

M. Saito associated to a complex algebraic variety X an abelian category $MHM(X)$, the category of algebraic *mixed Hodge modules* on X, together with a forgetful functor

$$rat : D^b MHM(X) \to D^b_c(X)$$

such that $rat(MHM(X)) \subset Perv(X)$ is faithful. For $\mathcal{M}^\bullet \in D^b MHM(X)$, $rat(\mathcal{M}^\bullet)$ is called the *underlying rational complex* of $\mathcal{M}^\bullet$.

Since $MHM(X)$ is an abelian category, the cohomology groups of every complex $\mathcal{M}^\bullet \in D^b MHM(X)$ are mixed Hodge modules. The usual truncation functors $\tau_{\leq}, \tau_{\geq}$ on $D^b MHM(X)$ correspond to the perverse truncations ${}^P\tau_{\leq}, {}^P\tau_{\geq}$ on $D^b_c(X)$, so the underlying rational complexes of the cohomology groups of a complex of mixed Hodge modules are the perverse cohomologies of the underlying rational complex, that is,

$$rat(H^j(\mathcal{M}^\bullet)) = {}^P\mathcal{H}^j(rat(\mathcal{M}^\bullet)). \tag{11.2}$$

Exercise 11.2.1 Show that $\mathcal{M}^\bullet \simeq 0$ in $D^b MHM(X)$ if, and only if, $rat(\mathcal{M}^\bullet) \simeq 0$ in $D^b_c(X)$. (Hint: use the faithfulness of the forgetful functor $rat : MHM(X) \to Perv(X)$.)

For a morphism $f : X \to Y$ of complex algebraic varieties, there are induced functors $f_*, f_! : D^b MHM(X) \to D^b MHM(Y)$ and $f^*, f^! : D^b MHM(Y) \to D^b MHM(X)$, which lift the analogous derived functors on the level of constructible complexes. Moreover, if f is proper, then $f_! = f_*$.

The Verdier duality functor $\mathcal{D}_X$ lifts to $MHM(X)$ as an involution (i.e., $\mathcal{D}^2_X = id$), in the sense that it commutes with the forgetful functor:

$$rat \circ \mathcal{D}_X = \mathcal{D}_X \circ rat. \tag{11.3}$$

Moreover, for a morphism $f : X \to Y$ of complex algebraic varieties one has $\mathcal{D}_X f_* = f_! \mathcal{D}_X$ and $\mathcal{D}_X f^* = f^! \mathcal{D}_X$.

Mixed Hodge modules are extensions in the singular context of variations of mixed Hodge structures, and can be regarded, informally, as sheaves of mixed Hodge structures. The precise definition of a mixed Hodge module is quite involved: it uses regular holonomic D-modules, perverse sheaves, and the theory of nearby and vanishing cycles. Since the actual definition of a regular holonomic D-module does not play a special role for the purpose of these notes, we refer the reader to [107] for a comprehensive reference on the theory of D-modules. Let us just mention that the equivalence between local systems and flat vector bundles on a complex manifold (cf. Remark 4.2.17) is replaced here by the *Riemann–Hilbert correspondence* between perverse sheaves with $\mathbb{C}$-coefficients and regular holonomic D-modules. More precisely, if $\mathcal{M}$ is a D-module on a nonsingular n-dimensional variety X (i.e., a left module over the sheaf of algebraic differential operators D_X), the *de Rham complex of* $\mathcal{M}$ is the $\mathbb{C}$-linear complex:

$$DR(\mathcal{M}) := \left[\mathcal{M} \longrightarrow \mathcal{M} \otimes \Omega^1_X \longrightarrow \cdots, \longrightarrow \mathcal{M} \otimes \Omega^n_X \right],$$

placed in degrees $-n, \cdots, 0$. If $\mathcal{M}$ is holonomic, then $DR(\mathcal{M})$ is constructible and it satisfies the axioms for a $\mathbb{C}$-perverse sheaf on X. In fact, the Riemann–Hilbert correspondence (cf. [118, 119, 172, 173]) establishes an equivalence between the category of regular holonomic D-modules and the category of $\mathbb{C}$-perverse sheaves on X, defined via the functor $\mathcal{M} \mapsto DR(\mathcal{M})$. We should also mention here that in the algebraic context the de Rham complex used for the Riemann–Hilbert correspondence is the associated analytic de Rham complex in the classical topology.

The objects of the category $MHM(X)$ can be roughly described as follows. For X nonsingular, $MHM(X)$ is a full subcategory of the category of objects

$$((\mathcal{M}, F_.), \mathcal{K}^\bullet, W_.)$$

such that:

1. $(\mathcal{M}, F_.)$ is an algebraic holonomic filtered D-module $\mathcal{M}$ on X, with an increasing "Hodge" filtration $F_.$ by coherent algebraic $\mathcal{O}_X$-modules;
2. $\mathcal{K}^\bullet \in Perv(X)$ is the underlying rational sheaf complex, and there is a quasi-isomorphism $\alpha : DR(\mathcal{M}) \simeq \mathbb{C} \otimes \mathcal{K}^\bullet$ of $\mathbb{C}$-perverse sheaves on X, where DR denotes as above the (shifted) de Rham functor, placed in degrees $-\dim_{\mathbb{C}} X, \ldots, 0$;
3. $W_.$ is a pair of increasing (weight) filtrations on $\mathcal{M}$ and $\mathcal{K}^\bullet$, compatible with α.

For a singular variety X, one works with local embeddings into manifolds and corresponding filtered D-modules with support on X. In addition, these objects have to satisfy a long list of very complicated properties that are beyond the scope of these notes. In the above notation, the functor rat is defined by

$$rat((\mathcal{M}, F_.), \mathcal{K}^\bullet, W_.) = \mathcal{K}^\bullet.$$

It follows from the definition of mixed Hodge modules that every $\mathcal{M} \in MHM(X)$ has a functorial increasing filtration $W_.$ in $MHM(X)$, called the *weight filtration* of $\mathcal{M}$, so that the functor $\mathcal{M} \to gr_k^W \mathcal{M}$ is exact. Moreover, every morphism of mixed Hodge modules is strictly compatible with the weight filtrations.

Definition 11.2.2 A mixed Hodge module $\mathcal{M} \in MHM(X)$ is said to be *pure of weight* k if $gr_i^W \mathcal{M} = 0$ for all $i \neq k$.

Definition 11.2.3 A complex $\mathcal{M}^\bullet \in D^b MHM(X)$ is *mixed of weight* $\leq k$ (resp., $\geq k$) if $gr_i^W H^j \mathcal{M}^\bullet = 0$ for all $i > j + k$ (resp., $i < j + k$), and it is *pure of weight* k if $gr_i^W H^j \mathcal{M}^\bullet = 0$ for all $i \neq j + k$.

Proposition 11.2.4 *If f is a map of algebraic varieties, then $f_!$ and f^* preserve weight $\leq k$, and f_* and $f^!$ preserve weight $\geq k$. If $\mathcal{M}^\bullet \in D^b MHM(X)$ is of weight $\leq k$ (resp. $\geq k$), then $H^j \mathcal{M}^\bullet$ has weight $\leq j + k$ (resp. $\geq j + k$).*

Example 11.2.5 If $\mathcal{M} \in MHM(X)$ is pure of weight k and $f : X \to Y$ is proper, then $H^i(f_* \mathcal{M})$ is pure of weight $i + k$.

Example 11.2.6 If $\mathcal{M}^\bullet \in D^b MHM(X)$ is pure and $f : X \to Y$ is proper, then $f_* \mathcal{M}^\bullet = f_! \mathcal{M}^\bullet \in D^b MHM(Y)$ is pure, of the same weight.

In the context of mixed Hodge modules the intermediate extension for an open inclusion $j : U \hookrightarrow X$ is defined as:

$$j_{!*} := \text{Image } (H^0 j_! \to H^0 j_*) : MHM(U) \longrightarrow MHM(X).$$

The following statement is then an immediate consequence of Proposition 11.2.4:

Proposition 11.2.7 *If $j : U \hookrightarrow X$ is a Zariski-open dense subset in X, the intermediate extension $j_{!*}$ preserves the weights.*

Proposition 11.2.8 *For every $\mathcal{M} \in MHM(X)$, $gr_i^W \mathcal{M}$ is a semi-simple object of $MHM(X)$.*

Lemma 11.2.9 $\text{Ext}^i(\mathcal{M}^\bullet, \mathcal{N}^\bullet) = 0$ *for $\mathcal{M}^\bullet$ of weight $\leq m$ and $\mathcal{N}^\bullet$ of weight $\geq n$, if $m < n + i$.*

Corollary 11.2.10 *If $\mathcal{M}^\bullet$ is pure of weight n, there is a non-canonical isomorphism in $D^b MHM(X)$:*

$$\mathcal{M}^\bullet \simeq \bigoplus_j H^j \mathcal{M}^\bullet[-j].$$

Proof Consider the following distinguished triangle in $D^b MHM(X)$:

$$\tau_{\leq j-1} \mathcal{M}^\bullet \longrightarrow \tau_{\leq j} \mathcal{M}^\bullet \longrightarrow \tau_{\leq j} \mathcal{M}^\bullet / \tau_{\leq j-1} \mathcal{M}^\bullet \overset{[1]}{\longrightarrow}$$

and note that $\tau_{\leq j} \mathcal{M}^\bullet / \tau_{\leq j-1} \mathcal{M}^\bullet \simeq H^j \mathcal{M}^\bullet[-j]$. Since $\mathcal{M}^\bullet$ has weight n, one gets that $\tau_{\leq j} \mathcal{M}^\bullet$ and $gr_j^\tau \mathcal{M}^\bullet := \tau_{\leq j} \mathcal{M}^\bullet / \tau_{\leq j-1} \mathcal{M}^\bullet$ have weight n as well, since $W_k(\tau_{\leq j} \mathcal{M}^\bullet) = \tau_{\leq j} W_k \mathcal{M}^\bullet$ and similarly for $gr_j^\tau \mathcal{M}^\bullet$. Since $\tau_{\leq j-1} \mathcal{M}^\bullet[1]$ has weight $n + 1$, one gets by Lemma 11.2.9 (or by strictness) that

$$\text{Ext}^1(gr_j^\tau \mathcal{M}^\bullet, \tau_{\leq j-1} \mathcal{M}^\bullet) = \text{Hom}(gr_j^\tau \mathcal{M}^\bullet, \tau_{\leq j-1} \mathcal{M}^\bullet[1]) = 0,$$

and hence there is a non-canonical splitting

$$\tau_{\leq j} \mathcal{M}^\bullet \simeq \tau_{\leq j-1} \mathcal{M}^\bullet \oplus gr_j^\tau \mathcal{M}^\bullet \simeq \tau_{\leq j-1} \mathcal{M}^\bullet \oplus H^j \mathcal{M}^\bullet[-j].$$

The result follows now by induction.   □

As a consequence, one has the following:

Theorem 11.2.11 *If $\mathcal{M}^\bullet$ is pure and $f : X \to Y$ is proper, there is a non-canonical isomorphism in $D^b MHM(Y)$:*

$$f_* \mathcal{M}^\bullet \simeq \bigoplus_j H^j f_* \mathcal{M}^\bullet [-j].$$

As it will be discussed below, Saito showed that the category of mixed Hodge modules supported on a point, $MHM(pt)$, coincides with the category mHs^p of (graded) polarizable rational mixed Hodge structures. (Here one has to switch the increasing D-module filtration $F_\bullet$ of the mixed Hodge module to the decreasing Hodge filtration of the mixed Hodge structure by $F^\bullet := F_{-\bullet}$, so that $gr_F^p \simeq gr_{-p}^F$.) In this case, the functor *rat* associates to a mixed Hodge structure the underlying rational vector space. By the identification $MHM(pt) \simeq mHs^p$, there exists a unique *Tate object* $\mathbb{Q}^H(k) \in MHM(pt)$ such that $rat(\mathbb{Q}^H(k)) = \mathbb{Q}(k)$, with $\mathbb{Q}(k)$ the (trivial) mixed Hodge structure of type $(-k, -k)$. For example,

$$\mathbb{Q}^H = \mathbb{Q}^H(0) = ((\mathbb{C}, F_\bullet), \mathbb{Q}, W_\bullet),$$

with $gr_i^F = 0 = gr_i^W$ for all $i \neq 0$, and $\alpha : \mathbb{C} \to \mathbb{C} \otimes \mathbb{Q}$ the obvious isomorphism. For a complex algebraic variety X with $a : X \to pt$ the map to a point, we define

$$\underline{\mathbb{Q}}_X^H(k) := a^* \mathbb{Q}^H(k) \in D^b MHM(X),$$

with

$$rat(\underline{\mathbb{Q}}_X^H(k)) = \underline{\mathbb{Q}}_X(k).$$

So tensoring with $\underline{\mathbb{Q}}_X^H(k)$ defines the *Tate twist operation* $\cdot(k)$ on mixed Hodge modules. To simplify the notations, we let $\underline{\mathbb{Q}}_X^H := \underline{\mathbb{Q}}_X^H(0)$. If X is *nonsingular* of pure dimension n, then $\underline{\mathbb{Q}}_X[n] \in Perv(X)$ and $\underline{\mathbb{Q}}_X^H[n] \in MHM(X)$ is a single mixed Hodge module (in degree 0), explicitly described by

$$\underline{\mathbb{Q}}_X^H[n] = ((\mathcal{O}_X, F_\bullet), \underline{\mathbb{Q}}_X[n], W_\bullet),$$

with $gr_i^F = 0 = gr_{i+n}^W$ for all $i \neq 0$. So if X is nonsingular of complex dimension n, then $\underline{\mathbb{Q}}_X^H[n]$ is a pure mixed Hodge module of weight n. By the stability of the intermediate extension functor, this shows that if X is a complex algebraic variety of pure dimension n and $j : U \hookrightarrow Z$ is the inclusion of a nonsingular Zariski-open dense subset, then the intersection cohomology module

$$IC_X^H := j_{!*}(\underline{\mathbb{Q}}_U^H[n])$$

is pure of weight n, with underlying perverse sheaf $rat(IC_X^H) = IC_X$. Note also that since IC_X is Verdier self-dual as a perverse sheaf, there is a corresponding isomorphism of mixed Hodge modules:

$$\mathcal{D}_X IC_X^H \simeq IC_X^H(n). \tag{11.4}$$

Remark 11.2.12 The BBDG decomposition theorem can now be deduced (after taking rat) by applying Theorem 11.2.11 to the pure object $\mathcal{M}^\bullet = IC_X$.

Definition 11.2.13 We say that $\mathcal{M} \in MHM(X)$ is supported on Z if and only if $rat(\mathcal{M})$ is supported on Z. We say that $\mathcal{M}$ has *strict support* Z if $\mathcal{M}$ has no sub-object and no quotient object supported in a proper subvariety of Z and $supp(rat(\mathcal{M})) = Z$.

The pure objects in Saito's theory are the *polarized Hodge modules*. The category $HM(X, k)^p$ of polarizable Hodge modules on X of weight k is a semi-simple abelian category, in the sense that every polarizable Hodge module on X can be written in a unique way as a direct sum of polarizable Hodge modules with strict support in irreducible closed subvarieties of X. This is what is called the *decomposition by strict support* of a pure Hodge module. If $HM_Z(X, k)^p$ denotes the category of pure Hodge modules of weight k and strict support Z, then $HM_Z(X, k)^p$ depends only on Z, and every $\mathcal{M} \in HM_Z(X, k)^p$ is *generically* a polarizable variation of Hodge structures $\mathcal{L}_U$ on a Zariski-open dense subset $U \subset Z$, with *quasi-unipotent monodromy at infinity*. Conversely, any such polarizable variation of Hodge structures can be extended uniquely to a pure Hodge module. In other words, there is an equivalence of categories:

$$HM_Z(X, k)^p \simeq VHS_{gen}(Z, k - dim(Z))^p, \tag{11.5}$$

where the right-hand side is the category of polarizable variations of Hodge structures of weight $k - dim(Z)$ defined on non-empty nonsingular subvarieties of Z, whose local monodromies are quasi-unipotent. Under this correspondence, if $\mathcal{M}$ is a pure Hodge module with strict support Z, then $rat(\mathcal{M}) = IC_Z(\mathcal{L})$, where $\mathcal{L}$ is the corresponding variation of Hodge structures.

Let us now consider in more detail the following example:

Example 11.2.14 Let X be a complex algebraic manifold of pure complex dimension n, with $\mathcal{L} := (\mathcal{L}, F^\bullet, W_\bullet)$ a *good* (i.e., admissible in the sense of Kashiwara [120][1], with quasi-unipotent monodromy at infinity) variation of (rational) mixed Hodge structures on X. Denote by $\mathcal{V} := \mathcal{L} \otimes_{\underline{\mathbb{Q}}_X} \mathcal{O}_X$ the flat bundle with its integrable connection ∇ associated to the local system $\mathcal{L}_{\mathbb{C}} = \mathcal{L} \otimes \mathbb{C}$. Since $\mathcal{L}$ underlies a variation of mixed Hodge structures, the bundle $\mathcal{V}$ comes equipped with its Hodge (decreasing) filtration by holomorphic sub-bundles $\mathcal{F}^p$, and these are required to satisfy the *Griffiths' transversality condition*

---

[1] Geometric variations of mixed Hodge structures are admissible, cf. [120].

$$\nabla(\mathcal{F}^p) \subset \Omega_X^1 \otimes \mathcal{F}^{p-1}.$$

(For the fact that ∇ and the holomorphic sub-bundles $\mathcal{F}^p$ are indeed algebraic, see [215, page 438].) The bundle $\mathcal{V}$ becomes a holonomic (left) D-module, with Hodge filtration given by

$$F_p(\mathcal{V}) := \mathcal{F}^{-p}\mathcal{V}.$$

Note that there is a quasi-isomorphism $\alpha : \mathrm{DR}(\mathcal{V}) \simeq \mathcal{L}[n]$, where we use the shifted de Rham complex

$$\mathrm{DR}(\mathcal{V}) := [\mathcal{V} \xrightarrow{\ \nabla\ } \cdots \xrightarrow{\ \nabla\ } \mathcal{V} \otimes_{\mathcal{O}_X} \Omega_X^n]$$

with $\mathcal{V}$ in degree $-n$, so that $DR(\mathcal{V}) \simeq \mathcal{L}[n]$ is a perverse sheaf on X. Moreover, α is compatible with the induced filtration $W_{.}$ defined by

$$W_i(\mathcal{L}[n]) := W_{i-n}\mathcal{L}[n] \quad \text{and} \quad W_i(\mathcal{V}) := (W_{i-n}\mathcal{L}) \otimes_{\underline{\mathbb{Q}}_X} \mathcal{O}_X.$$

This data defines a mixed Hodge module $\mathcal{L}^H[n]$ on X, with $rat(\mathcal{L}^H[n]) \simeq \mathcal{L}[n]$. Hence $rat(\mathcal{L}^H[n])[-n]$ is a local system on X.

Remark 11.2.15 A pure polarizable variation of weight k with quasi-unipotent monodromy at infinity yields a pure (polarizable) Hodge module of weight $k+n$ on X.

If X is nonsingular and of pure complex dimension n, an object $\mathcal{M} \in MHM(X)$ is called *smooth* if and only if $rat(\mathcal{M})[-n]$ is a local system on X. By associating to a good variation of mixed Hodge structures $\mathcal{L} = (\mathcal{L}, F^{\cdot}, W_{.})$ on X the mixed Hodge module $\mathcal{L}^H[n]$ as in Example 11.2.14, one obtains an equivalence between the category $MHM(X)_s$ of smooth mixed Hodge modules on X and good (i.e., admissible) variations of mixed Hodge structures on X, i.e., one has the following.

Theorem 11.2.16

$$MHM(X)_s \cong VMHS(X)_{ad}$$

If $X = pt$ is a point space, one obtains in particular an equivalence

$$MHM(pt) \cong mHs^p$$

between mixed Hodge modules over a point and the abelian category of (graded polarizable) mixed Hodge structures.

Note that, by the stability by the intermediate extension functor, it follows that if X is a complex algebraic variety of pure dimension n and $\mathcal{L}$ is an admissible variation of (pure) Hodge structures (of weight k) on a Zariski-open dense subset $U \subset X$, then $IC_X^H(\mathcal{L})$ is an algebraic mixed Hodge module (of pure weight $k + n$), so that $rat(IC_X^H(\mathcal{L})|_U) = \mathcal{L}[n]$.

We conclude this section with a brief discussion of the gluing procedure used by M. Saito to construct his mixed Hodge modules.

Let g be a function on a nonsingular variety X, with $Y = g^{-1}(0)_{red}$ and $U = X - Y$. The functors ${}^P\psi_g$ and ${}^P\varphi_g$ acting on perverse sheaves lift to functors

$$\psi_g^H : MHM(X) \to MHM(Y) \text{ and } \varphi_g^H : MHM(X) \to MHM(Y).$$

More precisely,

$$rat \circ \psi_g^H = {}^P\psi_g \circ rat \text{ and } rat \circ \varphi_g^H = {}^P\phi_g \circ rat. \tag{11.6}$$

Moreover, the morphisms *can*, *N*, *Var* of Section 10.7 and decompositions ${}^P\psi_g = {}^P\psi_{g,1} \oplus {}^P\psi_{g,\neq 1}$ (and similarly for ${}^P\varphi_g$) lift to the category of mixed Hodge modules.

Let X, g, Y, U be given as in the previous paragraph. Let $MHM(U, Y)_{gl}$ be the category whose objects are $(\mathcal{M}', \mathcal{M}'', u, v)$, with $\mathcal{M}' \in MHM(U)$, $\mathcal{M}'' \in MHM(Y)$, $u \in \text{Hom}(\psi_{g,1}^H \mathcal{M}', \mathcal{M}'')$, $v \in \text{Hom}(\mathcal{M}'', \psi_{g,1}^H \mathcal{M}'(1))$, and so that $v \circ u = N$. Then the following holds:

Theorem 11.2.17 *There is an equivalence of categories*

$$MHM(X) \cong MHM(U, Y)_{gl}$$

defined by:

$$\mathcal{M} \mapsto (\mathcal{M}|_U, \varphi_{g,1}^H, can, Var).$$

It follows that every object of $MHM(X)$ can be constructed by induction on the dimension of support by using Theorems 11.2.16 and 11.2.17.

We note in passing that if $f : X \to \mathbb{C}$ is a non-constant regular function on the complex algebraic variety X and $X_c = f^{-1}(c)$ is the fiber over c, then for each $x \in X_c$ one gets canonical mixed Hodge structures on the groups

$$H^j(F_x; \mathbb{Q}) = rat\left(H^j(i_x^* \psi_{f-c}^H \underline{\mathbb{Q}}_X^H[1])\right)$$

and

$$\tilde{H}^j(F_x; \mathbb{Q}) = rat\left(H^j(i_x^* \varphi_{f-c}^H \underline{\mathbb{Q}}_X^H[1])\right),$$

where F_x denotes the Milnor fiber of f at $x \in X_c$, and $i_x : \{x\} \hookrightarrow X_c$ is the inclusion of the point. Similarly, one obtains in this way the *limit mixed Hodge structure* on

$$\mathbb{H}^j(X_c; \psi_{f-c}\underline{\mathbb{Q}}_X) = rat\left(H^j(a_*\psi_{f-c}^H\underline{\mathbb{Q}}_X^H[1])\right)$$

with $a : X_c \to \{c\}$ the constant map.

Remark 11.2.18 A mixed Hodge module version of the Thom–Sebastiani theorem (Theorem 10.3.16) for vanishing cycles has been obtained by M. Saito in an unpublished preprint (see also [167]).

11.3 Hodge Theory on Intersection Cohomology Groups

In this section, we explain how to use Saito's mixed Hodge module theory to construct mixed Hodge structures on the intersection cohomology groups of complex algebraic varieties and, resp., of links of closed subvarieties. We also verify that the generalized Poincaré duality isomorphism is compatible with these mixed Hodge structures.

We start by noting that a graded polarizable mixed Hodge structure on the (compactly supported) rational cohomology of a complex algebraic variety X can be obtained by using the identifications:

$$H^i(X; \mathbb{Q}) = rat(H^i(a_*a^*\underline{\mathbb{Q}}_{pt}^H)) \tag{11.7}$$

and

$$H_c^i(X; \mathbb{Q}) = rat(H^i(a_!a^*\underline{\mathbb{Q}}_{pt}^H)), \tag{11.8}$$

with $a : X \to pt$ the constant map to a point space. Moreover, by a deep result of Saito [209], these structures coincide with the classical mixed Hodge structures constructed by Deligne.

More generally, we have the following:

Proposition 11.3.1 *Let X be a complex algebraic variety, and $a : X \to pt$ be the constant map to the point. For a bounded complex $\mathcal{M}^\bullet$ of mixed Hodge modules on X with underlying rational complex $\mathcal{K}^\bullet = rat(\mathcal{M}^\bullet)$, the vector spaces*

$$\mathbb{H}^j(X; \mathcal{K}^\bullet) = rat(H^j(a_*\mathcal{M}^\bullet)) \quad and \quad \mathbb{H}_c^j(X; \mathcal{K}^\bullet) = rat(H^j(a_!\mathcal{M}^\bullet))$$

are endowed with rational (graded polarizable) mixed Hodge structures.

Exercise 11.3.2 Show that if $\mathcal{K}^\bullet = rat(\mathcal{M}^\bullet) \in D_c^b(X)$ underlies a bounded complex of mixed Hodge modules, the perverse cohomology spectral sequence

$$E_2^{i,j} = \mathbb{H}_{(c)}^i(X; {}^p\mathcal{H}^j(\mathcal{K}^\bullet)) \Longrightarrow \mathbb{H}_{(c)}^{i+j}(X; \mathcal{K}^\bullet) \tag{11.9}$$

is a spectral sequence in the category of mixed Hodge structures. More generally, if F is a (left exact) functor that sends mixed Hodge modules to mixed Hodge modules, the corresponding spectral sequence

$$E_2^{i,j} = H^i F(H^j(\mathcal{M}^\bullet)) \Longrightarrow H^{i+j} F(\mathcal{M}^\bullet) \tag{11.10}$$

is a spectral sequences of mixed Hodge modules.

An immediate consequence of Proposition 11.3.1 is the following result, originally proved by Steenbrink–Zucker [223] for the curve case and by El Zein [70, 69] in the general situation:

Corollary 11.3.3 *Let X be a complex algebraic manifold and let $\mathcal{L}$ be an admissible variation of mixed Hodge structures on X with quasi-unipotent monodromy at infinity. Then the groups $H^j(X; \mathcal{L})$ and $H_c^j(X; \mathcal{L})$ get induced (graded polarizable) mixed Hodge structures. Moreover, these structures are pure if X is compact and $\mathcal{L}$ is a variation of pure Hodge structures.*

Similarly, we get the following:

Corollary 11.3.4 *If X is a complex algebraic variety of pure dimension n and $\mathcal{L}$ is an admissible variation of mixed Hodge structures on a Zariski-open dense subset $U \subset X$, then the intersection cohomology groups $IH^i(X; \mathcal{L})$ carry (graded polarizable) mixed Hodge structures. If, moreover, X is compact and $\mathcal{L}$ is pure of weight k, then $IH^i(X; \mathcal{L})$ carries a pure (polarizable) Hodge structure of weight $i + k$.*

For the constant variation $\mathcal{L} = \underline{\mathbb{Q}}_X$, Corollary 11.3.4 yields:

Corollary 11.3.5 *The intersection cohomology group $IH^i(X; \mathbb{Q})$ of a pure-dimensional complex projective (or compact) variety X admits a pure (polarizable) Hodge structure of weight i.*

By making use of Theorem 6.6.3, this further yields the following:

Corollary 11.3.6 *If X is a compact complex algebraic variety of pure dimension n that is also a rational homology manifold, its rational cohomology group $H^i(X; \mathbb{Q})$ carries a pure (polarizable) Hodge structure of weight i, for every $i \in \mathbb{Z}$.*

In view of Corollary 11.3.5, we denote by $Ih^{p,q}(X)$ the corresponding *intersection cohomology Hodge numbers* of a complex projective (or compact) variety X of pure dimension, with associated intersection cohomology Hodge polynomial

$$I\chi_y(X) := \sum_{p,q} (-1)^q Ih^{p,q}(X) \cdot y^p.$$

Then the following generalization of the Hodge index theorem (11.1) holds (see [164, Section 3.6]):

Theorem 11.3.7 *If X is a complex projective variety of even pure dimension, the Goresky–MacPherson signature $\sigma(X)$ of X can be computed from the intersection cohomology Hodge numbers of X as follows:*

$$\sigma(X) = I\chi_1(X) = \sum_{p,q}(-1)^q Ih^{p,q}(X). \tag{11.11}$$

Let us next note that by using the duality isomorphism (11.4), one obtains the following:

Proposition 11.3.8 *If X is a complex algebraic variety of complex pure dimension n, the generalized Poincaré duality isomorphism*

$$IH^k(X; \mathbb{Q}) \cong \left(IH_c^{2n-k}(X; \mathbb{Q})(n) \right)^{\vee}$$

is an isomorphism of mixed Hodge structures.

Proof Let $a : X \to pt$ be the constant map to a point space. For $\mathcal{M}^{\bullet} \in D^b MHM(X)$, one has the following identification in $MHM(pt)$:

$$H^j(a_*\mathcal{M}^{\bullet}) \simeq \mathcal{D}_{pt} H^{-j}(a_!\mathcal{D}_X\mathcal{M}^{\bullet}). \tag{11.12}$$

By taking $\mathcal{M}^{\bullet} = IC_X^H$ and $j = k - n$ in (11.12), and using (11.4), one gets the isomorphism

$$H^{k-n}(a_*IC_X^H) \simeq \mathcal{D}_{pt} H^{n-k}(a_!IC_X^H(n)) \tag{11.13}$$

in $MHM(pt)$. Applying the forgetful functor rat, one has the following sequence of mixed Hodge structure isomorphisms:

$$
\begin{aligned}
IH^k(X; \mathbb{Q}) &= \mathbb{H}^{k-n}(X; IC_X) \\
&= rat(H^{k-n}(a_*IC_X^H)) \\
&\overset{(11.13)}{\cong} \left(rat(H^{n-k}(a_!IC_X^H(n))) \right)^{\vee} \\
&= \left(\mathbb{H}_c^{n-k}(X; IC_X)(n) \right)^{\vee} \\
&= \left(IH_c^{2n-k}(X; \mathbb{Q})(n) \right)^{\vee},
\end{aligned}
$$

thus proving the claim.   □

Exercise 11.3.9 Let X be a complex algebraic variety. Show that the natural map $H^i(X; \mathbb{Q}) \to IH^i(X; \mathbb{Q})$ discussed in Exercise 6.7.1 and Remark 6.7.2 is a morphism of mixed Hodge structures. If X is projective, show that the kernel is the subspace of $H^i(X; \mathbb{Q})$ consisting of classes of Deligne weight $\leq i - 1$.

Remark 11.3.10 Recall that the standard t-structure τ on $D^b MHM(X)$ corresponds to the perverse t-structure ${}^p\tau$ on $D^b_c(X)$. However, in [207, Remark 4.6(2)], Saito constructed another t-structure ${}'\tau$ on $D^b MHM(X)$ that corresponds to the *standard* t-structure on $D^b_c(X)$ (cf. Example 8.1.12). By using the Deligne construction of intersection cohomology (see Chapter 6), it then follows that intersection cohomology complexes for *any* perversity $\overline{p}$ are complexes of mixed Hodge modules on the complex algebraic variety X. So, in particular, the intersection (co)homology groups $I^{(BM)}H_i^{\overline{p}}(X; \mathbb{Q})$ are endowed with mixed Hodge structures, for every perversity $\overline{p}$ and every integer i.

Exercise 11.3.11 Show that if $\mathcal{K}^\bullet = rat(\mathcal{M}^\bullet) \in D^b_c(X)$ underlies a bounded complex of mixed Hodge modules, the (compactly supported) hypercohomology spectral sequence

$$E_2^{i,j} = \mathbb{H}^i_{(c)}(X; \mathcal{H}^j(\mathcal{K}^\bullet)) \Longrightarrow \mathbb{H}^{i+j}_{(c)}(X; \mathcal{K}^\bullet) \tag{11.14}$$

is a spectral sequence in the category of mixed Hodge structures. Deduce that if $f : X \to Y$ is a morphism of complex algebraic varieties, the associated Leray spectral sequences

$$E_2^{i,j} = \mathbb{H}^i(Y; R^j f_* \underline{\mathbb{Q}}_X) \Longrightarrow \mathbb{H}^{i+j}(X; \mathbb{Q}) \tag{11.15}$$

and

$$E_2^{i,j} = \mathbb{H}^i_c(Y; R^j f_! \underline{\mathbb{Q}}_X) \Longrightarrow \mathbb{H}^{i+j}_c(X; \mathbb{Q}) \tag{11.16}$$

are spectral sequences in the category of mixed Hodge structures.

Another interesting application of the theory of mixed Hodge modules is the Durfee–Saito *semi-purity* result for the intersection cohomology groups of the link $L_X(Z)$ of a closed subvariety Z of X. Before stating the result, we need a few preparatory statements.

Let X be a complex algebraic variety of complex pure dimension n, and let $Z \subset X$ be a closed subvariety with link $L_X(Z)$ (see Section 7.3 and the definitions in [68]). The link $L_X(Z)$ of Z in X is of real dimension $2n - 1$, has only odd-dimensional strata, and it has an orientation induced from that of X. Moreover, one has the following:

Proposition 11.3.12 *The groups* $IH^k_c(L_X(Z); \mathbb{Q})$ *and* $IH^k(L_X(Z); \mathbb{Q})$ *are endowed with mixed Hodge structures.*

Proof As in Exercise 7.3.4, if $i : Z \hookrightarrow X$ and $j : U = X - Z \hookrightarrow X$ are the inclusion maps, then

$$IH_c^k(L_X(Z); \mathbb{Q}) \cong \mathbb{H}_c^{k-n}(Z; i^* Rj_* IC_U) = rat(H^{k-n}(a_! i^* j_* IC_U^H)) \qquad (11.17)$$

and

$$IH^k(L_X(Z); \mathbb{Q}) \cong \mathbb{H}^{k-n}(Z; i^* Rj_* IC_U) = rat(H^{k-n}(a_* i^* j_* IC_U^H)) \qquad (11.18)$$

with $a : Z \to pt$ the constant map to a point. An application of Proposition 11.3.1 then proves the claim. $\qquad \square$

Our next goal is to estimate the weights of the mixed Hodge structures on $IH_c^k(L_X(Z); \mathbb{Q})$ and $IH^k(L_X(Z); \mathbb{Q})$.

Lemma 11.3.13 *In the above notations, there is an isomorphism*

$$i^* j_* \simeq i^! j_! [1] \qquad (11.19)$$

in $D^b MHM(Z)$.

Proof For $\mathcal{M}^\bullet \in D^b MHM(X)$, apply the attaching triangle

$$j_! j^! \longrightarrow id \longrightarrow i_* i^* \overset{[1]}{\longrightarrow}$$

to $j_* \mathcal{M}^\bullet$ to get the following triangle in $D^b MHM(X)$:

$$j_! \mathcal{M}^\bullet \longrightarrow j_* \mathcal{M}^\bullet \longrightarrow i_* i^* j_* \mathcal{M}^\bullet \overset{[1]}{\longrightarrow} . \qquad (11.20)$$

Hence

$$i_* i^* j_* \mathcal{M}^\bullet \simeq cone(j_! \mathcal{M}^\bullet \to j_* \mathcal{M}^\bullet). \qquad (11.21)$$

Dualizing (11.20) and replacing $\mathcal{D}_X \mathcal{M}^\bullet$ with $\mathcal{M}^\bullet$, one gets the following triangle in $D^b MHM(X)$:

$$i_! i^! j_! \mathcal{M}^\bullet \longrightarrow j_! \mathcal{M}^\bullet \longrightarrow j_* \mathcal{M}^\bullet \overset{[1]}{\longrightarrow} . \qquad (11.22)$$

In particular,

$$i_! i^! j_! \mathcal{M}^\bullet [1] \simeq cone(j_! \mathcal{M}^\bullet \to j_* \mathcal{M}^\bullet). \qquad (11.23)$$

Therefore, (11.21) and (11.23) yield the isomorphism

$$i_* i^* j_* \mathcal{M}^\bullet \simeq i_! i^! j_! \mathcal{M}^\bullet [1] \qquad (11.24)$$

in $D^b MHM(X)$. The desired isomorphism (11.19) follows by applying i^* to (11.24). $\qquad \square$

The following result is the counterpart of Proposition 11.3.8 in the context of links (see [68, Proposition 3.3]):

Proposition 11.3.14 *Let X be a complex algebraic variety of complex pure dimension n, and let $Z \subset X$ be a closed subvariety with link $L_X(Z)$. The generalized Poincaré duality isomorphism*

$$I H^k(L_X(Z); \mathbb{Q}) \cong \left(I H_c^{2n-1-k}(L_X(Z); \mathbb{Q})(n)\right)^\vee$$

is an isomorphism of mixed Hodge structures.

Proof By Lemma 11.3.13 and using the above notations, with $a : Z \to pt$ the constant map, one has the following sequence of isomorphisms of mixed Hodge structures:

$$
\begin{aligned}
I H^k(L_X(Z); \mathbb{Q}) & \overset{(11.18)}{\cong} rat(H^{k-n}(a_* i^* j_* IC_U^H)) \\
& \overset{(11.19)}{\cong} rat(H^{k-n}(a_* i^! j_! IC_U^H[1])) \\
& = rat(H^{k+1-n}(a_* i^! j_! IC_U^H)) \\
& \overset{(11.12)}{\cong} \left(rat(H^{n-1-k}(a_! \mathcal{D}_Z(i^! j_! IC_U^H)))\right)^\vee \\
& \overset{(11.4)}{\cong} \left(rat(H^{n-1-k}(a_! i^* j_* IC_U^H(n)))\right)^\vee \\
& \cong \left(\mathbb{H}_c^{n-1-k}(Z; i^* j_* IC_U^H)(n)\right)^\vee \\
& \overset{(11.17)}{\cong} \left(I H_c^{2n-1-k}(L_X(Z); \mathbb{Q})(n)\right)^\vee.
\end{aligned}
$$

$\square$

We can now prove the following semi-purity result of Durfee–Saito, see [68, Theorem 4.1]:

Theorem 11.3.15 *Let X be a complex algebraic variety of pure dimension n, and let $Z \subset X$ be a closed subvariety with $\dim_{\mathbb{C}} Z \le d$. Let $L_X(Z)$ be the link of Z in X. Then $I H_c^k(L_X(Z); \mathbb{Q})$ carries a mixed Hodge structure of weight $\le k$ for $k < n - d$, and $I H^k(L_X(Z); \mathbb{Q})$ carries a mixed Hodge structure of weight $> k$ for $k \ge n + d$.*

Proof Recall from Proposition 11.2.4 that, if f is a map of complex algebraic varieties, then $f_!$ and f^* preserve weight $\le k$, and f_* and $f^!$ preserve weight $\ge k$. Moreover, if $\mathcal{M}^\bullet$ is a bounded complex of weight $\le k$ (resp., $\ge k$), then $H^j \mathcal{M}^\bullet$ has weight $\le j + k$ (resp., $\ge j + k$).

Let $U := X - Z$ with inclusion maps $j : U \hookrightarrow X$ and $i : Z \hookrightarrow X$. Since $IC_X = j_{!*} IC_U$, using properties of the intermediate extension (see also [12, 1.4.13, 1.4.23(ii)]) one can show that there is a distinguished triangle

$$IC_X \longrightarrow j_*ICU \longrightarrow i_*{}^P\tau_{>-1}i^*j_*ICU \longrightarrow .$$

Applying i^* gives a distinguished triangle

$$i^*ICX \longrightarrow i^*j_*ICU \longrightarrow {}^P\tau_{>-1}i^*j_*ICU \longrightarrow$$

underlying a corresponding triangle in $D^bMHM(X)$. Thus there is a long exact sequence of mixed Hodge structures

$$\cdots \longrightarrow \mathbb{H}_c^{k-n}(Z; i^*ICX) \longrightarrow \mathbb{H}_c^{k-n}(Z; i^*j_*ICU) \longrightarrow$$
$$\longrightarrow \mathbb{H}_c^{k-n}(Z; {}^P\tau_{>-1}i^*j_*ICU) \longrightarrow \cdots$$

The middle term in the above sequence is exactly $IH_c^k(L_X(Z); \mathbb{Q})$. The left term $\mathbb{H}_c^{k-n}(Z; i^*ICX) = rat(H^{k-n}(a_!i^*IC_X^H))$ has weight $\leq k$: indeed, since IC_X^H has weight n, it follows that $i^*IC_X^H$ has weight $\leq n$, and hence $a_!i^*IC_X^H$ has weight $\leq n$, with $a : Z \to pt$ the constant map to a point. The right term $\mathbb{H}_c^{k-n}(Z; {}^P\tau_{>-1}i^*j_*ICU)$ vanishes for $k - n < -d \leq -\dim_{\mathbb{C}} Z$, since $\mathcal{H}^i({}^P\tau_{>-1}i^*j_*ICU) = 0$ for $i < -d$ by the support condition for a perverse sheaf. This proves the first assertion.

The second assertion follows from the first by duality since, by Proposition 11.3.14, the duality isomorphism is an isomorphism of mixed Hodge structures.

$\square$

Remark 11.3.16 Such semi-purity statements preclude many closed odd-dimensional manifolds (e.g., tori) from being links of isolated singularities, see [68, Section 5] for more details.

11.4 Intersection Homology Betti Numbers, II

In this section, we use the theory of weights to refine the results of Section 6.7 on intersection homology Betti numbers. We follow Durfee's approach from [67]. All groups will be assumed to have the rational numbers as coefficients, unless otherwise indicated.

We begin by recalling the statement of Theorem 6.7.4 from Section 6.7:

Proposition 11.4.1 *Let X be a pure n-dimensional complex algebraic variety and U an open subvariety of X, with $Z = X - U$ a closed subvariety of dimension $\leq d$. Then (with $\mathbb{Q}$-coefficients):*

(a) $IH^k(X) \cong IH^k(U)$, *for $k < n - d$.*
(b) $IH^{n-d}(X) \hookrightarrow IH^{n-d}(U)$.
(c) $IH_c^{n+d}(X) \twoheadleftarrow IH_c^{n+d}(U)$.
(d) $IH_c^k(X) \cong IH_c^k(U)$, *for $k > n + d$.*

Recall from Corollary 11.3.4 that intersection cohomology groups of a complex algebraic variety have mixed Hodge structures. In particular, they have a weight filtration

$$\cdots \subseteq W_{m-1} \subseteq W_m \subseteq W_{m+1} \subseteq \cdots$$

with $gr_m^W = W_m / W_{m-1}$ an exact functor. One then has the following:

Proposition 11.4.2 *Let X be a pure n-dimensional complex algebraic variety and U an open subvariety of X. Then, for all integers k,*

(a) $gr_k^W IH_c^k(U) \hookrightarrow gr_k^W IH_c^k(X)$ *is a monomorphism.*
(b) $gr_k^W IH^k(X) \twoheadrightarrow gr_k^W IH^k(U)$ *is an epimorphism.*

Proof Let $a : X \to pt$ be the constant map to a point, and denote as before by $j : U \hookrightarrow X$ and $i : Z = X - U \hookrightarrow X$ the open and, respectively, closed inclusion. The attaching triangle

$$j_! j^* IC_X \longrightarrow IC_X \longrightarrow i_* i^* IC_X \longrightarrow \qquad (11.25)$$

underlies a distinguished triangle in $D^b MHM(X)$, and hence, upon applying

$$\mathbb{H}_c^{k-n}(X; -) = H^{k-n} a_!(-)$$

to it, one gets a corresponding long exact sequence of mixed Hodge structures:

$$\cdots \to IH_c^k(U) \to IH_c^k(X) \to \mathbb{H}_c^{k-n}(Z; i^* IC_X) \to \cdots \qquad (11.26)$$

Since IC_X^H is pure of weight n, it follows from Proposition 11.2.4 that, for every integer k, the mixed Hodge structures $IH_c^k(U)$, $IH_c^k(X)$, and $\mathbb{H}_c^{k-n}(Z; i^* IC_X)$ have weight $\leq k$. Taking gr_k^W in (11.26) proves (a).

Similarly, applying $\mathbb{H}^{k-n}(X; -) = H^{k-n} a_*(-)$ to the distinguished triangle

$$i_* i^! IC_X \longrightarrow IC_X \longrightarrow j_* j^! IC_X \longrightarrow$$

one gets a long exact sequence of mixed Hodge structures:

$$\cdots \to \mathbb{H}^{k-n}(Z; i^! IC_X) \to IH^k(X) \to IH^k(U) \to \cdots \qquad (11.27)$$

Moreover, Proposition 11.2.4 yields that for every integer k the mixed Hodge structures $\mathbb{H}^{k-n}(Z; i^! IC_X)$, $IH^k(X)$, and $IH^k(U)$ have weight $\geq k$. Taking gr_k^W in (11.27) proves (b).   □

An immediate consequence of Propositions 11.4.2 is the following estimate of the intersection homology Betti numbers:

Corollary 11.4.3 *Let X be a compact pure-dimensional complex algebraic variety, let $U \subset X$ be a nonsingular open subvariety, and let $\widetilde{X} \to X$ be a resolution of singularities. Then, for all integers k, one has that*

$$\max\{\dim gr_k^W H_c^k(U), \dim gr_k^W H^k(U)\} \le \dim IH^k(X) \le \dim H^k(\widetilde{X}).$$

Indeed, by Corollary 9.3.38 of the BBDG decomposition theorem, $IH^k(X; \mathbb{Q})$ is a direct summand of $H^k(\widetilde{X}; \mathbb{Q})$. In fact, the upper bound of the above estimate is realized when $\widetilde{X}$ is a small resolution (see Corollary 9.3.14), while the lower bound is realized, for example, if X has only isolated singularities.

As a consequence of Propositions 11.4.1 and 11.4.2 and of Corollary 11.3.5, one obtains the following:

Theorem 11.4.4 *Let X be a compact complex algebraic variety of pure complex dimension n, and let U be a nonsingular open subvariety with $Z = X - U$ of dimension $\le d$. Then (with $\mathbb{Q}$-coefficients):*

$$IH^k(X) \cong \begin{cases} H_c^k(U), & k > n+d, \\ gr_{n+d}^W H_c^{n+d}(U), & k = n+d, \\ gr_{n-d}^W H^{n-d}(U), & k = n-d, \\ H^k(U), & k < n-d. \end{cases}$$

In order to obtain a more explicit calculation of the outer intersection homology Betti numbers of an algebraic variety, Durfee used the notion of *weighted Euler characteristic*, defined as follows:

Definition 11.4.5 The *weighted Euler characteristic* of a complex algebraic variety Y is given by

$$\chi_m(Y) = \sum_i (-1)^i \cdot \dim gr_m^W H^i(Y),$$

where $W_\bullet$ denotes the weight filtration of the canonical mixed Hodge structure on $H^*(Y; \mathbb{Q})$. Similarly, one defines $\chi_m^c(Y)$ by using $H_c^*(Y)$.

One then has the following:

Theorem 11.4.6 *Let X be a compact complex algebraic variety of pure dimension n, and let Z be a closed subvariety of complex dimension $\le d$ containing the singular set of X. Let $\pi : \widetilde{X} \to X$ be an algebraic map with $\widetilde{X}$ smooth and $\pi^{-1}(Z) = E = E_1 \cup \cdots \cup E_r$ a divisor with simple normal crossings. Suppose that π is an analytic isomorphism of $\widetilde{U} = \widetilde{X} - E$ to $U = X - Z$. Then, for all $k \ge n+d$, one has (with $\mathbb{Q}$-coefficients):*

$$\dim I H^k(X) = \dim H^k(\widetilde{X}) - \sum_j (-1)^j \cdot \dim H^k(E_j). \tag{11.28}$$

Proof By Theorem 11.4.4, and the fact that $I H^k(X)$ is of pure weight k (Corollary 11.3.5) and the weights on $H_c^k(U) \cong \left(H^{2n-k}(U)(n) \right)^\vee$ are $\leq k$ for all k (by Theorem 11.1.19 and Poincaré duality), one gets that

$$(-1)^k \cdot \dim I H^k(X) = (-1)^k \cdot \dim gr_k^W H_c^k(U) = \chi_k^c(U),$$

for all $k \geq n + d$. The exact sequence of mixed Hodge structures

$$\cdots \longrightarrow H_c^i(\widetilde{U}) \longrightarrow H^i(\widetilde{X}) \longrightarrow H^i(E) \longrightarrow \cdots$$

yields

$$\chi_k^c(\widetilde{U}) = \chi_k^c(\widetilde{X}) - \chi_k^c(E).$$

Since $\widetilde{X}$ is nonsingular and compact, its cohomology is pure, so

$$\chi_k^c(\widetilde{X}) = (-1)^k \cdot \dim H^k(\widetilde{X}).$$

Finally, inclusion–exclusion yields that

$$\chi_k^c(E) = \sum_j (-1)^j \cdot \chi_k^c(E_j) = \sum_j (-1)^{j+k} \cdot \dim H^k(E_j),$$

which completes the proof.   □

Remark 11.4.7 By duality, (11.28) also gives the intersection homology Betti numbers for $k \leq n - d$. Furthermore, a similar argument gives the formula for the corresponding intersection cohomology Hodge numbers.

Chapter 12
Epilogue

In this last chapter, we provide a succinct summary of (and relevant references for) some of the recent applications (other than those already discussed) of intersection homology, perverse sheaves, and mixed Hodge modules in various fields such as topology, algebraic and enumerative geometry, representation theory, etc. This list of applications is by no means exhaustive, but rather reflects the author's own mathematical taste. While the discussion below is limited to a small fraction of the possible routes the interested reader might explore, it should nevertheless serve as a starting point for those interested in aspects of intersection homology, perverse sheaves and mixed Hodge modules in other areas than those already considered in the text.

12.1 Applications to Enumerative Geometry

There are recent applications of vanishing cycles and perverse sheaves in the context of enumerative geometry, more specifically, in *Donaldson–Thomas (DT) theory*.

Given a moduli space $\mathcal{M}$ of stable coherent sheaves on a Calabi–Yau threefold, the Donaldson–Thomas theory associates to it an integer $\chi_{\mathrm{vir}}(\mathcal{M})$ that is invariant under deformations of complex structures. In [10], Behrend showed that the Donaldson–Thomas invariant $\chi_{\mathrm{vir}}(\mathcal{M})$ can be computed as the weighted Euler characteristic over $\mathcal{M}$ of a certain constructible function, the *Behrend function* $v_{\mathcal{M}}$.

As many interesting constructible functions can be obtained by taking stalkwise Euler characteristics of constructible complexes of sheaves of vector spaces, one of the natural questions in DT theory concerns the *categorification* of the Donaldson–Thomas invariant. Specifically, one would like to find a constructible complex of vector spaces $\Phi_{\mathcal{M}} \in D^b_c(\mathcal{M})$ whose stalkwise Euler characteristic is the Behrend function $v_{\mathcal{M}}$ or, equivalently, whose Euler characteristic $\chi(\mathcal{M}, \Phi_{\mathcal{M}})$ computes the Donaldson–Thomas invariant $\chi_{\mathrm{vir}}(\mathcal{M})$. An intermediate level of categorification

© Springer Nature Switzerland AG 2019
L. G. Maxim, *Intersection Homology & Perverse Sheaves*, Graduate Texts
in Mathematics 281, https://doi.org/10.1007/978-3-030-27644-7_12

would associate to the moduli space $\mathcal{M}$ a *cohomological Donaldson–Thomas invariant*, i.e., a finite dimensional graded vector space $\mathcal{H}^*(\mathcal{M})$ such that

$$\chi_{\mathrm{vir}}(\mathcal{M}) = \sum_i (-1)^i \dim \mathcal{H}^i(\mathcal{M}).$$

Of course, once $\Phi_{\mathcal{M}}$ is known, a cohomological DT invariant can simply be defined as

$$\mathcal{H}^*(\mathcal{M}) := \mathbb{H}^*(\mathcal{M}; \Phi_{\mathcal{M}}),$$

the hypercohomology of $\Phi_{\mathcal{M}}$.

If $\mathcal{M}$ is nonsingular, Behrend's construction already implies that $\mathcal{H}^*(\mathcal{M})$ can be taken to be the hypercohomology $\mathbb{H}^*(\mathcal{M}; \underline{\mathbb{Q}}_{\mathcal{M}}[\dim \mathcal{M}])$ of the perverse sheaf $\Phi_{\mathcal{M}} := \underline{\mathbb{Q}}_{\mathcal{M}}[\dim \mathcal{M}]$ on $\mathcal{M}$.

Furthermore, if the moduli space $\mathcal{M}$ is the scheme-theoretic critical locus of some function $f : X \to \mathbb{C}$ defined on a smooth complex quasi-projective variety X,[1] a cohomological DT invariant can again be read off from Behrend's work, namely

$$\mathcal{H}^*(\mathcal{M}) = \mathbb{H}^*(\mathcal{M}; {}^P\varphi_f \underline{\mathbb{Q}}_{\mathcal{M}}[\dim \mathcal{M}]),$$

with $\Phi_{\mathcal{M}} := {}^P\varphi_f \underline{\mathbb{Q}}_{\mathcal{M}}[\dim \mathcal{M}] \in Perv(\mathcal{M})$ the self-dual complex of perverse vanishing cycles of f. It should be noted that the cohomological DT invariant $\mathcal{H}^*(\mathcal{M})$ carries in this case the action of the monodromy endomorphism induced from that of the vanishing cycles, and it is also endowed with a mixed Hodge structure since the perverse vanishing cycle complex underlies the mixed Hodge module $\varphi_f^H \underline{\mathbb{Q}}_{\mathcal{M}}^H[\dim \mathcal{M}]$.

More generally, it is known that a moduli space $\mathcal{M}$ of simple coherent sheaves on a Calabi–Yau threefold is, locally around every closed point, isomorphic to a critical locus. Then it can be shown [22] that the perverse sheaves of vanishing cycles on the critical charts glue to a self-dual global perverse sheaf $\Phi_{\mathcal{M}} \in Perv(\mathcal{M})$, the *DT sheaf* on $\mathcal{M}$, whose Euler characteristic $\chi(\mathcal{M}, \Phi_{\mathcal{M}})$ computes $\chi_{\mathrm{vir}}(\mathcal{M})$. Hence $\Phi_{\mathcal{M}}$ categorifies $\chi_{\mathrm{vir}}(\mathcal{M})$, and the graded vector space $\mathcal{H}^*(\mathcal{M}) = \mathbb{H}^*(\mathcal{M}; \Phi_{\mathcal{M}})$ is a cohomological Donaldson–Thomas invariant of $\mathcal{M}$. Furthermore, $\mathcal{H}^*(\mathcal{M})$ is endowed with a canonical mixed Hodge structure since $\Phi_{\mathcal{M}}$ underlies a mixed Hodge module, and it carries a monodromy action glued from the local monodromy endomorphisms.

We refer to [227] for a survey and for an extensive list of references.

---

[1]For example, this is the case for $\mathcal{M} = \mathrm{Hilb}^m_{\mathbb{C}^3}$, the Hilbert scheme of m points on $\mathbb{C}^3$.

12.2   Characteristic Classes of Complex Algebraic Varieties and Applications

There is an extensive theory of characteristic classes on stratified spaces, encompassing many different constructions and viewpoints. Part of the story (concerning L-classes) was already presented in Chapter 3. Here we focus on the more recent Hodge-theoretic characteristic classes, the *homology Hirzebruch classes* of singular complex algebraic varieties [21], and their various applications.

Characteristic classes of singular spaces are usually defined in (Borel–Moore) homology, as images of certain distinguished elements by a natural transformation on the Grothendieck group of suitable (e.g., constructible or coherent) sheaves, see e.g., [9, 31, 148].

Building on Saito's theory, Brasselet–Schürmann–Yokura defined in [21] the Hirzebruch class transformation

$$T_{y*} : K_0(MHM(X)) \longrightarrow H_*^{BM}(X) \otimes \mathbb{Q}[y, y^{-1}],$$

which assigns Borel–Moore homology classes to every Grothendieck class of a (complex of) mixed Hodge module(s) on a complex algebraic variety X. The value

$$T_{y*}(X) := T_{y*}([\underline{\mathbb{Q}}_X^H])$$

on the Grothendieck class of the constant Hodge module $\underline{\mathbb{Q}}_X^H$ is called the *homology Hirzebruch class of X*. These Hirzebruch classes have good functorial and normalization properties, e.g., for X nonsingular $T_{y*}(X)$ is Poincaré dual to the cohomology Hirzebruch class $T_y^*(TX)$ of the generalized Hirzebruch–Riemann–Roch theorem. Moreover, the Hirzebruch class transformation provides a functorial unification of the Chern class transformation of MacPherson [148], Todd class transformation of Baum–Fulton–MacPherson [9], and L-class transformation of Cappell–Shaneson [31] (as reformulated in [247]; see also [21, Section 4]), respectively, thus answering positively an old question of MacPherson about the existence of such a unifying theory. If the variety X is compact and $\mathcal{M}^\bullet \in D^b MHM(X)$ is a complex of mixed Hodge modules on X, the degree of the homology Hirzebruch class $T_{y*}([\mathcal{M}^\bullet])$ is the *Hodge polynomial* $\chi_y(X, \mathcal{M}^\bullet)$ corresponding to Saito's mixed Hodge structure on the hypercohomology $\mathbb{H}^*(X; rat(\mathcal{M}^\bullet))$.

Soon after their introduction, Hirzebruch classes have found a vast array of applications. For example, they were used in [161] to compute characteristic classes of toric varieties in terms of the Todd classes of closures of orbits, with applications to weighted lattice point counting in polytopes and generalized Pick-type formulae. Equivariant versions (for finite group actions) of homology Hirzebruch classes were developed in [36], and they were used to compute characteristic classes of symmetric products [38] as well as characteristic classes of Hilbert schemes of points [37], with applications to enumerative geometry (proof of a characteristic class version of the MNOP conjecture of Donaldson–Thomas theory [158]).

Homology Hirzebruch classes turn out to be extremely useful for understanding global invariants of complex hypersurfaces. Let $X = \{f = 0\}$ be the complex algebraic variety defined as the zero set (of complex codimension one) of an algebraic function $f : M \to \mathbb{C}$, for M a complex $(n + 1)$-dimensional algebraic manifold. The value

$$\mathcal{M}T_{y*}(X) := T_{y*}(\varphi_f^H \underline{\mathbb{Q}}_M^H[1])$$

of the Hirzebruch class transformation on the Grothendieck class of the vanishing cycle complex is supported on the singular locus of the hypersurface X, and hence it is a characteristic class measure of the "size" of singularities of X. The homology class $\mathcal{M}T_{y*}(X)$ is termed the *Hirzebruch–Milnor class of X*, as it is a far-reaching generalization of the Milnor number of isolated hypersurface singularities. For computations of these classes in terms of a stratification of the singular locus (providing, in particular, Hodge-theoretic generalizations of results from Section 10.4) see, e.g., [35] and [165]; see also [248] for a motivic treatment of such classes.

In [168], refined versions of Hirzebruch–Milnor classes of complex hypersurfaces are defined by taking into account the monodromy action on vanishing cycles. The obtained classes, termed *spectral Hirzebruch–Milnor classes* of X, may be viewed as a homology class version of the *Steenbrink spectrum*, and have surprising applications to the study of birational geometry invariants such as multiplier ideals, jumping coefficients, log canonical threshold, Du Bois singularities, etc.; see [168] for complete details.

For a comprehensive survey and more references about the many facets of Hirzebruch classes and their applications, see also [215] and [162].

12.3 Perverse Sheaves on Semi-Abelian Varieties: Cohomology Jump Loci, Propagation, Generic Vanishing

The study of perverse sheaves on (semi-)abelian varieties has seen a flurry of activity in recent years. Let us briefly describe some motivation and recent results.

Let X be a connected complex quasi-projective manifold. The character variety $Char(X) \cong (\mathbb{C}^*)^{b_1(X)}$, with $b_1(X) = \mathrm{rank}\,H_1(X; \mathbb{Z})$, is the connected component of $\mathrm{Hom}(\pi_1(X), \mathbb{C}^*)$ containing the identity character. Each character $\rho \in Char(X)$ defines a unique rank-one $\mathbb{C}$-local system $\mathcal{L}_\rho$ on X. The *cohomology jump loci of a constructible $\mathbb{C}$-complex* $\mathcal{F}^\bullet \in D_c^b(X)$ on X are defined as:

$$\mathcal{V}^i(X, \mathcal{F}^\bullet) := \{\rho \in Char(X) \mid \mathbb{H}^i(X; \mathcal{F}^\bullet \otimes_{\mathbb{C}} \mathcal{L}_\rho) \neq 0\} \subseteq Char(X).$$

These are generalizations of the (topological) cohomology jump loci

$$\mathcal{V}^i(X) := \mathcal{V}^i(X, \underline{\mathbb{C}}_X)$$

of X, which correspond to the constant sheaf $\underline{\mathbb{C}}_X$, and which are homotopy invariants of X. Cohomology jump loci provide a unifying framework for the study of a host of questions concerning homotopy types of complex algebraic varieties. In particular, they can be used to tackle Serre's problem concerning groups that can be realized as fundamental groups of complex quasi-projective manifolds.

It was recently shown in [28] that the irreducible components of the cohomology jump loci of bounded $\mathbb{C}$-constructible complexes on a complex algebraic manifold X are *linear* subvarieties. In particular, each $\mathcal{V}^i(X, \mathcal{F}^{\bullet})$ is a finite union of translated subtori of the character variety $Char(X)$. This fact vastly generalizes the classical monodromy theorem and imposes strong constraints on the topology of X.

By the classical Albanese map construction, cohomology jump loci of a complex quasi-projective manifold X are realized as cohomology jump loci of constructible complexes of sheaves (or, if the Albanese map is proper, of perverse sheaves) on the semi-abelian variety $\text{Alb}(X)$, the *Albanese variety* of X.[2] This motivates the investigation of cohomology jump loci of constructible complexes, and in particular of perverse sheaves, on complex semi-abelian varieties.[3]

Perverse sheaves on complex affine tori have been studied by Gabber–Loeser [78] via the *Mellin transformation*, whereas perverse sheaves on complex abelian varieties have been completely characterized by Schnell [213] in terms of certain codimension lower bounds of their cohomology jump loci.

Let $\mathcal{F}^{\bullet} \in D_c^b(\mathbb{G})$ be a bounded constructible complex of $\mathbb{C}$-sheaves on a semi-abelian variety $\mathbb{G}$, with cohomology jump loci $\mathcal{V}^i(\mathbb{G}, \mathcal{F}^{\bullet})$. By using the linear structure of irreducible components of the cohomology jump loci, in [145] we introduced refined notions of *(semi-)abelian codimensions*, $\text{codim}_{sa} \mathcal{V}^i(\mathbb{G}, \mathcal{F}^{\bullet})$, and $\text{codim}_a \mathcal{V}^i(\mathbb{G}, \mathcal{F}^{\bullet})$, and showed that the position of $\mathcal{F}^{\bullet}$ with respect to the perverse t-structure on $D_c^b(\mathbb{G})$ can be detected by the (semi-)abelian codimension of its cohomology jump loci. This result provides a complete characterization of $\mathbb{C}$-perverse sheaves on a semi-abelian variety $\mathbb{G}$ in terms of their cohomology jump loci, and generalizes Schnell's corresponding result [213, Theorem 7.4] for perverse sheaves on abelian varieties, as well as Gabber–Loeser's description [78] of perverse sheaves on complex affine tori (see also [144]).

---

[2]The Albanese map alb : $X \rightarrow \text{Alb}(X)$ induces an isomorphism on the free part of H_1. In particular, $Char(X) \cong Char(\text{Alb}(X))$.

[3]A complex *abelian variety* of dimension g is a compact complex torus $\mathbb{C}^g / \mathbb{Z}^{2g}$ that is also a complex projective variety. A *semi-abelian variety* $\mathbb{G}$ is an abelian complex algebraic group that is an extension

$$1 \rightarrow \mathbb{T} \rightarrow \mathbb{G} \rightarrow \mathbb{A} \rightarrow 1,$$

where $\mathbb{A}$ is an abelian variety and $\mathbb{T}$ is a complex affine torus.

Furthermore, the cohomology jump loci of perverse sheaves on semi-abelian varieties satisfy the following list of properties, collectively termed the *propagation package*, see [145]:

Theorem 12.3.1 *Let* $\mathbb{G}$ *be a semi-abelian variety defined as the extension*

$$1 \to \mathbb{T} \to \mathbb{G} \to \mathbb{A} \to 1,$$

where $\mathbb{A}$ *is an abelian variety of complex dimension g and* $\mathbb{T} = (\mathbb{C}^*)^m$ *is a complex affine torus of dimension m. The cohomology jump loci of a* $\mathbb{C}$*-perverse sheaf* $\mathcal{F}^\bullet$ *on* $\mathbb{G}$ *satisfy the following:*

(i) Propagation property:

$$\mathcal{V}^{-m-g}(\mathbb{G}, \mathcal{F}^\bullet) \subseteq \cdots \subseteq \mathcal{V}^{-1}(\mathbb{G}, \mathcal{F}^\bullet) \subseteq \mathcal{V}^0(\mathbb{G}, \mathcal{F}^\bullet),$$

$$\mathcal{V}^0(\mathbb{G}, \mathcal{F}^\bullet) \supseteq \mathcal{V}^1(\mathbb{G}, \mathcal{F}^\bullet) \supseteq \cdots \supseteq \mathcal{V}^g(\mathbb{G}, \mathcal{F}^\bullet).$$

Furthermore, $\mathcal{V}^i(\mathbb{G}, \mathcal{F}^\bullet) = \emptyset$ *if* $i \notin [-m - g, g]$.
(ii) Generic vanishing: there exists a non-empty Zariski-open subset U *of* $Char(\mathbb{G})$ *such that, for every closed point* $\rho \in U$, $\mathbb{H}^i(\mathbb{G}; \mathcal{F}^\bullet \otimes_{\mathbb{C}} \mathcal{L}_\rho) = 0$ *for all* $i \neq 0$.
(iii) Signed Euler characteristic property:

$$\chi(\mathbb{G}, \mathcal{F}^\bullet) \geq 0.$$

Moreover, the equality holds if and only if $\mathcal{V}^0(\mathbb{G}, \mathcal{F}^\bullet) \neq Char(\mathbb{G})$.

An equivalent (and perhaps more suggestive) formulation of the propagation property can be given as follows. Let $\mathcal{F}^\bullet$ be a $\mathbb{C}$-perverse sheaf on a semi-abelian variety $\mathbb{G}$ so that not all $\mathbb{H}^j(\mathbb{G}; \mathcal{F}^\bullet)$ are zero. Let

$$k_+ := \max\{j \mid \mathbb{H}^j(\mathbb{G}; \mathcal{F}^\bullet) \neq 0\} \text{ and } k_- := \min\{j \mid \mathbb{H}^j(\mathbb{G}; \mathcal{F}^\bullet) \neq 0\}.$$

Then the propagation property (i) is equivalent to the following assertions: $k_+ \geq 0$, $k_- \leq 0$ and

$$\mathbb{H}^j(\mathbb{G}; \mathcal{F}^\bullet) \neq 0 \iff k_- \leq j \leq k_+.$$

Some of the properties of Theorem 12.3.1 have been also obtained by other authors by different methods. For example, the generic vanishing property (ii) for perverse sheaves on complex affine tori, abelian varieties and, respectively, semi-abelian varieties was proved in various settings, see, e.g., [130, Theorem 2.1], [131, Theorem 1.1], [213, Corollary 7.5], [239, Vanishing Theorem], [13, 144, Theorem 1.1]. The signed Euler characteristic property (iii) is originally due to Franecki and Kapranov [73, Corollary 1.4].

The results mentioned in this section have a wide range of applications, including to the study of cohomology jump loci (and hence of the homotopy type) of complex quasi-projective manifolds, for understanding the topology of the Albanese map, as well as in the context of homological duality properties of complex algebraic varieties. Let us only mention here the following corollary of Theorem 12.3.1; for more details the reader may consult [143, 144, 145].

Corollary 12.3.2 *Let X be a complex quasi-projective manifold of complex dimension n, with Albanese map* alb : $X \to \mathrm{Alb}(X)$. *Assume that* $R\mathrm{alb}_* \underline{\mathbb{C}}_X[n]$ *is a perverse sheaf on* $\mathrm{Alb}(X)$ *(e.g.,* alb *is proper and semi-small). Then the cohomology jump loci* $\mathcal{V}^i(X)$ *of X satisfy the following properties:*

(1) Propagation:

$$\mathcal{V}^n(X) \supseteq \mathcal{V}^{n-1}(X) \supseteq \cdots \supseteq \mathcal{V}^0(X) = \{1\};$$

$$\mathcal{V}^n(X) \supseteq \mathcal{V}^{n+1}(X) \supseteq \cdots \supseteq \mathcal{V}^{2n}(X).$$

(2) Codimension lower bound: for all $i \geq 0$,

$$\mathrm{codim}\, \mathcal{V}^{n-i}(X) \geq i \ \text{ and } \ \mathrm{codim}\, \mathcal{V}^{n+i}(X) \geq 2i.$$

(3) Generic vanishing: $H^i(X; \mathcal{L}_\rho) = 0$ *for generic* $\rho \in \mathrm{Char}(X)$ *and all* $i \neq n$.
(4) Signed Euler characteristic property: $(-1)^n \cdot \chi(X) \geq 0$.
(5) Betti property: if $b_i(X)$ denotes the i-th Betti number of X, then $b_i(X) > 0$ for every $i \in [0, n]$, and $b_1(X) \geq n$.

12.4 Generic Vanishing Theory via Mixed Hodge Modules

There is a corresponding propagation package in the coherent setting, which describes the behavior of the cohomology of a line bundle as it varies over the Picard torus. Such results originate from the work of Green–Lazarsfeld [90], who showed that on a complex projective manifold X, the cohomology of a generic line bundle $L \in Pic^0(X)$ vanishes in degrees below $\dim \mathrm{alb}(X)$, where alb : $X \to \mathrm{Alb}(X)$ is the Albanese map of X. This is a consequence of a more general result involving the *algebraic cohomology jump loci* of the structure sheaf $\mathcal{O}_X$, namely

$$\mathrm{codim}\, V^i(\mathcal{O}_X) \geq \dim \mathrm{alb}(X) - i, \tag{12.1}$$

where for a coherent sheaf $\mathcal{F}$ on X one sets

$$V^i(\mathcal{F}) := \{L \in Pic^0(X) \mid H^i(X; \mathcal{F} \otimes L) \neq 0\} \subset Pic^0(X).$$

The Green–Lazarsfeld generic vanishing theorem has been recently extended by Popa–Schnell in [200] to coherent sheaves of a Hodge-theoretic nature. More precisely, one has the following result:

Theorem 12.4.1 *Let $\mathbb{A}$ be a complex abelian variety, and let $\mathcal{M}$ be a mixed Hodge module on $\mathbb{A}$ with underlying filtered D-module $(\mathcal{M}, F_\bullet)$. Then for each $k \in \mathbb{Z}$, the algebraic cohomology jump loci of the coherent sheaf $gr_k^F \mathcal{M}$ satisfy the following codimension lower bound:*

$$\mathrm{codim}\, V^i(gr_k^F \mathcal{M}) \geq i, \tag{12.2}$$

for all $i > 0$. In particular, each $gr_k^F \mathcal{M}$ is a GV-sheaf on $\mathbb{A}$ (i.e., it satisfies generic vanishing).

The use of mixed Hodge module theory allows one to obtain a broader picture, and to extract much more information than previously known. One important example is a formula for the codimension of the cohomology jump loci $V^i(\Omega_X^p)$ for *all* bundles of holomorphic forms on a projective manifold X. This can be obtained by applying Theorem 12.4.1 to the direct image $R\mathrm{alb}_* \underline{\mathbb{Q}}_X^H[\dim X]$ and using the decomposition theorem for Hodge modules (see Theorem 11.2.11):

Theorem 12.4.2 *Let X be a complex projective manifold of dimension n. Then*

$$\mathrm{codim}\, V^q(\Omega_X^p) \geq |p + q - n| - r(\mathrm{alb}), \tag{12.3}$$

for every $p, q \in \mathbb{N}$, where $r(\mathrm{alb})$ is the defect of semi-smallness of the Albanese map of X. Moreover, the Albanese map is semi-small if and only if X satisfies the generic Nakano vanishing theorem, i.e.,

$$\mathrm{codim}\, V^q(\Omega_X^p) \geq |p + q - n|$$

for every $p, q \in \mathbb{N}$.

This also shows that if the Albanese map of X is semi-small, then

$$(-1)^{n-p}\chi(X, \Omega_X^p) \geq 0,$$

which leads to interesting new bounds for the topological Euler characteristic of irregular varieties.

Generic vanishing theorems for Ω_X^p have also been obtained in [131] by different methods.

A coherent analogue of the propagation property (i) of Theorem 12.3.1 was proved in [192] for the algebraic cohomology jump loci of GV-sheaves on abelian varieties.

For more details on cohomology jump loci (both algebraic and topological) as well as additional references, the reader may benefit from consulting the comprehensive surveys [199] and [29]. In particular, [199] gives an overview of various vanishing and positivity theorems for Hodge modules.

12.5 Applications to Representation Theory

The importance of intersection cohomology in representation theory was already acknowledged in Lusztig's work, e.g., see his ICM address [146] for an excellent survey. We also mention here the *geometric Satake equivalence* of Mirković–Vilonen [181], which identifies equivariant perverse sheaves on the affine Grassmannian Gr_G with the Langlands dual group of a reductive group G. For more applications of the BBDG decomposition theorem in geometric representation theory of reductive algebraic groups, see [43]. More recently, striking applications of intersection homology, perverse sheaves, and the decomposition theorem in representation theory (e.g., in relation to the Kazhdan–Lusztig conjecture) were obtained by Williamson and his collaborators, see [244] and [71] for nice overviews of recent developments.

12.6 Alexander-Type Invariants of Complex Hypersurface Complements

As already indicated in Section 10.2, intersection homology and perverse sheaves play an important role in understanding the topology of abelian covers of complex hypersurface complements in terms of local invariants of singularities. For more details, the reader may consult [62, 63, 141, 142, 160, 163], and also [61, Section 6.4].

Bibliography

1. A'Campo, N.: Le nombre de Lefschetz d'une monodromie. Nederl. Akad. Wetensch. Proc. Ser. A **76**, Indag. Math. **35**, 113–118 (1973)
2. Andreotti, A., Frankel, T.: The Lefschetz theorem on hyperplane sections. Ann. Math. (2) **69**, 713–717 (1959)
3. Ardila, F., Boocher, A.: The closure of a linear space in a product of lines. J. Algebraic, Combin. **43**(1), 199–235 (2016)
4. Banagl, M.: Extending Intersection Homology Type Invariants to Non-Witt Spaces. Memoirs of the American Mathematical Society, vol. 160, no. 760. American Mathematical Society, Providence, RI (2002)
5. Banagl, M.: The L-class of non-Witt spaces. Ann. Math. (2) **163**(3), 743–766 (2006)
6. Banagl, M.: Topological Invariants of Stratified Spaces. Springer Monographs in Mathematics. Springer, Berlin (2007)
7. Barthel, G., Brasselet, J.-P., Fieseler, K.-H., Gabber, O., Kaup, L.: Relèvement de cycles algébriques et homomorphismes associés en homologie d'intersection. Ann. Math. (2) **141**(1), 147–179 (1995)
8. Barthel, G., Brasselet, J.-P., Fieseler, K.-H., Kaup, L.: Combinatorial intersection cohomology for fans. Tohoku Math. J. (2) **54**(1), 1–41 (2002)
9. Baum, P., Fulton, W., MacPherson, R.: Riemann-Roch for singular varieties. Inst. Hautes Études Sci. Publ. Math. **45**, 101–145 (1975)
10. Behrend, K.: Donaldson-Thomas type invariants via microlocal geometry. Ann. Math. (2) **170**(3), 1307–1338 (2009)
11. Beĭlinson, A.A.: On the derived category of perverse sheaves. In: K-Theory, Arithmetic and Geometry (Moscow, 1984–1986). Lecture Notes in Mathematics, vol. 1289, pp. 27–41. Springer, Berlin (1987)
12. Beĭlinson, A.A., Bernstein, J.N., Deligne, P.: Faisceaux pervers. In: Analysis and Topology on Singular Spaces, I (Luminy, 1981). Astérisque, vol. 100, pp. 5–171. Soc. Math. France, Paris (1982)
13. Bhatt, B., Schnell, C., Scholze, P.: Vanishing theorems for perverse sheaves on abelian varieties, revisited. Selecta Math. (N.S.) **24**(1), 63–84 (2018)
14. Billera, L.J., Lee, C.W.: Sufficiency of McMullen's conditions for f-vectors of simplicial polytopes. Bull. Amer. Math. Soc. (N.S.) **2**(1), 181–185 (1980)
15. Borel, A.: Intersection cohomology. Modern Birkhäuser Classics. Birkhäuser Boston, Inc., Boston, MA (2008)
16. Borel, A., Moore, J.C.: Homology theory for locally compact spaces. Michigan Math. J. **7**, 137–159 (1960)

© Springer Nature Switzerland AG 2019
L. G. Maxim, *Intersection Homology & Perverse Sheaves*, Graduate Texts in Mathematics 281, https://doi.org/10.1007/978-3-030-27644-7

17. Borho, W., MacPherson, R.: Représentations des groupes de Weyl et homologie d'intersection pour les variétés nilpotentes. C. R. Acad. Sci. Paris Sér. I Math. **292**(15), 707–710 (1981)

18. Borho, W., MacPherson, R.: Partial resolutions of nilpotent varieties. In: Analysis and Topology on Singular Spaces, II, III (Luminy, 1981). Astérisque, vol. 101, pp. 23–74. Soc. Math. France, Paris (1983)

19. Bott, R., Tu, L.W.: Differential Forms in Algebraic Topology. Graduate Texts in Mathematics, vol. 82. Springer, New York/Berlin (1982)

20. Braden, T.: Remarks on the combinatorial intersection cohomology of fans. Pure Appl. Math. Q. **2**(4, Special Issue: In honor of Robert D. MacPherson. Part 2), 1149–1186 (2006)

21. Brasselet, J.-P., Schürmann, J., Yokura, S.: Hirzebruch classes and motivic Chern classes for singular spaces. J. Topol. Anal. **2**(1), 1–55 (2010)

22. Brav, C., Bussi, V., Dupont, D., Joyce, D.D., Szendrői, B.: Symmetries and stabilization for sheaves of vanishing cycles. J. Singul. **11**, 85–151 (2015). With an appendix by Jörg Schürmann

23. Bredon, G.E.: Sheaf Theory. Graduate Texts in Mathematics, vol. 170, 2nd edn. Springer, New York (1997)

24. Bressler, P., Lunts, V.A.: Intersection cohomology on nonrational polytopes. Compos. Math. **135**(3), 245–278 (2003)

25. Brieskorn, E.: Beispiele zur Differentialtopologie von Singularitäten. Invent. Math. **2**, 1–14 (1966)

26. Brieskorn, E.: Die Monodromie der isolierten Singularitäten von Hyperflächen. Manuscripta Math. **2**, 103–161 (1970)

27. Browder, W.: Surgery on Simply-Connected Manifolds. Springer, New York/Heidelberg (1972). Ergebnisse der Mathematik und ihrer Grenzgebiete, Band 65

28. Budur, N., Wang, B.: Absolute sets and the Decomposition Theorem. Ann Sci. École Norm. Sup. (to appear). arXiv e-prints. art. arXiv:1702.06267, Feb 2017

29. Budur, N., Wang, B.: Recent results on cohomology jump loci. In: Hodge Theory and L^2-Analysis. Adv. Lect. Math. (ALM), vol. 39, pp. 207–243. Int. Press, Somerville, MA (2017)

30. Budur, N., Liu, Y, Wang, B.: The monodromy theorem for compact Kähler manifolds and smooth quasi-projective varieties. Math. Ann. **371**(3–4), 1069–1086 (2018)

31. Cappell, S.E., Shaneson, J.L.: Singular spaces, characteristic classes, and intersection homology. Ann. Math. (2) **134**(2), 325–374 (1991)

32. Cappell, S.E., Maxim, L., Shaneson, J.L.: Hodge genera of algebraic varieties. I. Commun. Pure Appl. Math. **61**(3), 422–449 (2008)

33. Cappell, S.E., Maxim, L., Shaneson, J.L.: Intersection cohomology invariants of complex algebraic varieties. In: Singularities I. Contemporary Mathematics, vol. 474, pp. 15–24. American Mathematical Society, Providence, RI (2008)

34. Cappell, S.E., Maxim, L., Shaneson, J.L.: Euler characteristics of algebraic varieties. Commun. Pure Appl. Math. **61**(3), 409–421 (2008)

35. Cappell, S.E., Maxim, L., Schürmann, J., Shaneson, J.L.: Characteristic classes of complex hypersurfaces. Adv. Math. **225**(5), 2616–2647 (2010)

36. Cappell, S.E., Maxim, L., Schürmann, J., Shaneson, J.L.: Equivariant characteristic classes of singular complex algebraic varieties. Commun. Pure Appl. Math. **65**(12), 1722–1769 (2012)

37. Cappell, S., Maxim, L., Ohmoto, T., Schürmann, J., Yokura, S.: Characteristic classes of Hilbert schemes of points via symmetric products. Geom. Topol. **17**(2), 1165–1198 (2013)

38. Cappell, S.E., Maxim, L., Schürmann, J., Shaneson, J.L., Yokura, S.: Characteristic classes of symmetric products of complex quasi-projective varieties. J. Reine Angew. Math. **728**, 35–63 (2017)

39. Cheeger, J., Goresky, M., MacPherson, R.: L^2-cohomology and intersection homology of singular algebraic varieties. In: Seminar on Differential Geometry. Annals of Mathematics Studies, vol. 102, pp. 303–340. Princeton Univ. Press, Princeton, N.J. (1982)

40. Chern, S.-S.: A simple intrinsic proof of the Gauss-Bonnet formula for closed Riemannian manifolds. Ann. Math. (2) **45**, 747–752 (1944)

41. Chern, S.-S.: Characteristic classes of Hermitian manifolds. Ann. Math. (2) **47**, 85–121 (1946)

42. Chern, S.-S.: On the multiplication in the characteristic ring of a sphere bundle. Ann. Math. (2) **49**, 362–372 (1948)

43. Chriss, N., Ginzburg, V.: Representation Theory and Complex Geometry. Birkhäuser Boston, Inc., Boston, MA (1997)

44. Clemens, H.: Picard-Lefschetz theorem for families of nonsingular algebraic varieties acquiring ordinary singularities. Trans. Am. Math. Soc. **136**, 93–108 (1969)

45. Cox, D.A., Little, J.B., Schenck, H.K.: Toric Varieties. Graduate Studies in Mathematics, vol. 124. American Mathematical Society, Providence, RI (2011)

46. de Bruijn, N.G., Erdös, P.: On a combinatorial problem. Nederl. Akad. Wetensch. Proc. **51**, 1277–1279, Indag. Math. **10**, 421–423 (1948)

47. de Cataldo, M.A.A.: Perverse sheaves and the topology of algebraic varieties. In: Geometry of Moduli Spaces and Representation Theory. IAS/Park City Math. Ser., vol. 24, pp. 1–58. American Mathematical Society, Providence, RI (2017)

48. de Cataldo, M.A.A., Migliorini, L.: The hard lefschetz theorem and the topology of semismall maps. Ann. Sci. École Norm. Sup. (4) **35**(5), 759–772 (2002)

49. de Cataldo, M.A.A., Migliorini, L.: The Hodge theory of algebraic maps. Ann. Sci. École Norm. Sup. (4) **38**(5), 693–750 (2005)

50. de Cataldo, M.A.A., Migliorini, L.: Intersection forms, topology of maps and motivic decomposition for resolutions of threefolds. In: Algebraic Cycles and Motives, vol. 1. London Mathematical Society Lecture Note Series, vol. 343, pp. 102–137. Cambridge University Press, Cambridge (2007)

51. de Cataldo, M.A.A., Migliorini, L.: The decomposition theorem, perverse sheaves and the topology of algebraic maps. Bull. Amer. Math. Soc. (N.S.) **46**(4), 535–633 (2009)

52. de Cataldo, M.A.A., Migliorini, L.: What is... a perverse sheaf? Notices Amer. Math. Soc. **57**(5), 632–634 (2010)

53. de Cataldo, M.A., Migliorini, L., Mustaţă, M.: Combinatorics and topology of proper toric maps. J. Reine Angew. Math. **744**, 133–163 (2018)

54. Deligne, P.: Théorème de Lefschetz et critères de dégénérescence de suites spectrales. Inst. Hautes Études Sci. Publ. Math. **35**, 259–278 (1968)

55. Deligne, P.: Théorie de Hodge. II. Inst. Hautes Études Sci. Publ. Math. **40**, 5–57 (1971)

56. Deligne, P.: Théorie de Hodge. III. Inst. Hautes Études Sci. Publ. Math. **44**, 5–77 (1974)

57. Deligne, P.: Décompositions dans la catégorie dérivée. In: Motives (Seattle, WA, 1991). Proceedings of Symposia in Pure Mathematics, vol. 55, pp. 115–128. American Mathematical Society, Providence, RI (1994)

58. Dieudonné, J.: A History of Algebraic and Differential Topology. 1900–1960. Birkhäuser Boston, Inc., Boston, MA (1989)

59. Dimca, A.: Topics on Real and Complex Singularities. Advanced Lectures in Mathematics. Friedr. Vieweg & Sohn, Braunschweig (1987)

60. Dimca, A.: Singularities and Topology of Hypersurfaces. Universitext. Springer, New York (1992)

61. Dimca, A.: Sheaves in Topology. Universitext. Springer, Berlin (2004)

62. Dimca, A., Libgober, A.: Regular functions transversal at infinity. Tohoku Math. J. (2) **58**(4), 549–564 (2006)

63. Dimca, A., Maxim, L.: Multivariable Alexander invariants of hypersurface complements. Trans. Am. Math. Soc. **359**(7), 3505–3528 (2007)

64. Dimca, A., Némethi, A.: Hypersurface complements, Alexander modules and monodromy. In: Real and Complex Singularities. Contemporary Mathematics, vol. 354, pp. 19–43. American Mathematical Society, Providence, RI (2004)

65. Dowling, T.A., Wilson, R.M.: The slimmest geometric lattices. Trans. Am. Math. Soc. **196**, 203–215 (1974)

66. Dowling, T.A., Wilson, R.M.: Whitney number inequalities for geometric lattices. Proc. Am. Math. Soc. **47**, 504–512 (1975)

67. Durfee, A.H.: Intersection homology Betti numbers. Proc. Am. Math. Soc. **123**(4), 989–993 (1995)

68. Durfee, A.H., Saito, M.: Mixed Hodge structures on the intersection cohomology of links. Compos. Math. **76**(1–2), 49–67 (1990)

69. El Zein, F.: Dégénérescence diagonale. I. C. R. Acad. Sci. Paris Sér. I Math. **296**(1), 51–54 (1983)

70. El Zein, F.: Dégénérescence diagonale. II. C. R. Acad. Sci. Paris Sér. I Math. **296**(4), 199–202 (1983)

71. Elias, B., Williamson, G.: Kazhdan-Lusztig conjectures and shadows of Hodge theory. In: Arbeitstagung Bonn 2013. Progr. Math., vol. 319, pp. 105–126. Birkhäuser/Springer, Cham (2016)

72. Fieseler, K.-H.: Rational intersection cohomology of projective toric varieties. J. Reine Angew. Math. **413**, 88–98 (1991)

73. Franecki, J., Kapranov, M.M.: The Gauss map and a noncompact Riemann-Roch formula for constructible sheaves on semiabelian varieties. Duke Math. J. **104**(1), 171–180 (2000)

74. Fried, D.: Monodromy and dynamical systems. Topology **25**(4), 443–453 (1986)

75. Friedman, G.: Singular Intersection Homology. http://faculty.tcu.edu/gfriedman/IHbook.pdf

76. Friedman, G.: An introduction to intersection homology with general perversity functions. In: Topology of Stratified Spaces. Mathematical Sciences Research Institute Publications, vol. 58, pp. 177–222. Cambridge University Press, Cambridge (2011)

77. Fulton, W.: Introduction to Toric Varieties. Annals of Mathematics Studies, vol. 131. Princeton University Press, Princeton, NJ (1993). The William H. Roever Lectures in Geometry

78. Gabber, O., Loeser, F.: Faisceaux pervers *l*-adiques sur un tore. Duke Math. J. **83**(3), 501–606 (1996)

79. Godement, R.: Topologie algébrique et théorie des faisceaux. Hermann, Paris (1973) Troisième édition revue et corrigée, Publications de l'Institut de Mathématique de l'Université de Strasbourg, XIII, Actualités Scientifiques et Industrielles, No. 1252

80. Goresky, M.: Triangulation of stratified objects. Proc. Am. Math. Soc. **72**(1), 193–200 (1978)

81. Goresky, M., MacPherson, R.: Intersection homology theory. Topology **19**(2), 135–162 (1980)

82. Goresky, M., MacPherson, R.: On the topology of complex algebraic maps. In: Algebraic Geometry (La Rábida, 1981). Lecture Notes in Math., vol. 961, pp. 119–129. Springer, Berlin (1982)

83. Goresky, M., MacPherson, R.: Intersection homology. II. Invent. Math. **72**(1), 77–129 (1983)

84. Goresky, M., MacPherson, R.: Morse theory and intersection homology theory. In: Analysis and Topology on Singular Spaces, II, III (Luminy, 1981). Astérisque, vol. 101, pp. 135–192. Soc. Math. France, Paris (1983)

85. Goresky, M., MacPherson, R.: Stratified Morse Theory. Ergebnisse der Mathematik und ihrer Grenzgebiete (3) [Results in Mathematics and Related Areas (3)], vol. 14. Springer, Berlin (1988)

86. Goresky, M., Siegel, P.: Linking pairings on singular spaces. Comment. Math. Helv. **58**(1), 96–110 (1983)

87. Göttsche, L.: The Betti numbers of the Hilbert scheme of points on a smooth projective surface. Math. Ann. **286**(1–3), 193–207 (1990)

88. Göttsche, L., Soergel, W.: Perverse sheaves and the cohomology of Hilbert schemes of smooth algebraic surfaces. Math. Ann. **296**(2), 235–245 (1993)

89. Grauert, H.: Über Modifikationen und exzeptionelle analytische Mengen. Math. Ann. **146**, 331–368 (1962)

90. Green, M., Lazarsfeld, R.: Deformation theory, generic vanishing theorems, and some conjectures of Enriques, Catanese and Beauville. Invent. Math. **90**(2), 389–407 (1987)

91. Griffiths, P., Harris, J.: Principles of Algebraic Geometry. Wiley Classics Library. Wiley, New York (1994) Reprint of the 1978 original

92. Groupes de monodromie en géométrie algébrique. I. Lecture Notes in Mathematics, vol. 288. Springer, Berlin/New York (1972). Séminaire de Géométrie Algébrique du Bois-Marie 1967–1969 (SGA 7 I), Dirigé par A. Grothendieck. Avec la collaboration de M. Raynaud et D. S. Rim

93. Groupes de monodromie en géométrie algébrique. II. Lecture Notes in Mathematics, vol. 340. Springer, Berlin/New York (1973). Séminaire de Géométrie Algébrique du Bois-Marie 1967–1969 (SGA 7 II), Dirigé par P. Deligne et N. Katz

94. Hamm, H.: Lokale topologische Eigenschaften komplexer Räume. Math. Ann. **191**, 235–252 (1971)

95. Hamm, H.: Lefschetz theorems for singular varieties. In: Singularities, Part 1 (Arcata, Calif., 1981), Proc. Sympos. Pure Math., vol. 40, pp. 547–557. Amer. Math. Soc., Providence, RI (1983)

96. Hamm, H.: Zum Homotopietyp q-vollständiger Räume. J. Reine Angew. Math. **364**, 1–9 (1986)

97. Hartshorne, H. Algebraic Geometry. Graduate Texts in Mathematics, No. 52 Springer, New York/Heidelberg (1977)

98. Hatcher, A.: Algebraic Topology. Cambridge University Press, Cambridge (2002)

99. Hironaka, H.: Resolution of singularities of an algebraic variety over a field of characteristic zero. I, II. Ann. Math. (2) **79**, 109-203 (1964); ibid. (2) **79**, 205–326 (1964)

100. Hirzebruch, F.: On Steenrod's reduced powers, the index of inertia, and the Todd genus. Proc. Nat. Acad. Sci. USA **39**, 951–956 (1953)

101. Hirzebruch, F.: Arithmetic genera and the theorem of Riemann-Roch for algebraic varieties. Proc. Nat. Acad. Sci. USA **40**, 110–114 (1954)

102. Hirzebruch, F.: Topological Methods in Algebraic Geometry. Die Grundlehren der Mathematischen Wissenschaften, Band 131. Springer, New York (1966)

103. Hirzebruch, F.: The signature theorem: reminiscences and recreation. Ann. Math. Stud. **70**, 3–31 (1971)

104. Hirzebruch, F.: Singularities and exotic spheres. In: Séminaire Bourbaki, vol. 10, Exp. No. 314, pp. 13–32. Soc. Math. France, Paris (1995)

105. Hodge, W.V.D.: The Theory and Applications of Harmonic Integrals. Cambridge University Press, Cambridge; Macmillan Company, New York (1941)

106. Hodge, W.V.D.: The topological invariants of algebraic varieties. In: Proceedings of the International Congress of Mathematicians, Cambridge, MA, 1950, vol. 1, pp. 182–192. American Mathematical Society, Providence, RI (1952)

107. Hotta, R., Takeuchi, K., Tanisaki, T.: D-modules, Perverse Sheaves, and Representation Theory. Progress in Mathematics, vol. 236. Birkhäuser Boston, Inc., Boston, MA (2008). Translated from the 1995 Japanese edition by Takeuchi

108. Hudson, J.F.P.: Piecewise Linear Topology. University of Chicago Lecture Notes Prepared with the Assistance of J. L. Shaneson and J. Lees. W. A. Benjamin, Inc., New York/Amsterdam (1969)

109. Huh, J., Wang, B.: Enumeration of points, lines, planes, etc. Acta Math. **218**(2), 297–317 (2017)

110. Hurewicz, W., Wallman, H.: Dimension Theory. Princeton Mathematical Series, v. 4. Princeton University Press, Princeton, N. J. (1941)

111. Hurwitz, A.: Über Riemann'sche Flächen mit gegebenen Verzweigungspunkten. Math. Ann. **39**(1), 1–60 (1891)

112. Iversen, B.: Critical points of an algebraic function. Invent. Math. **12**, 210–224 (1971)

113. Iversen, B.: Cohomology of sheaves. Universitext. Springer, Berlin (1986)

114. Kaledin, D.: Symplectic singularities from the Poisson point of view. J. Reine Angew. Math. **600**, 135–156 (2006)

115. Karčjauskas, K.: A generalized Lefschetz theorem. Funkcional. Anal. i Priložen. **11**(4), 80–81 (1977)

116. Karu, K.: Hard Lefschetz theorem for nonrational polytopes. Invent. Math. **157**(2), 419–447 (2004)

117. Kashiwara, M.: On the maximally overdetermined system of linear differential equations. I. Publ. Res. Inst. Math. Sci. **10**, 563–579 (1974/75)

118. Kashiwara, M.: Faisceaux constructibles et systèmes holonômes d'équations aux dérivées partielles linéaires à points singuliers réguliers. In: Séminaire Goulaouic-Schwartz, 1979–1980 (French) pages Exp. No. 19, 7. École Polytech., Palaiseau (1980)

119. Kashiwara, M.: The Riemann-Hilbert problem for holonomic systems. Publ. Res. Inst. Math. Sci. **20**(2), 319–365 (1984)

120. Kashiwara, M.: A study of variation of mixed Hodge structure. Publ. Res. Inst. Math. Sci. **22**(5), 991–1024 (1986)

121. Kashiwara, M.: Semisimple holonomic D-modules. In: Topological Field Theory, Primitive Forms and Related Topics (Kyoto, 1996). Progr. Math., vol. 160, pp. 267–271. Birkhäuser Boston, Boston, MA (1998)

122. Kashiwara, M., Schapira, P.: Sheaves on manifolds. Grundlehren der Mathematischen Wissenschaften [Fundamental Principles of Mathematical Sciences], vol. 292. Springer, Berlin (1994)

123. Kato, M., Matsumoto, Y.: On the connectivity of the Milnor fiber of a holomorphic function at a critical point. In: Manifolds—Tokyo 1973 (Proc. Internat. Conf., Tokyo, 1973), pp. 131–136. Univ. Tokyo Press, Tokyo (1975)

124. Katz, N.M.: Nilpotent connections and the monodromy theorem: applications of a result of Turrittin. Inst. Hautes Études Sci. Publ. Math. **39**, 175–232 (1970)

125. Kaup, L., Kaup, B.: Holomorphic Functions of Several Variables. De Gruyter Studies in Mathematics, vol. 3. Walter de Gruyter & Co., Berlin (1983) An introduction to the fundamental theory, With the assistance of Gottfried Barthel, Translated from the German by Michael Bridgland

126. Kervaire, M.A., Milnor, J.W.: Groups of homotopy spheres. I. Ann. Math. (2) **77**, 504–537 (1963)

127. King, H.C.: Topological invariance of intersection homology without sheaves. Topology Appl. **20**(2), 149–160 (1985)

128. Kirwan, F., Woolf, J.: An Introduction to Intersection Homology Theory, 2nd edn. Chapman & Hall/CRC, Boca Raton, FL (2006)

129. Kleiman, S.L.: The development of intersection homology theory. Pure Appl. Math. Q. **3**(1, Special Issue: In honor of Robert D. MacPherson. Part 3), 225–282 (2007)

130. Krämer, T.: Perverse sheaves on semiabelian varieties. Rend. Semin. Mat. Univ. Padova **132**, 83–102 (2014)

131. Krämer, T., Weissauer, R.: Vanishing theorems for constructible sheaves on abelian varieties. J. Algebraic Geom. **24**(3), 531–568 (2015)

132. Landman, A.: On the Picard-Lefschetz transformation for algebraic manifolds acquiring general singularities. Trans. Am. Math. Soc. **181**, 89–126 (1973)

133. Lazarsfeld, R.: Positivity in Algebraic Geometry. I. Ergebnisse der Mathematik und ihrer Grenzgebiete. 3. Folge. A Series of Modern Surveys in Mathematics, vol. 28. Springer, Berlin (2004). Classical setting: line bundles and linear series

134. Lê, D.T.: Topologie des singularités des hypersurfaces complexes. In: Singularités à Cargèse (Rencontre Singularités Géom. Anal., Inst. Études Sci., Cargèse, 1972), pp. 171–182. Astérisque, Nos. 7 et 8. Soc. Math. France, Paris (1973)

135. Lê, D.T.: Calcul du nombre de cycles évanouissants d'une hypersurface complexe. Ann. Inst. Fourier (Grenoble) **23**(4), 261–270 (1973)

136. Lê, D.T.: Some remarks on relative monodromy. In: Real and Complex Singularities (Proc. Ninth Nordic Summer School/NAVF Sympos. Math., Oslo, 1976), pp. 397–403. Sijthoff and Noordhoff, Alphen aan den Rijn (1977)

137. Lê, D.T.: The geometry of the monodromy theorem. In: C. P. Ramanujam—a tribute. Tata Inst. Fund. Res. Studies in Math., vol. 8, pp. 157–173. Springer, Berlin/New York (1978)

138. Lê, D.T.: Sur les cycles évanouissants des espaces analytiques. C. R. Acad. Sci. Paris Sér. A-B **288**(4), A283–A285 (1979)

139. Lefschetz, S.: L'analysis situs et la géométrie algébrique. Gauthier-Villars, Paris (1950)

140. Libgober, A.: Homotopy groups of the complements to singular hypersurfaces. II. Ann. Math. (2) **139**(1), 117–144 (1994)

141. Liu, Y.: Nearby cycles and Alexander modules of hypersurface complements. Adv. Math. **291**, 330–361 (2016)

142. Liu, Y., Maxim, L.: Characteristic varieties of hypersurface complements. Adv. Math. **306**, 451–493 (2017)

143. Liu, Y., Maxim, L., Wang, B.: Generic vanishing for semi-abelian varieties and integral alexander modules. Math. Z. **293**(1–2), 629–645 (2019).

144. Liu, Y., Maxim, L., Wang, B.: Mellin transformation, propagation, and abelian duality spaces. Adv. Math. **335**, 231–260 (2018)

145. Liu, Y., Maxim, L., Wang, B.: Perverse Sheaves on Semi-Abelian Varieties. arXiv e-prints. art. arXiv:1804.05014, Apr 2018

146. Lusztig, G. Intersection cohomology methods in representation theory. In: Proceedings of the International Congress of Mathematicians (Kyoto, 1990), vols. I, II, pp. 155–174. Math. Soc. Japan, Tokyo (1991)

147. Macdonald, I.G.: The Poincaré polynomial of a symmetric product. Proc. Cambridge Philos. Soc. **58**, 563–568 (1962)

148. MacPherson, R.: Chern classes for singular algebraic varieties. Ann. Math. (2) **100**, 423–432 (1974)

149. MacPherson, R.: Global questions in the topology of singular spaces. In: Proceedings of the International Congress of Mathematicians (Warsaw, 1983), vol. 1, 2, pp. 213–235. PWN, Warsaw (1984)

150. MacPherson, R.: Intersection Homology and Perverse Sheaves. Unpublished Colloquium Lectures, 1990

151. MacPherson, R. Vilonen, K.: Elementary construction of perverse sheaves. Invent. Math. **84**(2), 403–435 (1986)

152. Massey, D.B.: Critical points of functions on singular spaces. Topol. Appl. **103**(1), 55–93 (2000)

153. Massey, D.B.: The Sebastiani-Thom isomorphism in the derived category. Compos. Math. **125**(3), 353–362 (2001)

154. Massey, D.B.: Numerical Control Over Complex Analytic Singularities. Memoirs of the American Mathematical Society, vol. 163, no. 778. American Mathematical Society, Providence, RI (2003)

155. Massey, D.B.: Natural commuting of vanishing cycles and the Verdier dual. Pac. J. Math. **284**(2), 431–437 (2016)

156. Mather, J.: Notes on topological stability. Bull. Amer. Math. Soc. (N.S.) **49**(4), 475–506 (2012)

157. Maulik, D., Yun, Z.: Macdonald formula for curves with planar singularities. J. Reine Angew. Math. **694**, 27–48 (2014)

158. Maulik, D., Nekrasov, N.A., Okounkov, A., Pandharipande, R.: Gromov-Witten theory and Donaldson-Thomas theory. I. Compos. Math. **142**(5), 1263–1285 (2006)

159. Maxim, L.: A decomposition theorem for the peripheral complex associated with hypersurfaces. Int. Math. Res. Not. **43**, 2627–2656 (2005)

160. Maxim, L.: Intersection homology and Alexander modules of hypersurface complements. Comment. Math. Helv. **81**(1), 123–155 (2006)

161. Maxim, L., Schürmann, J.: Characteristic classes of singular toric varieties. Commun. Pure Appl. Math. **68**(12), 2177–2236 (2015)

162. Maxim, L., Schürmann, J.: Characteristic classes of mixed Hodge modules and applications. In: Schubert varieties, equivariant cohomology and characteristic classes—IMPANGA 15, EMS Ser. Congr. Rep., pp. 159–202. Eur. Math. Soc., Zürich (2018)

163. Maxim, L., Wong, K.T.: Twisted Alexander invariants of complex hypersurface complements. Proc. R. Soc. Edinb. Sect. A **148**(5), 1049–1073 (2018)

164. Maxim, L., Saito, M., Schürmann, J.: Symmetric products of mixed Hodge modules. J. Math. Pures Appl. (9) **96**(5), 462–483 (2011)

165. Maxim, L., Saito, M., Schürmann, J.: Hirzebruch-Milnor classes of complete intersections. Adv. Math. **241**, 220–245 (2013)

166. Maxim, L., Israel Rodriguez, J., Wang, B.: Euclidean distance degree of the multiview variety. SIAM J. Appl. Algebra Geom. (to appear). arXiv e-prints. art. arXiv:1812.05648, Dec 2018
167. Maxim, L., Saito, M., Schürmann, J.: Thom–Sebastiani theorems for filtered D-modules and for multiplier ideals. International Mathematics Research Notices, rny032 (2018). https://doi.org/10.1093/imrn/rny032
168. Maxim, L., Saito, M., Schürmann, J.: Spectral Hirzebruch-Milnor classes of singular hypersurfaces. Math. Ann. 1–35 (2018) https://doi.org/10.1007/s00208-018-1750-4
169. Maxim, L., Israel Rodriguez, J., Wang, B.: Defect of Euclidean Distance Degree. arXiv e-prints art. arXiv:1905.06758, May 2019
170. McCrory, C.: Cone complexes and PL transversality. Trans. Am. Math. Soc. **207**, 269–291 (1975)
171. McMullen, P.: The numbers of faces of simplicial polytopes. Isr. J. Math. **9**, 559–570 (1971)
172. Mebkhout, Z.: Sur le problème de Hilbert-Riemann. In: Complex analysis, microlocal calculus and relativistic quantum theory (Proc. Internat. Colloq., Centre Phys., Les Houches, 1979). Lecture Notes in Physics, vol. 126, pp. 90–110. Springer, Berlin/New York (1980)
173. Mebkhout, Z.: Une autre équivalence de catégories. Compos. Math. **51**(1), 63–88 (1984)
174. Migliorini, L., Shende, V.: A support theorem for Hilbert schemes of planar curves. J. Eur. Math. Soc. (JEMS) **15**(6), 2353–2367 (2013)
175. Migliorini, L., Shende, V.: Higher discriminants and the topology of algebraic maps. Algebr. Geom. **5**(1), 114–130 (2018)
176. Milnor, J.W.: Construction of universal bundles. II. Ann. Math. (2) **63**, 430–436 (1956)
177. Milnor, J.W.: On the cobordism ring Ω^* and a complex analogue. I. Amer. J. Math. **82**, 505–521 (1960)
178. Milnor, J.W.: Morse Theory. Annals of Mathematics Studies, No. 51. Princeton University Press, Princeton, NJ (1963)
179. Milnor, J.W.: Singular Points of Complex Hypersurfaces. Annals of Mathematics Studies, No. 61. Princeton University Press, Princeton, NJ; University of Tokyo Press, Tokyo (1968)
180. Milnor, J.W., Stasheff, J.D.: Characteristic Classes. Annals of Mathematics Studies, No. 76. Princeton University Press, Princeton, NJ; University of Tokyo Press, Tokyo (1974)
181. Mirković, I., Vilonen, K.: Geometric Langlands duality and representations of algebraic groups over commutative rings. Ann. Math. (2) **166**(1), 95–143 (2007)
182. Mochizuki, T.: Asymptotic behaviour of tame harmonic bundles and an application to pure twistor D-modules. I. Mem. Amer. Math. Soc. **185**(869) (2007)
183. Mochizuki, T.: Asymptotic behaviour of tame harmonic bundles and an application to pure twistor D-modules. II. Mem. Amer. Math. Soc. **185**(870) (2007)
184. Motzkin, T.S.: The lines and planes connecting the points of a finite set. Trans. Am. Math. Soc. **70**, 451–464 (1951)
185. Nakajima, H.: Lectures on Hilbert Schemes of Points on Surfaces. University Lecture Series, vol. 18. American Mathematical Society, Providence, RI (1999)
186. Némethi, A.: Generalized local and global Sebastiani-Thom type theorems. Compos. Math. **80**(1), 1–14 (1991)
187. Ngô, B.C.: Le lemme fondamental pour les algèbres de Lie. Publ. Math. Inst. Hautes Études Sci. **111**, 1–169 (2010)
188. Nori, M.V.: Constructible sheaves. In: Algebra, Arithmetic and Geometry, Part I, II (Mumbai, 2000). Tata Inst. Fund. Res. Stud. Math., vol. 16, pp. 471–491. Tata Inst. Fund. Res., Bombay (2002)
189. Novikov, S.P.: Some problems in the topology of manifolds connected with the theory of Thom spaces. Sov. Math. Dokl. **1**, 717–720 (1960)
190. Novikov, S.P.: Topological invariance of rational classes of Pontrjagin. Dokl. Akad. Nauk SSSR **163**, 298–300 (1965)
191. Oka, M.: On the homotopy types of hypersurfaces defined by weighted homogeneous polynomials. Topology **12**, 19–32 (1973)
192. Pareschi, G., Popa, M.: GV-sheaves, Fourier-Mukai transform, and generic vanishing. Amer. J. Math. **133**(1), 235–271 (2011)

193. Parusiński, A., Pragacz, P.: A formula for the Euler characteristic of singular hypersurfaces. J. Algebraic Geom. **4**(2), 337–351 (1995)
194. Parusiński, A., Pragacz, P.: Characteristic classes of hypersurfaces and characteristic cycles. J. Algebraic Geom. **10**(1), 63–79 (2001)
195. Peters, C.A.M., Steenbrink, J.H.M.: Mixed Hodge structures. Ergebnisse der Mathematik und ihrer Grenzgebiete. 3. Folge. A Series of Modern Surveys in Mathematics [Results in Mathematics and Related Areas. 3rd Series. A Series of Modern Surveys in Mathematics], vol. 52. Springer-Verlag, Berlin (2008)
196. Poincaré, H.: Complément à l'Analysis Situs. Rendiconti del Circolo Matematico di Palermo **13**, 285–343 (1899)
197. Poincaré, H.: Second Complément à l'Analysis Situs. Proc. Lond. Math. Soc. **32**, 277–308 (1900)
198. Pontryagin, L.S.: Characteristic cycles on differentiable manifolds. Mat. Sbornik N. S. **21**(63), 233–284 (1947)
199. Popa, M.: Positivity for Hodge modules and geometric applications. In: Algebraic geometry: Salt Lake City 2015. Proc. Sympos. Pure Math., vol. 97, pp. 555–584. American Mathematical Society, Providence, RI (2018)
200. Popa, M., Schnell, C.: Generic vanishing theory via mixed Hodge modules. Forum Math. Sigma **1**, e1, 60 pp. (2013)
201. Rietsch, K.: An introduction to perverse sheaves. In: Representations of Finite Dimensional Algebras and Related Topics in Lie Theory and Geometry. Fields Inst. Commun., vol. 40, pp. 391–429. American Mathematical Society, Providence, RI (2004)
202. Rota, G.-C.: Combinatorial theory, old and new. In: Actes du Congrès International des Mathématiciens (Nice, 1970), Tome 3, pp. 229–233. Gauthier-Villars, Paris (1971)
203. Rota, G.-C., Harper, L.W.: Matching theory, an introduction. In: Advances in Probability and Related Topics, vol. 1, pp 169–215. Dekker, New York (1971)
204. Sabbah, C.: Polarizable twistor D-modules. Astérisque, vol. 300 (2005)
205. Saito, M.: Modules de Hodge polarisables. Publ. Res. Inst. Math. Sci. **24**(6), 849–995 (1988)
206. Saito, M.: Introduction to mixed Hodge modules. Astérisque, vol. 179–180, pp. 10, 145–162 (1989). Actes du Colloque de Théorie de Hodge (Luminy, 1987)
207. Saito, M.: Mixed Hodge modules. Publ. Res. Inst. Math. Sci. **26**(2), 221–333 (1990)
208. Saito, M.: Decomposition theorem for proper Kähler morphisms. Tohoku Math. J. (2) **42**(2), 127–147 (1990)
209. Saito, M.: Mixed Hodge complexes on algebraic varieties. Math. Ann. **316**(2), 283–331 (2000)
210. Saito, M.: A Young Person's Guide to Mixed Hodge Modules. arXiv e-prints art. arXiv:1605.00435, May 2016
211. Sakamoto, K.: Milnor fiberings and their characteristic maps. In: Manifolds—Tokyo 1973 (Proc. Internat. Conf., Tokyo, 1973), pp. 145–150. Univ. Tokyo Press, Tokyo (1975)
212. Schmid, W.: Variation of Hodge structure: the singularities of the period mapping. Invent. Math. **22**, 211–319 (1973)
213. Schnell, C.: Holonomic D-modules on abelian varieties. Publ. Math. Inst. Hautes Études Sci. **121**, 1–55 (2015)
214. Schürmann, J.: Topology of Singular Spaces and Constructible Sheaves. Monografie Matematyczne, vol. 63. Birkhäuser Verlag, Basel (2003)
215. Schürmann, J.: Characteristic classes of mixed Hodge modules. In: Topology of stratified spaces. Math. Sci. Res. Inst. Publ., vol. 58, pp. 419–470. Cambridge Univ. Press, Cambridge (2011)
216. Schürmann, J., Yokura, S.: A survey of characteristic classes of singular spaces. In: Singularity Theory, pp. 865–952. World Sci. Publ., Hackensack, NJ (2007)
217. Sebastiani, M., Thom, R.: Un résultat sur la monodromie. Invent. Math. **13**, 90–96 (1971)
218. Siegel, P.H.: Witt spaces: a geometric cycle theory for KO-homology at odd primes. Amer. J. Math. **105**(5), 1067–1105 (1983)

219. Spanier, E.H.: Algebraic topology. McGraw-Hill Book Co., New York-Toronto, Ont.-London (1966)
220. Stanley, R.P.: Generalized H-vectors, intersection cohomology of toric varieties, and related results. In: Commutative Algebra and Combinatorics. Advanced Studies in Pure Mathematics, vol. 11, pp. 187–213. Mathematical Society of Japan, Kyoto (1985)
221. Stanley, R.P.: The number of faces of a simplicial convex polytope. Adv. Math. **35**(3), 236–238 (1980)
222. Stanley, R.P.: Combinatorial applications of the hard Lefschetz theorem. In: Proceedings of the International Congress of Mathematicians, vols. 1, 2 (Warsaw, 1983), pp. 447–453. PWN, Warsaw (1984)
223. Steenbrink, J., Zucker, S.: Variation of mixed Hodge structure. I. Invent. Math. **80**(3), 489–542 (1985)
224. Stein, K.: Analytische Zerlegungen komplexer Räume. Math. Ann. **132**, 63–93 (1956)
225. Stiefel, E.: Richtungsfelder und Fernparallelismus in n-dimensionalen Mannigfaltigkeiten. Comment. Math. Helv. **8**(1), 305–353 (1935)
226. Sullivan, D.: Combinatorial invariants of analytic spaces. In: Proceedings of Liverpool Singularities—Symposium, I (1969/70), pp. 165–168. Springer, Berlin (1971)
227. Szendrői, B.: Cohomological Donaldson-Thomas theory. In: String-Math 2014. Proc. Sympos. Pure Math., vol. 93, pp. 363–396. American Mathematical Society, Providence, RI (2016)
228. Théorie des topos et cohomologie étale des schémas. Tome 3. Lecture Notes in Mathematics, vol. 305. Springer, Berlin/New York (1973). Séminaire de Géométrie Algébrique du Bois-Marie 1963–1964 (SGA 4), Dirigé par M. Artin, A. Grothendieck et J. L. Verdier. Avec la collaboration de P. Deligne et B. Saint-Donat
229. Thom, R.: Espaces fibrés en sphères et carrés de Steenrod. Ann. Sci. Ecole Norm. Sup. (3) **69**, 109–182 (1952)
230. Thom, R.: Quelques propriétés globales des variétés différentiables. Comment. Math. Helv. **28**, 17–86 (1954)
231. Thom, R.: Les classes caractéristiques de Pontrjagin des variétés triangulées. In: Symposium internacional de topología algebraica (International Symposium on Algebraic Topology), pp. 54–67. Universidad Nacional Autónoma de México and UNESCO, Mexico City (1958)
232. Thom, R.: Ensembles et morphismes stratifiés. Bull. Amer. Math. Soc. **75**, 240–284 (1969)
233. Verdier, J.-L.: Stratifications de Whitney et théorème de Bertini-Sard. Invent. Math. **36**, 295–312 (1976)
234. Verdier, J.-L.: Catégories dérivées: quelques résultats (état 0). In: Cohomologie étale. Lecture Notes in Mathematics, vol. 569, pp. 262–311. Springer, Berlin (1977)
235. Verdier, J.-L.: Dualité dans la cohomologie des espaces localement compacts. In: Séminaire Bourbaki, vol. 9, Exp. No. 300, pp. 337–349. Soc. Math. France, Paris (1995)
236. Verdier, J.-L.: Des catégories dérivées des catégories abéliennes. Astérisque, vol. 239, 1996. With a preface by Luc Illusie, Edited and with a note by Georges Maltsiniotis
237. Wall, C.T.C.: Determination of the cobordism ring. Ann. Math. (2) **72**, 292–311 (1960)
238. Weber, A.: A morphism of intersection homology induced by an algebraic map. Proc. Am. Math. Soc. **127**(12), 3513–3516 (1999)
239. Weissauer, R.: Vanishing theorems for constructible sheaves on abelian varieties over finite fields. Math. Ann. **365**(1–2), 559–578 (2016)
240. Whitney, H.: On the theory of sphere-bundles. Proc. Nat. Acad. Sci. USA **26**, 148-153 (1940)
241. Whitney, H.: On the topology of differentiable manifolds. In: Lectures in Topology, pp. 101–141. University of Michigan Press, Ann Arbor, MI (1941)
242. Whitney, H.: Tangents to an analytic variety. Ann. Math. (2) **81**, 496–549 (1965)
243. Whitney, H.: Local properties of analytic varieties. In: Differential and Combinatorial Topology (A Symposium in Honor of Marston Morse), pp. 205–244. Princeton University Press, Princeton, NJ (1965)
244. Williamson, G.: Algebraic representations and constructible sheaves. Jpn. J. Math. **12**(2), 211–259 (2017)

245. Williamson, G.: The Hodge theory of the decomposition theorem. Astérisque, vol. 390, Exp. No. 1115, pp. 335–367 (2017) Séminaire Bourbaki, vol. 2015/2016. Exposés 1104–1119
246. Woolf, J.: Witt groups of sheaves on topological spaces. Comment. Math. Helv. **83**(2), 289–326 (2008)
247. Yokura, S.: On Cappell-Shaneson's homology L-classes of singular algebraic varieties. Trans. Am. Math. Soc. **347**(3), 1005–1012 (1995)
248. Yokura, S. Motivic Milnor classes. J. Singul. **1**, 39–59 (2010)

Index

© Springer Nature Switzerland AG 2019
L. G. Maxim, *Intersection Homology & Perverse Sheaves*, Graduate Texts
in Mathematics 281, https://doi.org/10.1007/978-3-030-27644-7

Printed in the United States
By Bookmasters